IEEE Recommended Practice for the Design of Reliable Industrial and Commercial Power Systems

Published by
The Institute of Electrical and Electronics Engineers, Inc.

ISBN 1-55937-066-1

Recognized as an
American National Standard (ANSI)

IEEE
Std 493-1990
(Revision of IEEE
Std 493-1980)

IEEE Recommended Practice for the Design of Reliable Industrial and Commercial Power Systems

Sponsor

Power Systems Reliability Subcommittee
of the
Power Systems Engineering Committee
of the
IEEE Industry Applications Society

Approved September 28, 1990

IEEE Standards Board

Approved February 27, 1991

American National Standards Institute

Abstract: The fundamentals of reliability analysis as it applies to the planning and design of industrial and commercial electric power distribution systems are presented. Included are basic concepts of reliability analysis by probability methods, fundamentals of power system reliability evaluation, economic evaluation of reliability, cost of power outage data, equipment reliability data, examples of reliability analysis. Emergency and standby power, electrical preventive maintenance, and evaluating and improving reliability of the existing plant are also addressed. The presentation is self-contained and should enable trade-off studies during the design of industrial and commercial power systems.
Keywords: Designing reliable industrial and commercial power systems, equipment reliability data, industrial and commercial power systems reliability analysis, reliability analysis.

Second Printing
May 1995

The Institute of Electrical and Electronics Engineers, Inc.
345 East 47th Street, New York, NY 10017-2394, USA

© 1991 by Institute of Electrical and Electronics Engineers, Inc.
All rights reserved. Published 1991
Printed in the United States of America

ISBN 1-55937-066-1

Library of Congress Catalog Number 91-055199

May 3, 1991

SH13730

Foreword

(This Foreword is not a part of IEEE Std 493-1990, IEEE Recommended Practice for the Design of Reliable Industrial and Commercial Power Systems.)

The design of reliable industrial and commercial power systems is of considerable interest to many people. Prior to 1962, a qualitative viewpoint was taken when attempting to achieve this objective. The need for a quantitative approach was first recognized in the early 1960's when a small group of pioneers led by W. H. Dickinson organized an extensive AIEE survey of the reliability of electrical equipment in industrial plants. The AIEE survey that was taken in 1962 was followed by several IEEE reliability surveys, which were published in 1973 through 1979. These surveys from the the 1970's were the basis for the reliability data contained in IEEE Std 493-1980. Six additional IEEE reliability surveys have been conducted and published during the 1980's and have been used in this revision of IEEE Std 493-1980. IEEE Std 493-1990 also includes pertinent tutorial reliability material and the cost of power interruptions data.

Tutorial reliability sessions on the design of industrial and commercial power systems were conducted at technical conferences of the IEEE Industry Applications Society in 1971, 1976, and 1980.

This recommended practice was prepared by a working group of the Power Systems Reliability Subcommittee, Power Systems Engineering Committee, Industrial and Commercial Power Systems Department of the IEEE Industry Application Society.

At the time it recommended these practices, the working group of the Power Systems Reliability Subcommittee had the following members:

C. R. Heising, *Chair*

J. W. Aquilino	E. Golpashin	A. D. Patton
C. E. Becker	D. O. Koval	C. Singh
W. F. Braun	A. T. Norris	H. P. Stickley
B. G. Douglas	P. O'Donnell	S. J. Wells
P. E. Gannon		D. W. McWilliams*

*Deceased

This recommended practice was reviewed and approved by the Power Systems Reliability Subcommittee which had the following members:

C. R. Heising, *Chair*   **D. O. Koval,** *Vice Chair*

J. W. Aquilino	B. G. Douglas	A. T. Norris
P. F. Albrecht	R. H. Gauger	P. O'Donnell
C. E. Becker	P. E. Gannon	A. D. Patton
W. F. Braun	E. Golpashin	C. Singh
L. E. Conrad	S. C. Kapoor	H. P. Stickley
W. H. Dickinson	R. H. McFadden	S. J. Wells
	L. D. Monaghan	

When the IEEE Standards Board approved this standard on September 28, 1990, it had the following membership:

Acknowledgment is given to the members of the working groups that compiled the 14 IEEE committee reports which are contained in the Appendixes. Acknowledgment is also given for the three equipment reliability surveys conducted by others which are summarized in Appendixes I, J, and L.

IEEE Recommended Practice for the Design of Reliable Industrial and Commercial Power Systems

Working Group Members and Contributors

C. R. Heising, *Working Group Chair*

Chapter 1 — Introduction: C. R. Heising

Chapter 2 — Planning and Design: A. D. Patton, P. E. Gannon, C. R. Heising, D. O. Koval, C. E. Becker, E. Golpashin

Chapter 3 — Summary of Equipment Reliability Data: D. O. Koval, *Chair*; J. W. Aquilino, W. F. Braun, B. G. Douglas, C. R. Heising, A. T. Norris, P. O'Donnell, H. P. Stickley

Chapter 4 — Evaluating and Improving the Reliability of an Existing Plant: C. E. Becker

Chapter 5 — Electrical Preventive Maintenance: S. J. Wells, C. R. Heising

Chapter 6 — Emergency and Standby Power: D. W. McWilliams

Chapter 7 — Examples of Reliability Analysis and Cost Evaluation: C. R. Heising, P. E. Gannon

Chapter 8 — Basic Concepts of Reliability Analysis by Probability Methods: C. Singh

Contents

Chapter 1
Introduction

1.1 Objectives and Scope. The objective of this book is to present the fundamentals of reliability analysis as it applies to the planning and design of industrial and commercial electric power distribution systems. The text material is primarily directed toward consulting engineers and plant electrical engineers.

The design of reliable industrial and commercial power distribution systems is important because of the high costs associated with power outages. There is a need to be able to consider the cost of power outages when making design decisions for new power distribution systems and to be able to make quantitative cost versus reliability trade-off studies. The lack of credible data concerning equipment reliability and the cost of power outages has hindered engineers in making such studies.

The authors of this book have attempted to provide sufficient information so that reliability analyses can be performed on power systems without the need for cross references to other texts. Included are:
  (1) Basic concepts of reliability analysis by probability methods
  (2) Fundamentals of power system reliability evaluation
  (3) Economic evaluation of reliability
  (4) Cost of power outage data
  (5) Equipment reliability data
  (6) Examples of reliability analysis
In addition, discussion and information are provided on:
  (1) Emergency and standby power
  (2) Electrical preventive maintenance
  (3) Evaluating and improving reliability of the existing plant
A quantitative reliability analysis includes making a disciplined evaluation of alternate power distribution system design choices. When costs of power outages at the various building and plant locations are factored into the evaluation, the decisions can be based upon *total owning cost over the useful life of the equipment* rather than simply *first cost* of the system. The material in this book should enable engineers to make more use of quantitative cost versus reliability trade-off studies during the design of industrial and commercial power systems.

1.2 IEEE Surveys of Industrial Plants. From 1973 through 1989, the Power Systems Reliability Subcommittee of the Power Systems Engineering Committee of the IEEE Industry Applications Society conducted and published the results of extensive surveys of the reliability of electrical equipment in industrial plants and also the cost of power outages for both industrial plants and commercial buildings. This included motors, motor starters, generators, power transformers, rectifier transformers, circuit breakers, disconnect switches, bus duct, switchgear bus-bare, switchgear bus-insulated, open wire, cable, cable joints, cable terminations, and electric utility power supplies. The results from these surveys have been published in 15 IEEE committee reports, 14 of which are included in this book in Appendixes A, B, C, D, E, G, H, and K. Appendix F gives the procedure used for conducting these surveys. It has been considered important that the "reasons for conducting the survey" be written down at the beginning of each new survey. It has also been considered important that the final report receive both oral and written discussion at the end of each survey. Some of the IEEE surveys have also included the cost of power interruptions, critical service loss duration time, and plant restart time. The most important results from these 15 surveys have been summarized in Chapters 2, 3, and 5. Table 10 in Chapter 3 contains a summary of the latest equipment reliability data from these surveys; and these values are suggested for use in the absence of better data which may be available from the reader's own experience. Table 9 presents a guide of where to look in this book for additional reliability data for each of several equipment categories.

Three important equipment reliability surveys conducted by others have been summarized and included as Appendixes I, J, and L; these appendixes supplement the IEEE equipment reliability surveys in some categories in which there has been little or no data and in other categories in which the data is more recent and/or much more extensive. These three equipment reliability surveys include:

(1) Cable, cable splices, and cable terminations
(2) High-voltage circuit breakers above 63 kV
(3) Diesel and gas turbine generating units

An extensive bibliography on electrical service interruption costs is presented in Appendix M.

The reliability survey data contained in this book provide historical experience to those who have not been able to collect their own data. Such data can be an aid in analyzing, designing, or redesigning an industrial or commercial system and can provide a basis for the quantitative comparison of alternate designs.

1.3 How to Use This Book. This book is primarily directed toward consulting engineers and plant electrical engineers and covers the fundamentals of reliability analysis as it applies to the planning and design of industrial and commercial electric power distribution systems. The methods of reliability analysis are based upon probability and statistics. Some users of this book may wish to read Chapter 8 on basic probability concepts before reading Chapter 2 on planning and design. Other users may wish to start with Chapter 2 and not wish to attempt to fully understand the derivation of the statistical formulas given in 2.1.9, and 2.1.11.1.

The most important parts of planning and design are covered in 2.1 and 2.2 on fundamentals of power system reliability evaluation and on the economic evaluation of reliability. Chapter 7 gives six examples using these methods of analysis. These examples cover some of the most common decisions that engineers are faced with when designing a power distribution system. Some discussion on the limitations of reliability and availability predictions are given in the latter part of 7.1.

Those wishing to obtain equipment reliability data should go to Chapter 3. Those wishing to obtain data on the cost of electrical interruptions to industrial plants or commercial buildings should consult 2.2. Any data on costs may need to be updated to take into account the effects of inflation.

The importance of electrical preventive maintenance in planning and design is covered in 2.3 and 2.4. Chapter 5 discusses the subject in further detail and contains data showing the effect of maintenance quality on equipment failure rates.

Many reliability studies need to be followed up by considerations for emergency and standby power. This subject is covered in Chapter 6 and may also be considered part of planning and design.

An approach to evaluating and upgrading the reliability of an existing plant is presented in Chapter 4. Some users of this book may wish to start with this chapter.

1.4 Definitions. The following definitions should be used in conjunction with this recommended practice:

availability. The steady-state probability that a component is in service.

component. A piece of electrical or mechanical equipment, a line or circuit, or a section of a line or circuit, or a group of items that is viewed as an entity for the purposes of reliability evaluation.

electrical equipment. A general term including materials, fittings, devices, appliances, fixtures, apparatus, machines, etc., used as a part of, or in connection with, an electric installation.

electrical preventive maintenance. A system of planned inspection, testing, cleaning, drying, monitoring, adjusting, corrective modification, and minor repair of electrical equipment to minimize or forestall future equipment operating problems or failures; which, depending upon equipment type, may require exercising or proof testing.

expected failure duration. The expected or long-term average duration of a single failure event.

expected interruption duration. The expected, or average, duration of a single load interruption event.

exposure time. The time during which a component is performing its intended function and is subject to failure.

failure. Any trouble with a power system component that causes any of the following events to occur:
  (1) Partial or complete plant shutdown, or below-standard plant operation
  (2) Unacceptable performance of user's equipment
  (3) Operation of the electrical protective relaying or emergency operation of the plant electrical system

(4) De-energization of any electric circuit or equipment

A failure on a public utility supply system may cause the user to have either of the following:

(1) A power interruption or loss of service
(2) A deviation from normal voltage or frequency outside the normal utility profile

A failure on an in-plant component causes a forced outage of the component, that is, the component is unable to perform its intended function until it is repaired or replaced. The terms "failure" and "forced outage" are often used synonymously.

failure rate (forced outage rate). The mean number of failures of a component per unit exposure time. Usually exposure time is expressed in years and failure rate is expressed in failures per year.

forced outage. An outage (failure) that cannot be deferred.

forced unavailability. The long-term average fraction of time that a component or system is out of service due to a forced outage (failure).

interruption. The loss of electric power supply to one or more loads.

interruption frequency. The expected (average) number of power interruptions to a load per unit time, usually expressed as interruptions per year.

mean time between failures (MTBF). The mean exposure time between consecutive failures of a component. It can be estimated by dividing the exposure time by the number of failures in that period, provided that a sufficient number of failures has occurred in that period.

mean time to repair (MTTR). The mean time to repair a failed component. It can be estimated by dividing the summation of repair times by the number of repairs, and, therefore, it is practically the average repair time.

minimal cut-set. A set of components that, if removed from the system, results in loss of continuity to the load point being investigated and that does not contain as a subset any set of components that is itself a cut-set of the system.

offline system. A system that is dormant until it is called upon to operate, such as a diesel generator that is started up when a power failure occurs.

online system. A system that is operating at all times, such as an inverter supplied by dc power via the primary power source through a battery charger.

outage. The state of a component or system when it is not available to properly perform its intended function.

repair time. The repair time of a failed component or the duration of a failure is the clock time from the occurrence of the failure to the time when the component is restored to service, either by repair of the failed component or by substitution of a spare component for the failed component. It includes time for diagnosing the trouble, locating the failed component, waiting for parts, repairing or replacing, testing, and restoring the component to service. The terms "repair time" and "forced outage duration" are often used synonymously. It is not the time required to restore service to a load by putting alternate circuits into operation.

scheduled outage. An outage that results when a component is deliberately taken out of service at a selected time, usually for purposes of construction, maintenance, or repair.

scheduled outage duration. The period from the initiation of a scheduled outage until construction, preventive maintenance, or repair work is completed and the affected component is made available to perform its intended function.

scheduled outage rate. The mean number of scheduled outages of a component per unit exposure time.

switching time. The period from the time a switching operation is required due to a component failure until that switching operation is completed. Switching operations include such operations as: throwover to an alternate circuit, opening or closing a sectionalizing switch or circuit breaker, reclosing a circuit breaker following a tripout due to a temporary fault, etc.

system. A group of components connected or associated in a fixed configuration to perform a specified function of distributing power.

unavailability. The long-term average fraction of time that a component or system is out of service due to failures or scheduled outages. An alternative definition is the steady-state probability that a component or system is out of service due to failures or scheduled outages.

Chapter 2
Planning and Design

2.1 Fundamentals of Power System Reliability Evaluation

2.1.1 Reliability Evaluation Fundamentals. Fundamentals necessary for a quantitative reliability evaluation of electric power systems include definitions of basic terms, discussions of useful measures of system reliability and the basic data needed to compute these indexes, and a description of the procedure for system reliability analysis including computation of quantitative reliability indexes.

2.1.2 Power System Design Considerations. An important aspect of power system design involves consideration of the service reliability requirements of loads that are to be supplied and the service reliability that will be provided by any proposed system. System reliability assessment and evaluation methods based on probability theory that allow the reliability of a proposed system to be assessed quantitatively are finding wide application today. Such methods permit consistent, defensible, and unbiased assessments of system reliability that are not otherwise possible.

The quantitative reliability evaluation methods presented here permit reliability indexes for any electric power system to be computed from knowledge of the reliability performance of the constituent components of the system. Thus, alternative system designs can be studied to evaluate the impact on service reliability and cost of changes in component reliability, system configuration, protection and switching scheme, or system operating policy including maintenance practice.

2.1.3 Definitions. The definitions presented here include those used in the survey of the reliability of electrical equipment in industrial plants (see Reference [16]).[1] The definitions presented here are not exhaustive, but do provide much of the required nomenclature for discussions of power system reliability.

availability. This term may apply either to the performance of individual components or to that of a system. Availability is defined to be the long-term average fraction of time that a component or system is in service satisfactorily performing its intended function. An alternative and equivalent definition for availability is the steady-state probability that a component or system is in service.

component. A piece of electrical or mechanical equipment, a line or circuit, or a section of a line or circuit, or a group of items that is viewed as an entity for the purposes of reliability evaluation.

---

[1] The numbers in brackets correspond to those in the references at the end of this chapter.

expected interruption duration. The expected, or average, duration of a single load interruption event.

exposure time. The time during which a component is performing its intended function and is subject to failure.

failure. Any trouble with a power system component that causes any of the following events to occur:
(1) Partial or complete plant shutdown, or below-standard plant operation
(2) Unacceptable performance of user's equipment
(3) Operation of the electrical protective relaying or emergency operation of the plant electrical system
(4) De-energization of any electric circuit or equipment

A failure on a public utility supply system may cause the user to have either of the following:
(1) A power interruption or loss of service
(2) A deviation from normal voltage or frequency outside the normal utility profile

A failure on an in-plant component causes a forced outage of the component, that is, the component is unable to perform its intended function until it is repaired or replaced. The terms "failure" and "forced outage" are often used synonymously.

failure rate (forced outage rate). The mean number of failures of a component per unit exposure time. Usually exposure time is expressed in years and failure rate is expressed in failures per year.

forced outage. An outage (failure) that cannot be deferred.

forced unavailability. The long-term average fraction of time that a component or system is out of service due to a forced outage (failure).

interruption. The loss of electric power supply to one or more loads.

interruption frequency. The expected (average) number of power interruptions to a load per unit time, usually expressed as interruptions per year.

outage. The state of a component or system when it is not available to properly perform its intended function.

repair time. The repair time of a failed component or the duration of a failure is the clock time from the occurrence of the failure to the time when the component is restored to service, either by repair of the failed component or by substitution of a spare component for the failed component. It includes time for diagnosing the trouble, locating the failed component, waiting for parts, repairing or replacing, testing, and restoring the component to service. The terms "repair time" and "forced outage duration" are often used synonymously. It is not the time required to restore service to a load by putting alternate circuits into operation.

scheduled outage. An outage that results when a component is deliberately taken out of service at a selected time, usually for purposes of construction, maintenance, or repair.

scheduled outage duration. The period from the initiation of a scheduled outage until construction, preventive maintenance, or repair work is completed and the affected component is made available to perform its intended function.

scheduled outage rate. The mean number of scheduled outages of a component per unit exposure time.

switching time. The period from the time a switching operation is required due to a component failure until that switching operation is completed. Switching operations include such operations as: throwover to an alternate circuit, opening or closing a sectionalizing switch or circuit breaker, reclosing a circuit breaker following a tripout due to a temporary fault, etc.

system. A group of components connected or associated in a fixed configuration to perform a specified function of distributing power.

unavailability. The long-term average fraction of time that a component or system is out of service due to failures or scheduled outages. An alternative definition is the steady-state probability that a component or system is out of service. Mathematically, unavailability = (1 – availability).

2.1.4 System Reliability Indexes. The basic system reliability indexes (see References [2], [9], [10], and [20]) that have proven most useful and meaningful in power distribution system design are:
(1) Load interruption frequency
(2) Expected duration of load interruption events
These indexes can be readily computed using methods that will be described later. The two basic indexes of interruption frequency and expected interruption duration can be used to compute other indexes that are also useful:
(1) Total expected (average) interruption time per year (or other time period)
(2) System availability or unavailability as measured at the load supply point in question
(3) Expected, demanded, but unsupplied, energy per year
It should be noted here that the disruptive effect of power interruptions is often nonlinearly related to the duration of the interruption. Thus, it is often desirable to compute not only an overall interruption frequency but also frequencies of interruptions categorized by the appropriate durations.

2.1.5 Data Needed for System Reliability Evaluations. The data needed for quantitative evaluations of system reliability depend to some extent on the nature of the system being studied and the detail of the study. In general, however, data on the performance of individual components together with the times required to perform various switching operations are required.
System component data that are generally required are summarized as follows:
(1) Failure rates (forced outage rates) associated with different modes of component failure
(2) Expected (average) time to repair or replace failed component
(3) Scheduled (maintenance) outage rate of component
(4) Expected (average) duration of a scheduled outage event

If possible, component data should be based on the historical performance of components in the same environment as those in the proposed system being studied. The reliability surveys conducted by the Power Systems Reliability Subcommittee (see References [16] and [17]) provide a source of component data when such specific data are not available. These data and the data from later surveys have been summarized in Chapter 3.

The needed switching time data include:

(1) Expected times to open and close a circuit breaker
(2) Expected times to open and close a disconnect or throwover switch
(3) Expected time to replace a fuse link
(4) Expected times to perform such emergency operations as cutting in clear, installing jumpers, etc.

Switching times should be estimated for the system being studied based on experience, engineering judgment, and anticipated operating practices.

2.1.6 Method for System Reliability Evaluation. The method for system reliability evaluation that is recommended and presented here has evolved over a number of years (see References [2], [3], [9], [10], and [12]). The method, called the "minimal cut-set method," is believed to be particularly well suited to the study and analysis of electric power distribution systems as found in industrial plants and commercial buildings. The method is systematic and straightforward and lends itself to either manual or computer computation. An important feature of the method is that system weak points can be readily identified, both numerically and nonnumerically, thereby focusing design attention on those sections of the system that contribute most to service unreliability. See Chapter 8 for a derivation of the minimal cut-set method.

The procedure for system reliability evaluation is outlined as follows:

(1) Assess the service reliability requirements of the loads and processes that are to be supplied and determine the appropriate service interruption definition or definitions.
(2) Perform a failure modes and effects analysis identifying and listing those component failures and combinations of component failures that result in service interruptions and that constitute minimal cut-sets of the system.
(3) Compute the interruption frequency contribution, the expected interruption duration, and the probability of each of the minimal cut-sets of step (2).
(4) Combine the results of step (3) to produce system reliability indexes.

These steps will be discussed in more detail in the sections that follow.

2.1.7 Service Interruption Definition. The first step in any electric power system reliability study should be a careful assessment of the power supply quality (e.g., sags, surges, harmonics, etc.) and continuity required by the loads that are to be served. This assessment should be summarized and expressed in a service interruption definition that can be used in the succeeding steps of the reliability evaluation procedure. The interruption definition specifies, in general, the reduced voltage level (voltage dip) together with the minimum duration of such a reduced voltage period that results in substantial degradation or complete loss of function of the load or process being served. Frequently, reliability studies are conducted on a continuity basis, in which case, interruption definitions reduce to a minimum duration specification with voltage assumed to be zero during the interruption.

A further discussion of interruption definitions together with examples of such definitions is given in 7.1.2.

2.1.8 Failure Modes and Effects Analysis (FMEA). The FMEA for power distribution systems amounts to the determination and listing of those component outage events or combinations of component outages that result in an interruption of service at the load point being studied according to the interruption definition that has been adopted. This analysis must be made in consideration of the different types and modes of outages that components may exhibit and the reaction of the system's protection scheme to these events.

The primary result of the FMEA as far as quantitative reliability evaluation is concerned is the list of minimal cut-sets it produces. The use of the minimal cut-sets in the calculation of system reliability indexes is described in Chapter 8 of this book. A minimal cut-set is defined to be "a set of components that, if removed from the system, results in loss of continuity to the load point being investigated and that does not contain as a subset any set of components that is itself a cut-set of the system." In the present context, the components in a cut-set are just those components whose overlapping outage results in an interruption according to the interruption definition adopted.

An important nonquantitative benefit of the FMEA is the thorough and systematic thought process and investigation that it requires. Often weak points in system design will be identified before any quantitative reliability indexes are computed. Thus, the FMEA is a useful reliability design tool even in the absence of the data needed for quantitative evaluation.

The FMEA and the determination of minimal cut-sets is most efficiently conducted by considering first the effects of outages of single components and then the effects of overlapping outages of increasing numbers of components. Those cut-sets containing a single component are termed "first-order cut-sets." Similarly, cut-sets containing two components are termed "second-order cut-sets," etc. In theory, the FMEA should continue until all the minimal cut-sets of the system have been found. In practice, however, the FMEA can be terminated earlier since high-order cut-sets have low probability compared to lower order cut-sets. A good rule of thumb is to determine minimal cut-sets up to order $n + 1$, where n is the lowest-order minimal cut-set of the system. Since most power distribution systems have at least some first-order minimal cut-sets, the analysis can usually be terminated after the second-order minimal cut-sets have been found.

2.1.9 Computation of System Reliability Indexes. The list of minimal cut-sets obtained from FMEA is used to compute system reliability indexes. Since the occurrence of any cut-set will result in system failure, these cut-sets can be regarded as acting in series. The failure frequency and average outage duration can therefore be computed using Eqs 1 and 2.

f_s = System interruption frequency = $\sum_i f_{cs_i}$ (Eq 1)

r_s = System expected interruption duration = $\sum_i f_{cs_i} r_{cs_i}/f_s$ (Eq 2)

NOTE: These are approximate formulas and should only be used when $f_{cs_i} \times r_{cs_i}$ are less than 0.01.

where

f_{cs_i} = Frequency of cut-set event i.
r_{cs_i} = Expected duration of cut-set event i.

It can be seen from Eqs 1 and 2 that, once the frequency and duration of the various cut-sets are known, the load point interruption frequency and duration can be easily computed. Since the various cut-set events are not mutually exclusive, Eq 1 is an upper bound on the frequency of system failure. Assuming, however, that the time a component spends on outage is very small compared to the time it is operating satisfactorily, Eqs 1 and 2 give results close to the exact values. A later section gives equations for computing the frequency and duration for various types of outage events.

2.1.10 Component Failure Modes. Distribution system components, such as lines, transformers, and circuit breakers, are subject to a variety of failure modes that, in general, have different impacts on system reliability performance. For system reliability evaluation purposes, it is useful to categorize system components as switching devices or nonswitching devices. First, consider nonswitching devices, such as lines or transformers. The important modes of failure are those events that cause the component to be unable to fulfill its current-carrying function, generally due to a fault and subsequent isolation of the faulted component by a protective device. Such failure modes can be modeled in system reliability calculations through the use of permanent forced outage rates and transient forced outage rates, where

λ = Permanent forced outage rate of the component = Rate of occurrence of forced outages in which the component is damaged and cannot be restored to service until repair or replacement has been completed.

λ' = Transient forced outage rate of component = Rate of occurrence of forced outages in which the component is undamaged and can be immediately restored to service.

NOTE: A forced outage is defined as "an outage (failure) that cannot be deferred."

Now consider the failure modes of protection systems and of switching devices, such as circuit breakers. In contrast to the components described above whose only function is carrying current (a continuously required function), protection systems and switching devices generally have both continuously required as well as response functions. The inability to perform a continuously required function, such as current carrying, will immediately impact system performance while the inability to perform a response function, such as tripping open on command, will be manifested only when the response is required. Some of the more important failure modes of protection systems and switching devices and the parameters used to model these failure modes in reliability calculations are summarized as follows:

Continuous Functions

(1) Component short circuit resulting in operation of backup protective devices. The modeling parameter is λ, which is the rate of occurrence of such short-circuit events.

(2) Switching device opening without the proper command. The modeling parameter is λ_{FT}, which is the rate of occurrence of such events given that the device is closed.

(3) Switching device closing without the proper command. The modeling parameter is λ_{FC}, which is the rate of occurrence of such events given that the device is open.

Response Functions

(1) Switching device failure to open on command. The modeling parameter is p_s, which is the probability that the device will not open on command.

(2) Switching device failure to close on command. The modeling parameter is p_c, which is the probability that the device will not close on command.

(3) Protection system trips incorrectly due to a fault outside of the protection zone. The modeling parameter is p_o, which is the probability of an incorrect trip, given a fault outside the protection zone.

2.1.11 Expressions for Outage Events. Expressions for computing the frequency, f_{cs}, and the expected durations, r_{cs}, of a cut-set event are summarized in this section. These expressions are generally approximate, but are sufficiently accurate for practical calculations in typical situations. The given expressions presume that all physically parallel paths in a distribution system are fully redundant, that is, it is presumed that any one path of parallel set is fully capable of carrying the highest load that may be experienced. Further, the failure bunching effects of storms and other common-mode or common-cause failures are not considered in the given expressions. These issues are fully described elsewhere (see References [2] and [10]) and are usually not numerically important in industrial and commercial distribution systems whose reliability performance is dominated by series components that yield first-order cut-sets.

2.1.11.1 Forced Outages of Current-Carrying Components. Now we will consider events of cessation of the continuous current-carrying function of any component. The following notations are used:

f_{cs} = Frequency of cut-set event.

r_{cs} = Expected duration of cut-set event = Expected duration of system failure event due to occurrence of the cut-set event.

λ_i = Permanent forced outage rate of component i.

λ'_i = Transient forced outage of component i.

r_i = Expected repair or replacement time of component i.

t = Time to perform an appropriate switching operation.

First, consider cut-sets associated with permanent forced outages.

First-Order (Single-Component) Cut-Sets

$$f_{cs} = \lambda_i \qquad \text{(Eq 3)}$$

$$r_{cs} = \min{(r_i, t)} = \text{Minimum of } r_i \text{ or } t \qquad \text{(Eq 4)}$$

Second-Order (Dual-Component) Cut-Sets

$$f_{cs} = \lambda_i \lambda_j (r_i + r_j) \qquad \text{(Eq 5)}$$

$$r_{cs} = \min{(r_i r_j / (r_i + r_j), t)} \qquad \text{(Eq 6)}$$

NOTE: Equations 5 and 6 are approximate formulas and should only be used when both $\lambda_i \times r_i$ and $\lambda_j \times r_j$ are less than 0.01.

Note that the above expressions for f_{cs} are approximate and assume that λ is much less than $1/r$. This is usually a reasonable assumption, but exact expressions are given in Chapter 8 and should be used if needed. Also note that, particularly in the above expressions for r_{cs}, system interruption durations may be determined by component repair or replacement times or by the time to restore service to interrupted loads through a switching operation. Thus, r_{cs} is very much a function of system topology and switching arrangements. It should also be noted that f_{cs} for second-order cut-sets may, in certain circumstances, also be a function of switching times rather than repair and replacement times. In such cases, the times r_i and r_j should be viewed as the appropriate switching times.

Next, consider cut-sets associated with transient forced outages or transient forced outage events overlapping permanent forced outage events. The likelihood of overlapping transient forced outages is considered remote and is not considered in this book.

First-Order (Single-Component) Cut-Sets

$$f_{cs} = \lambda'_i \qquad \text{(Eq 7)}$$

$$r_{cs} = t \qquad \text{(Eq 8)}$$

Second-Order (Dual-Component) Cut-Sets

$$f_{cs} = \lambda_j \lambda'_i r_j \qquad \text{(Eq 9)}$$

$$r_{cs} = t \qquad \text{(Eq 10)}$$

NOTE: Equation 9 is approximate and should only be used when $\lambda_j \times r_j$ is less than 0.01.

2.1.11.2 Failures of Switching Devices or Protection Systems. Now consider failure events of switching devices or protection systems. The frequency and duration of cut-set events associated with the short-circuit failure mode of switching devices can be calculated using Eqs 3 through 10 as appropriate. Similarly, if a switching device is normally operated closed, the effects of false trip events having a rate of λ_{FT} can be calculated using the approaches of Eqs 3 through 10. The event of switching device closure without proper command is not viewed as generally important from a distribution system reliability point of view (though it certainly is important from a safety viewpoint) and will not be treated in this book.

Switching device or protection system failures that render the device or system unable to respond properly to some other event may occur at the instant of required action or more probably represent undetected prior failures. Such latent failures are only revealed by the event calling for the device or system action. It follows, therefore, that response function failures of switching devices or protection systems never constitute first-order cut-sets since such failures do not in and of themselves result in load interruptions. Expressions for f_{cs} and r_{cs} for each of the response function failure modes follow. In these expressions, λ is the rate of occurrence of the event requiring a response.

Failure to open on command

$$f_{cs} = \lambda p_s \qquad\qquad\qquad\qquad \text{(Eq 11)}$$

$$r_{cs} = r \text{ or } t \text{ as appropriate} \qquad\qquad \text{(Eq 12)}$$

Failure to close on command

$$f_{cs} = \lambda p_c \qquad\qquad\qquad\qquad \text{(Eq 13)}$$

$$r_{cs} = r \text{ or } t \text{ as appropriate} \qquad\qquad \text{(Eq 14)}$$

Incorrect trip due to fault outside protection zone

$$f_{cs} = \lambda p_o \qquad\qquad\qquad\qquad \text{(Eq 15)}$$

$$r_{cs} = r \text{ or } t \text{ as appropriate} \qquad\qquad \text{(Eq 16)}$$

The probabilities p_s and p_c are very much influenced by inspection, maintenance, and testing policies. This follows since p_s and p_c largely reflect undetected prior failures at the time of a required response.

2.1.11.3 Scheduled Outages of Components. We will now consider system interruptions and their related cut-sets, which are associated with scheduled outages of components. A scheduled outage is defined as "an outage that results when a component is deliberately taken out of service at a selected time, usually for purposes of construction, maintenance, or repair." The distinction between a forced outage and a scheduled outage is the degree to which the outage can be postponed; a forced outage cannot be postponed, while a scheduled outage can be postponed, if necessary, to avoid consumer interruptions. Clearly, a scheduled outage of a component that constitutes a first-order cut-set will result in a consumer interruption regardless of the degree to which it can be postponed. However, the timing of such an outage is entirely controllable and can, therefore, be taken at times of minimum inconvenience and with forewarning. Therefore, such interruptions may not have the same impact as interruptions that occur at a random time and without warning. The frequency and duration of first-order cut-sets associated with scheduled outages are:

$$f_{cs} = \lambda''_i \qquad\qquad\qquad\qquad \text{(Eq 17)}$$

$$r_{cs} = r''_i \qquad\qquad\qquad\qquad \text{(Eq 18)}$$

where λ''_i and r''_i are the scheduled outage rate and average scheduled outage duration of the ith component.

In systems possessing redundant supply paths, consumer interruption should never occur due to overlapping scheduled outages of components. However, a component forced outage may overlap a preexisting component scheduled outage, thereby producing a consumer interruption and a second-order cut-set. The frequency and duration of a cut-set in which a forced outage of component j overlaps a scheduled outage of component i are given as follows:

$$f_{\text{cs}} = \lambda''_i \lambda_j r''_i \qquad\qquad\qquad\qquad\qquad\qquad \text{(Eq 19)}$$

$$r_{\text{cs}} = \frac{r''_i r_j}{r''_i + r_j} \quad \text{or } t \text{ as appropriate} \qquad\qquad\qquad \text{(Eq 20)}$$

Again, the assumption is that λ is much greater than $1/r$.

Table 1
Data for Example Systems

Line Sections

$\quad \lambda = 0.20/\text{yr}$
$\quad r = 3 \text{ h}$

Breakers and Switches

$\quad \lambda = 0.01/\text{yr}$
$\quad \lambda_{\text{FT}} = 0.003/\text{yr}$
$\quad p_s = 0.001$
$\quad p_o = 0.01$
$\quad r = 5 \text{ h}$

Switching Times

$\quad t_s =$ Normal manual switching time $= 0.5 \text{ h}$
$\quad t_B =$ Time to isolate breaker or switch or to repair noncatastrophic failure $= 1 \text{ h}$

2.1.12 Example. A simple example will now be used to illustrate the application of the reliability evaluation concepts that have been presented to the evaluation of alternative system protection and sectionalizing schemes. The alternatives to be studied are shown in Figs 1, 2, and 3. More detailed examples using typical data are given in Chapter 7. In these examples, only the labeled line sections and circuit breakers or switches are considered fallible. Furthermore, in the interest of simplifying the example, scheduled outages and transient forced outages of components are not considered. Assumed numerical data for the example systems are shown in Table 1. In every case, the reliability performance indexes desired are the interruption rate and expected duration that would be experienced by a load served from line section L_1.

Here an interruption is defined to be "the loss of continuity from the source to the load point for a time longer than that required for an automatic or remotely controlled switching operation."

The analysis of each system is shown in the tables within Figs 1, 2, and 3. In the analysis, it is assumed that breakers are operated automatically or remotely, while

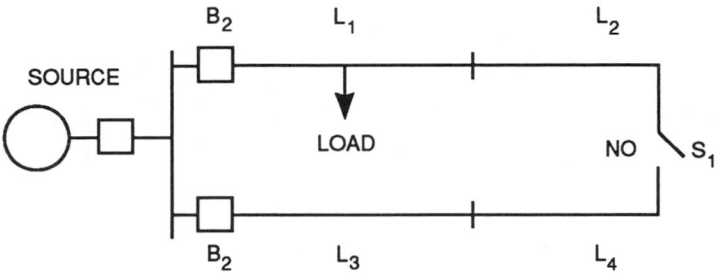

Cut-Set	(1) Frequency (failures/yr)	(2) Duration (h/failure)	(1) × (2)
Line Failures			
L_1	$\lambda = 0.20$	$r = 3$	0.20×3
L_2	$\lambda = 0.20$	$r = 3$	0.20×3
Breaker/Switch Failures			
Type 1: B_1	$\lambda = 0.01$	$t_B = 1$	0.01×1
Type 1: B_2	$\lambda = 0.01$	$t_B = 1$	0.01×1
Type 1: S_1	$\lambda = 0.01$	$t_B = 1$	0.01×1
Type 2: B_1	$\lambda_{FT} = 0.003$	$t_B = 1$	0.003×1
	$\Sigma = 0.433$		$\Sigma = 1.233$

where
$f_s = 0.433$ interruptions/yr
$r_s = 1.233 / 0.433 = 2.85$ h/interruption

Fig 1
Example System — No Line Sectionalizing

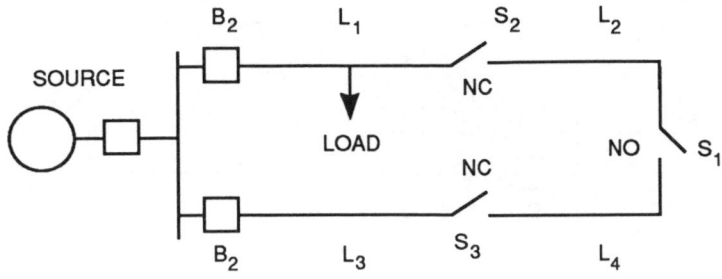

Cut-Set	(1) Frequency (failures/yr)	(2) Duration (h/failure)	$(1) \times (2)$
Line Failures			
L_1	$\lambda = 0.20$	$r = 3$	0.20×3
L_2	$\lambda = 0.20$	$t_s = 0.5$	0.20×0.5
Breaker/Switch Failures			
Type 1: B_1	$\lambda = 0.01$	$t_B = 1$	0.01×1
Type 1: B_2	$\lambda = 0.01$	$t_B = 1$	0.01×1
Type 1: S_1	$\lambda = 0.01$	$t_B = 1$	0.01×1
Type 1: S_2	$\lambda = 0.01$	$t_B = 1$	0.01×1
Type 2: B_1	$\lambda_{FT} = 0.003$	$t_B = 1$	0.003×1
Type 4: B_2	$p_s(\lambda_{L_3} + \lambda_{L_4} + \lambda_{S_3})$	$t_B = 1$	0.00041×1
	$= 0.00041$		$\Sigma = 0.74341$
	$\Sigma = 0.44341$		

where
  $f_s = 0.44341$ interruptions/yr
  $r_s = 0.74341 / 0.44341 = 1.68$ h/interruption

Fig 2
Example System — Lines Sectionalized with Switches

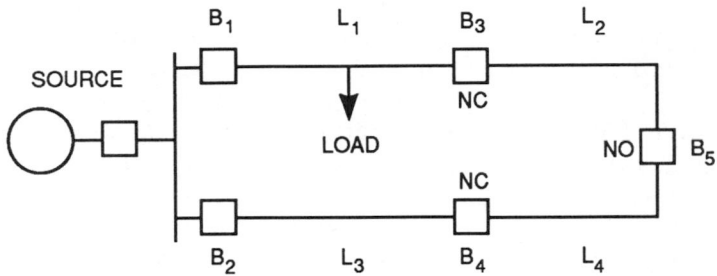

Cut-Set	(1) Frequency (failures/yr)	(2) Duration (h/failure)	$(1) \times (2)$
Line Failures			
L_1	$\lambda = 0.20$	$r = 3$	0.20×3
Breaker/Switch Failures			
Type 1: B_1	$\lambda = 0.01$	$t_B = 1$	0.01×1
Type 1: B_2	$\lambda = 0.01$	$t_B = 1$	0.01×1
Type 1: B_3	$\lambda = 0.01$	$t_B = 1$	0.01×1
Type 4: B_3	$p_s(\lambda_{L_2} + \lambda_{B_5}) = 0.00021$	$t_B = 1$	0.00021×1
Type 4: B_2	$p_s(\lambda_{L_3} + \lambda_{B_4}) = 0.00021$	$t_B = 1$	0.00021×1
Type 6: B_1	$p_o(\lambda_{L_2} + \lambda_{B_5}) = 0.0021$	$t_B = 1$	0.0021×1
	$\Sigma = 0.23252$		$\Sigma = 0.63252$

where
$f_s = 0.23$ interruptions/yr
$r_s = 0.63252 / 0.23252 = 2.72$ h/interruption

Fig 3
Example System — Lines Sectionalized with Circuit Breakers

switches are operated manually. The results of the analyses are in agreement with intuition:

(1) Sectionalizing circuits with noninterrupting devices reduces average interruption duration but has a minimal effect on the interruption rate.

(2) Sectionalizing circuits with fault-interrupting devices cuts the interruption rate.

Note, however, that the average interruption duration of Case 3 is close to that of Case 1 and higher than that of Case 2. This points out that f_s and r_s may not move in the same direction as changes are made in the protection scheme and that the indexes f_s and r_s should be viewed as a complementary pair in reliability analysis.

2.1.13 Incomplete Redundancy. A common method of improving the reliability performance of a system is through component redundancy, for example, more than one transformer in a substation. Typically, each component of the redundant set has sufficient capacity, perhaps based on an emergency rating, to carry the peak load that the system may be asked to deliver. Such full redundancy is effective in improving system reliability performance but is usually quite expensive. If the load of the system is variable, the opportunity exists to cut costs by reducing the capacity of redundant components to levels less than that required to carry system peak load. Such component capacity reductions admit the possibility that one component of the redundant set might be called upon to carry system peak load and would thereby suffer an overload outage. An overload outage might result in an actual interruption of load or perhaps only some loss of life in the overloaded component, depending on the protection scheme in service.

A method exists (see References [1] and [7]) for computing the frequency, average duration, and probability of overload outage events as a function of component capacities and load characteristics. This method, which is compatible with the general reliability evaluation procedure outlined earlier, can be used to evaluate the cost/reliability trade-offs of incomplete redundancy. The method is briefly presented hereafter.

Consider a system possessing incomplete redundancy, and consider the forced outage of some set i of the components of this system. Let the frequency and probability of this forced outage event be f_i and P_i. Then the frequency, probability, and average duration of overloading events that are precipitated by loss of the components in set i are given approximately by:

$$f_{OL_i} = f_i \times P \text{ (load} \geq \text{capacity of remaining components)}$$
$$+ P_i \times f \text{ (load} \geq \text{capacity of remaining components)}$$
$$P_{OL_i} = P_i \times P \text{ (load} \geq \text{capacity of remaining components)}$$
$$D_{OL_i} = P_{OL_i} / f_{OL_i}$$

In the above expressions, $P(\text{load} \geq X)$ is called the "load-duration characteristic" and is simply the probability or proportion of time that the load is greater than or equal to X. A typical load-duration characteristic for utility load is shown in Fig 4. Similarly, $f(\text{load} \geq X)$ is called the "load-frequency characteristic" and is the rate with which events (load $\geq X$) occur. A typical load-frequency characteristic is shown in Fig 5. The reader is referred to Reference [1] for additional discussion of the load-duration and load-frequency characteristics.

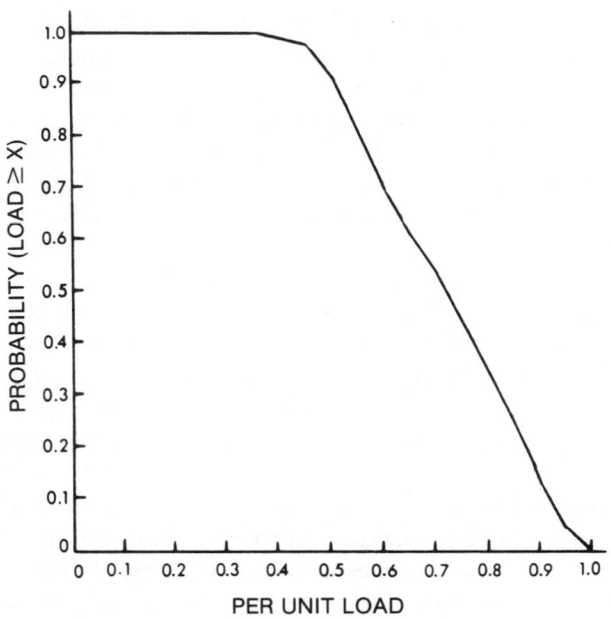

Fig 4
Typical Load-Duration Characteristic

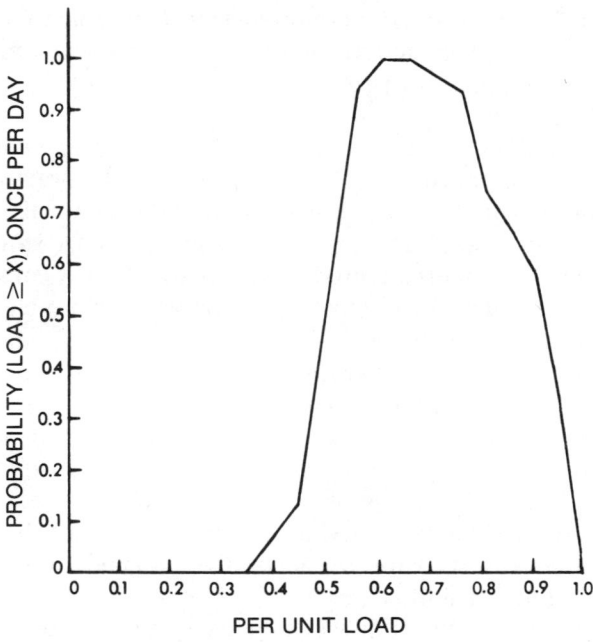

Fig 5
Typical Load-Frequency Characteristic

2.2 Costs of Interruptions — Economic Evaluation of Reliability

2.2.1 Cost of Interruptions versus Capital Cost. The type and extent of new or rehabilitated electric systems for industrial plants or commercial buildings must carefully balance the costs of anticipated interruptions to electrical service against the capital costs of the systems involved. Each instance requires a separate analysis taking into account special production and occupancy needs. Because of the many variables involved, one of the most difficult items to obtain is the cost of the electrical interruptions.

2.2.1.1 What Is an Interruption? Economic evaluation of reliability begins with the establishment of an interruption definition. Such a definition specifies the magnitude of the voltage dip and the minimum duration of such a reduced-voltage period that results in a loss of production or other function for the plant, process, or building in question. Frequently, interruption definitions are given only in terms of a minimum duration and assume that the voltage is zero during that period.

IEEE surveys (see References [8], [18], and [21]) have revealed a wide variation in the minimum or *critical* service loss duration. Table 2 summarizes results for industrial plants, and Table 3 gives results for commercial buildings. It is clear from these tables that careful attention must be paid to choosing the proper interruption definition in any specific reliability evaluation.

Another important consideration in the economic evaluation of reliability is the time required to restart a plant or process following a power interruption. An IEEE survey (see Reference [8] and is also shown in Table 4) indicates that industrial plant restart time following a complete plant shutdown due to a power interruption averages 17.4 hours. The median plant restart time was found to be 4 hours. Clearly, specific data on plant or process restart time should be used if possible in any particular evaluation.

Many industrial plants reported that 1 to 10 cycles were considered critical interruption time, as compared to 1.39 hours, required for start up (plant outage time being considered equal to plant start-up time). This indicates that the critical factor must be carefully explored prior to assigning a cost to the interruption. Fifteen percent of the commercial buildings reported the critical service loss duration time to be 1 second or less, which is probably attributable to the fact that computer installations were involved.

Further data from Reference [8] graphically illustrates the time required to start an industrial plant after an interruption.

Thus, step 1 of the cost analysis becomes the selection of the *critical duration time* of the outage and the *plant start-up time*, including equipment repair or replacement time required because of the interruption.

2.2.1.2 Cost of an Electrical Service Interruption. With the establishment of expected downtime per interruption, costs are assigned to all individual items involved, including but not limited to:

(1) Value of lost production time less expenses saved (expected restart time is used along with the repair or replacement time)
(2) Damaged plant equipment
(3) Spoiled or off-specification product

Table 2
Critical Service Loss Duration for Industrial Plants*
(Maximum length of time an interruption of electrical service will not stop plant production.)

25th Percentile	Median	75th Percentile	Average Plant Outage Time for Equipment Failure Between 1- and 10-Cycle Duration
10 cycles	10 s	15 min	1.39 h

*Fifty-five plants in the United States and Canada reporting; all industry.

Table 3
Critical Service Loss Duration for Commercial Buildings*
(Maximum length of time before an interruption to electrical service is considered critical.)

Service Loss Duration Time							
1 cycle (%)	2 cycles (%)	8 cycles (%)	1 s (%)	5 min (%)	30 min (%)	1 h (%)	12 h (%)
3	6	9	15	36	64	74	100

*Fifty-four buildings reporting; percentage of buildings with critical service loss for duration less than or equal to time indicated.

Table 4
Plant Restart Time*
(After service is restored following a failure that has caused a complete plant shutdown.)

Average (h)	Median (h)
17.4	4.0

*Forty-three plants in the United States and Canada reporting; all industry.

(4) Extra maintenance costs

(5) Cost for repair of failed component

If possible, the cost for each interruption of service should be expressed in a short interruption plus an amount of dollars per hour for the total outage time in order to utilize the reliability data and analysis presented.

2.2.1.3 Economic Evaluation of Reliability. There are many methods of varying degrees of complexity for accomplishing economic evaluations. For quick *order of magnitude* or *Is it worth further investigation?* type of evaluations, cost data from References [8] and [21] can be used. Caution must be exercised, however, since these data are very general in nature, and wide variations are possible in individual cases. Some of the more commonly accepted methods for economic analyses are:

(1) Revenue requirements (RR)
(2) Return on investment (ROI)
(3) Life cycle costing (LCC)

It is not the intent of this chapter to stipulate the method to be used nor the depth to which each analysis is to be made. These are considered to be the prerogative of the engineer and will depend heavily on management choice and the time available for the analysis. The revenue requirements (RR) method is given in this chapter as an example.

2.2.2 "Order of Magnitude" Cost of Interruptions. IEEE surveys (see References [9], [15], and [21]) present general data on the cost of interruptions to industrial plants and commercial buildings in the United States and Canada. Additional cost of interruption data is presented in various IEEE-IAS and IEEE-PES publications. A bibliography of these data is listed in Reference [5]. Other data are listed in References [4], [6], [13], and [19] and are primarily for areas in the middle of Canada and the Province of Ontario. The reader is again cautioned that such general data should be used only for "order of magnitude" evaluations where data specific to the system being studied are not available. A review of the reliability data can probably best be used in adjudicating the type of utility company service that should be provided.

The costs based on the kW interrupted and the kWh not delivered to industrial plants are presented in Tables 5 and 6.

Interruption costs based on kWh not delivered and reflecting the relationship to duration of interruptions for commercial buildings are presented in Tables 7 and 8.

Interruption costs as they are related to interruption time from Table 7 and from Reference [19] are graphically represented in Fig 6. Small industrials are

Table 5
Average Cost of Power Interruptions for Industrial Plants*

All plants	$4.69/kW + $6.65/kWh
Plants > 1000 kW max demand	$2.60/kW + $2.33/kWh
Plants < 1000 kW max demand	$11.38/kW + $20.11/kWh

*Forty-one plants in the United States and Canada reporting; published in 1973 with costs updated to July 1987.

Table 6
Median Cost of Power Interruptions for Industrial Plants*

All plants	$1.71/kW + $2.06/kWh
Plants > 1000 kW max demand	$0.79/kW + $0.89/kWh
Plants < 1000 kW max demand	$9.13/kW + $10.96/kWh

*Forty-one plants in the United States and Canada reporting; published in 1973 with costs updated to July 1987.

Table 7
Average Cost of Power Interruptions for Commercial Buildings

All Commercial Buildings*	$14.65/kWh not delivered
Office Buildings Only	$17.99/kWh not delivered

*Fifty-four buildings in the United States reporting; published in 1975 with costs updated to July 1987.

Table 8
Cost of Power Interruptions as a Function of Duration
for Office Buildings (with Computers)*

Power Interruptions	Sample Size	Maximum	Cost/Peak kWh Not Delivered Minimum	Average
15 min Duration	14	$45.11	$3.82	$18.05
1 h Duration	16	$50.61	$3.82	$16.85
Duration > 1 h	10	$137.35	$0.32	$19.91

*Published in 1975 with costs updated to July 1987.

Fig 6
Cost of Interruption versus Duration

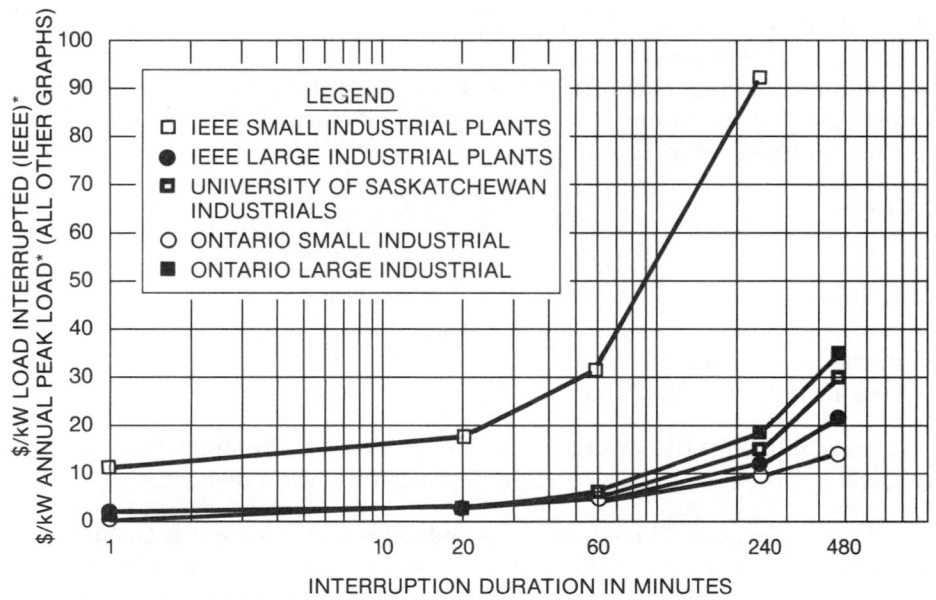

*All costs are updated to July 1987 and are in U.S. dollars.

considered to be those with a maximum demand of less than 1000 kW and large industrials are considered to be those with a demand of greater than 1000 kW.

2.2.3 Economic Analysis of Reliability in Electrical Systems. There are several acceptable methods for accomplishing an economic analysis of the reliability in electric systems. The examples of reliability analysis included in this chapter and Chapter 7 utilize the revenue requirements (RR) method. The application of this method as it applies to the analyses of the reliability in industrial plant electrical systems was presented in part 6 of Reference [9]. Applicable excerpts from this reference are included herein.

2.2.3.1 The Revenue Requirements (RR) Method. Although there are many ways in use to compare alternatives, some of these have defects and weaknesses, especially when comparing design alternatives in contrast to overall projects. The RR method is "mathematically rigorous and quantitatively correct to the extent permitted by accuracy with which items of cost can be forecast" (see References [9] and [18]).

The essence of the RR method is that, for each alternative plan being considered, the minimum revenue requirements (MRR) are determined. This means that we find out how much product we must sell to achieve minimum acceptable earnings on the investment involved plus all expenses associated with that investment. These minimum revenue requirements for alternative plans may be compared directly. The plan having the lowest MRR is the economic choice.

Minimum revenue requirements are made up of and equal to

(1) Variable operating expenses
(2) Minimum acceptable earnings
(3) Depreciation
(4) Income taxes
(5) Fixed operating expenses

These minimum revenue requirements may be separated into two main parts, one proportional and the other not proportional to investment in the alternative. This may be expressed in an equation

$$G = X + CF \qquad \text{(Eq 21)}$$

where

G = Minimum revenue requirements to achieve minimum acceptable earnings.
X = Nonfixed or variable operating expenses.
C = Capital investment.
F = Fixed investment charge factor.

The last term in Eq 21, the product of C and F includes the items (2), (3), (4), and (5) listed in the preceding paragraph. Equation 21 will now be discussed.

X (Variable Expenses) — The effect of the failure of a component is to cause an increase in variable expenses. How serious this increase is depends to a great extent on the location of the component in the system and on the type of power distribution system employed. The quality of a component as installed can have a significant effect on the number of failures experienced. A poor quality component installed with poor workmanship and with poor application engineering may greatly increase the number of failures that occur as compared with a high quality component installed with excellent workmanship and sound application engineering.

When a failure does occur, variable expenses are increased in two ways. In the first way, the increase is the result of the failure itself. In the second way, the increase is proportional to the duration of the failure.

Considering the first way, the increased expense due to the failure includes the following:

(1) Damaged plant equipment
(2) Spoiled or off-specification product
(3) Extra maintenance costs
(4) Costs for repair of the failed component

Considering the second way, plant downtime resulting from failures is made up of the time required to restart the plant, if necessary, plus the time to

(1) Effect repairs, if it is a radial system, or
(2) Effect a transfer from the source on which the failure occurred to an energized source

During plant downtime, production is lost. This lost production is not available for sale, so revenues are lost. However, during plant downtime, some expenses may be saved, such as expenses for material, labor, power, and fuel costs. Therefore, the value of the lost production is the revenues lost, because production stopped less the expenses saved. Some of the variable expenses may vary depending on the duration of plant downtime. For example, if plant downtime is only 1 hour, perhaps no labor costs are saved. But, if plant downtime exceeds 8 hours, labor costs may be saved.

If we assume that the value/hour of variable expenses does not vary with the duration of plant downtime, then the value of lost production can be expressed on a per hour basis, and the total value of lost production is the product of plant downtime in hours and the value of lost production per hour.

It should be noted that both the value of lost production and expenses incurred are proportional to the failure rate. The total effect on variable expenses, if the value of lost production is a constant on a per hour basis, may be expressed in the following equation:

$$X = \lambda \left[x_i + (g_p - x_p)(r + s) \right] \tag{Eq 22}$$

where
X = Variable expenses ($ per year).
λ = Failures per year or failure rate.
x_i = Extra expenses incurred per failure ($ per failure).
g_p = Revenues lost per hour of plant downtime ($ per hour).
x_p = Variable expenses saved per hour of plant downtime ($ per hour).
r = Repair or replacement time after a failure (or transfer time if not radial system), in hours.
s = Plant start-up time after a failure, in hours.

Assume that

λ = 0.1 failure per year.
x_i = \$40 000 per failure, extra expenses incurred.
g_p = \$16 000 per hour, revenues lost.
x_p = \$12 000 per hour, expenses saved.
r = Ten hours per failure.
s = Twenty hours per failure.

Then, variable expenses affected would be

$$X = (0.1)\,[\$40\,000 + (\$16\,000 - \$12\,000)(10 + 20)] = \$16\,000 \text{ per year}$$

The term g_p represents revenues lost, and it is not really an expense. However, it is a negative revenue, and as such has the same effect on the economics as a positive expense item. It is convenient to treat it as though it were an expense.

A failure rate of 0.1 failure per year is equivalent to a mean time between failures of 10 years. Since we are dealing with probability, this is what we can expect, but in a specific case, we might have two failures in one 10 year period and no failures in another 10 year period. But considering many similar cases, we expect to have an average of 0.1 failure per year, with each failure costing an average of \$160 000. This gives an equal average amount per year in the above example of \$16 000.

The point is that even though the actual failures cost \$160 000 each and occur once every 10 years, a given failure is just as likely to occur in any of the 10 years. The equivalent equal annual amount of \$16 000 per year is the average value of one failure in 10 years.

C (Investment) — Each different alternative in an industrial plant power distribution system involves different investments. The system requiring the least investment will usually be some form of radial system. By varying the type of construction and the quality of the components in the system, the investment in radial systems can vary widely.

The best method is to find one total investment in each alternative plan. Another common method is to find the incremental investment in all alternatives over a base or least expensive plan. The main reason that the total investment method is preferable is that, in comparing alternatives, the investment is multiplied by an F factor (which will be explained later). This factor is usually the same for alternative plans of the sort being considered here, but this is not necessarily the case.

Using the incremental investment may thus introduce a slight error into the economic comparisons.

F (Investment Charge Factor) — This discussion of the investment charge factor is taken from Reference [9].

The F factor includes the following items, which are constant in relation to the investment:

(1) Minimum acceptable rate of return on investment, allowing for risk
(2) Income taxes
(3) Depreciation
(4) Fixed expenses

An equation to calculate the F factor is:

$$F = \left[\frac{(S_c a_L / f_r) - t \bar{d}_t}{1 - t} \right] + e \qquad \text{(Eq 23)}$$

This may also take the following form:

$$F = \bar{r} + \bar{d} + \bar{t} + e \qquad \text{(Eq 24)}$$

where

a_n = $R + d_n$, amortization factor or leveling factor.
d_n = $R / (S_n - 1)$, sinking fund factor.
S_n = $(1 + R)^n$, growth factor or future value factor.
n = Period of years, such as c or L.
c = Years prior to start up that an investment is made.
L = Life of investment years.
R = Minimum acceptable earnings per \$ of C (investment).
f_r = Probability of success or risk adjustment factor.
t = Income taxes per \$ of C (investment).
$\bar{d}_t$ = Income tax depreciation, levelized per \$ of C (investment) = $1 / L$, $\bar{d}_t = 1L$.
e = Fixed expenses per \$ of C (investment).
$\bar{r}$ = Levelized return on investment per \$ of C (investment).
$\bar{d}$ = Levelized depreciation on investment per \$ of C (investment).
$\bar{t}$ = Levelized income taxes on investment per \$ of C (investment).

Assume

L = Twenty years (life of investment).
c = One year.
R = 0.15 (minimum acceptable rate of return).
f_r = 1 (risk adjustment factor).
t = 0.5 (income tax rate).
$\bar{d}_t$ = $1 / L$ = 0.05.
e = 0.0825.

then

S_c = $(1 + R)^c = (1 + 0.15)^1 = 1.15$
S_L = $(1 + R)^L = (1 + 0.15)^{20} = 16.37$
d_L = $(R / S_L - 1) = 0.15 / (16.37 - 1) = 0.0098$
a_L = $(R + d_L) = 0.15 + 0.0098 = 0.1598$

Substituting into Eq 23 to calculate the F factor, we get

$$F = \left\{ \left[\frac{\left(\frac{(1.15)(0.1598)}{1.0} \right) - (0.5)(0.05)}{1 - 0.5} \right] \right\} + 0.0825 = 0.04 \qquad \text{(Eq 25)}$$

All the assumed values are believed to be typical for the average electrical distribution system, except the value of e = 0.0825. This latter value was arbitrarily assumed in order to make R round out to 0.4. The term e covers such items as insurance, property taxes, and fixed maintenance costs. A typical value for e is probably less than 0.0825.

It is believed that a typical value for minimum acceptable return on investment in many industrial plants is 15%, that is, $R = 0.15$. The company's average rate of return, based on either past history or anticipated results, is a measure of what R should be. In plants of higher risk than the average, the risk adjustment factor, f_r, should probably be less than 1. However, company management determines what the value of R should be.

The value of F can be calculated from Eq 23. In Reference [8], tabular values are given for the factors S_n and a_n for various rates of return and plant lifes.

2.2.3.2 Steps for Economic Comparisons

(1) Prepare single-line diagrams of alternative plans and assign failure rates, repair times, and investment in each component, and then determine the total investment C in each plan.
(2) Determine X, the increased variable expense for each plan as the sum of the value of lost production and the extra variable expenses incurred.
(3) Determine F, the fixed investment charge factor F from Eq 23.
(4) Calculate $G = X + CF$, the minimum revenue requirements G of each plan from Eq 21.
(5) Select as the economic choice the plan having the lowest value of G.

2.2.3.3 Conclusions. A technique has been presented for the economic evaluation of power system reliability. The method of determining the failure rates and repair times of different alternatives is not covered here. Additional information relative to the RR method is included in Reference [18].

2.2.4 Examples. Examples of electrical systems with varying degrees of reliability (availability), together with fixed and variable costs are given in Chapter 7.

2.2.5 Worth of Improved Reliability in Electrical Components. All of the data and examples presented in this chapter utilize failure rates and average repair time data for standard electrical components. Unfortunately, industry and commercial standards for recording failure history are very unsophisticated and do not allow differentiation between various grades of equipment or between different manufacturers.

A summary of several studies on the worth of improved reliability and of reduced maintenance costs for high-voltage circuit breakers is given in Reference [14], which is included as Appendix J in this book. These studies were made by a working group in CIGRE (Conference Internationale des Grand Reseaux Electriques a² Haute Tension), which is a technical arm of the International Electrotechnical Commission (IEC). In addition, this CIGRE working group has made a worldwide survey that collected and published all of the necessary reliability data and maintenance cost data that are needed in order to make studies on the worth of improved reliability and reduced maintenance costs of high-voltage circuit breakers. A summary of this data is given in Reference [14].

2.3 Cost of Scheduled Electrical Preventive Maintenance. In the economic evaluation of reliability, it is always appropriate to consider the costs of scheduled electrical preventive maintenance. Sometimes these costs are large enough to make it desirable to analyze them separately when comparing alternative designs of industrial power systems. The revenue requirements (RR) method described in 2.2.3.1 includes a term called the "investment charge factor (F)" which is given by Eq 23 in 2.2.3.1 and includes e (the fixed yearly expenses) as a percentage of the capital investment. Both F and e are attributed to scheduled electrical preventive maintenance, insurance, property taxes, etc. Since the yearly average costs for scheduled electrical preventive maintenance may not be the same percentage of investment for every component within the industrial power system, a separate more detailed look is often taken at these costs for each component.

Scheduled electrical preventive maintenance has two major cost elements: labor effort and spare parts consumed. These costs are often expressed on an average yearly basis so as to be usable with the RR method when an economic evaluation is made. These data are needed for each different type of component used in the industrial power system and can be compiled for each component as follows:

(1) Labor costs in manhours per component per year
(2) Cost of spare parts consumed in dollars per component per year
(3) Labor rate in dollars per manhour

If, for example, a component is only maintained once every 3 years, then its maintenance costs should be divided by 3 in order to determine the average yearly maintenance cost. The labor rate used should probably only include the overhead costs associated with the storage of spare parts, direct supervision of maintenance, and the costs for the necessary test equipment. The labor costs in dollars per component per year can be calculated by multiplying items (1) and (3) together; the result can then be added to item (2) to get the total average yearly costs that are attributable to scheduled electrical preventive maintenance.

Data thus collected can become obsolete at a later date due to inflation that can result in changing the labor rate used and also the average yearly cost of spare parts consumed. But the data for labor in manhours per component per year does not become obsolete due to inflation. Some engineers have chosen to use their labor rate to convert their average yearly cost data for spare parts consumed into average yearly "equivalent manhours" data. This is then added to the labor manhours data to get total equivalent manhours per component per year that includes both the labor cost and the cost of spare parts consumed. The use of equivalent manhours for cost data instead of dollars has two advantages

(1) The equivalent manhours data do not become obsolete due to inflation.
(2) The equivalent manhours data can be considered an international currency. The data are not affected by changing exchange rates between different currencies or countries. This enables the cost data to be easily compared with studies from other countries.

Component data on the cost of scheduled electrical preventive maintenance are not included in this book except for the data on high-voltage circuit breakers above 63 kV collected by a CIGRE working group (see Reference [14]), which is included in this book as Appendix J. It would be desirable to have such data for all of the electrical equipment categories listed in Table 10 of Chapter 3. It would then be possible to consider the cost of scheduled electrical preventive maintenance in design decisions of the industrial power system by adding this into the MRR method.

2.4 Effect of Scheduled Electrical Preventive Maintenance on Failure Rate. One of the important total operating cost decisions made by the management of an industrial plant is how much money to spend for scheduled electrical preventive maintenance. The amount of maintenance performed on a component can affect its failure rate. Very little quantitative data have been collected and published on this subject. Yet this is an important factor when attempting to study the total owning costs of a complete power system. If the maintenance effort is reduced, the maintenance costs go down. This may increase the failure rate of the components in the power system and raise the costs associated with failures. There is an optimum amount of maintenance for minimum total owning cost of a complete power system.

The subject of electrical preventive maintenance is discussed in Chapter 5. Some data are shown in Tables 43 and 44 on the effect of the frequency and quality of scheduled electrical preventive maintenance. These data have been used to calculate the effect of maintenance quality on the failure rate of transformers, circuit breakers, and motors shown in Table 45. Unfortunately, the data do not relate the amount or cost of component maintenance to the failure rate.

The effect of the cost of component scheduled electrical preventive maintenance on failure rate has not been included in this book. More industry studies and published data are needed on this subject, like the example described next.

2.4.1 Example. A paper containing quantitative data and analysis of optimum maintenance intervals has been published (see Reference [23]). This work was based upon 10 000 failures collected at the author's company over a period of 7 years for 23 categories of electrical equipment. Included in this paper was a description of just what failures could be prevented by maintenance. Actual data were used to determine how this failure rate varied with the maintenance interval. The optimum maintenance interval was then determined based upon the maintenance cost and the cost of failures/power outages. Failures that could be prevented by diagnostic testing were then studied in a similar manner to those that could be prevented by maintenance. The optimum diagnostic interval was then calculated for 15 equipment categories based upon the cost of diagnostic testing and the cost of failures/power outages.

It was reported that 25% of the failures could have been prevented by maintenance, and additional failures could have been prevented by diagnostic testing.

2.5 References. The information in this chapter shall be used in conjunction with the following publications:

NOTE: References [16], [17], [15], and [21], respectively, are reprinted in Appendixes A, B, C, and D. References [5] and [19] are reprinted in Appendix M. Reference [14] is reprinted in Appendix J.

[1] Ayoub, A. K. and Patton, A. D. "A Frequency and Duration Method for Generating System Reliability Evaluation," *IEEE Transactions on Power Apparatus and Systems*, Nov./Dec. 1976, pp. 1929–1933.

[2] Billinton, R. and Allan, R. N. "Reliability Evaluation of Power Systems," Plenum Publishing Corp., 1983.

[3] Billinton, R. and Grover, M. S. "A Sequential Method for Reliability Analysis of Distribution and Transmission Systems," *Proceedings of the 1975 Annual Reliability and Maintainability Symposium*, Jan. 1975, pp. 460–469.

[4] Billinton, R. and Wacker, G. "Cost of Electrical Service Interruptions to Industrial and Commercial Consumers," *IEEE IAS Conference Record*, Oct. 7–11, 1985.

[5] Billinton, R., Wacker, G., and Wojczynski, E. "Comprehensive Bibliography on Electrical Service Interruption Costs," *IEEE Transactions on Power Apparatus and Systems*, vol. PAS-102, no. 6, Jun. 1983, pp. 1831–1837.

[6] Billinton, R., Wacker, G., and Wojczynski, E. "Interruption Cost Methodology and Results — A Canadian Commercial and Small Industry Survey," *IEEE Transactions on Power Apparatus and Systems*, vol. PAS-103, no. 2, Feb. 1984, pp. 437–443.

[7] Christiaanse, W. R. "Reliability Calculations including the Effects of Overloads and Maintenance," *IEEE Transactions on Power Apparatus and Systems*, Jul./Aug. 1971, pp 1664-1676.

[8] Dickinson, W. H. "Economic Evaluation of Industrial Power Systems Reliability," *Transactions of the AIEE (Industry Applications)*, vol. 76, Nov. 1957, pp. 264–272.

[9] Dickinson, W. H. et al. "Fundamentals of Reliability Techniques as Applied to Industrial Power Systems," *Conference Record of the 1971 IEEE I&CPS Technical Conference*, pp. 10–31.

[10] Endrenyi, J. "Reliability Modeling in Electric Power Systems," John Wiley & Sons, 1978.

[11] Endrenyi, J., Maenhaut, P. C., and Payne, L. C. "Reliability Evaluation of Transmission Systems with Switching After Faults — Approximations and a Computer Program," *IEEE Transactions on Power Apparatus and Systems*, Nov./Dec. 1973, pp. 1863–1875.

[12] Gaver, D. P., Montmeat, F. E., and Patton, A. D. "Power System Reliability, I — Measures of Reliability and Methods of Calculation," *IEEE Transactions on Power Apparatus and Systems*, Jul. 1964, pp. 727-737.

[13] Goushleff, D. C. "Use of Interruption Costs in Regional Supply Planning," Ontario Hydro Research Division, presented at the IEEE IAS Conference, Oct. 7–11, 1985.

[14] Heising, C. R. "Summary of CIGRE 13-06 Working Group Worldwide Reliability Data, Maintenance Cost Data, and Studies on the Worth of Improved Reliability of High Voltage Circuit Breakers," *IEEE I&CPS Conference Record*, May 5–8, 1986.

[15] IEEE Committee Report. "Reliability of Electric Utility Supplies to Industrial Plants," *Conference Record of the 1975 IEEE I&CPS Technical Conference*, May 5–8, 1975, pp. 131–133.

[16] IEEE Committee Report. "Report on Reliability Survey of Industrial Plants," *IEEE Transactions on Industry Applications*, Mar./Apr. 1974, pp. 213–235.

[17] IEEE Committee Report. "Report on Reliability Survey of Industrial Plants," *IEEE Transactions on Industry Applications*, Jul./Aug. 1975, pp. 456–476; Sep./Oct. 1975, p. 681.

[18] Jeynes, P. H. and Van Nemwegen, L. "The Criterion of Economic Choice," *Transactions of the AIEE (Power Apparatus and Systems)*, vol. 77, Aug. 1958, pp. 606–635.

[19] Koval, D. O. and Billinton, R. "Statistical and Analytical Evaluation of the Duration and Cost of Consumers' Interruptions," Paper no. A79-057-1, IEEE 1979 Winter Power Meeting, New York City, NY.

[20] Patton, A. D. and Ayoub, A. K. "Reliability Evaluation, Systems Engineering for Power: Status and Prospects," U.S. Energy Research and Development Administration, publication CONF-750867 1975, pp. 275-289.

[21] Patton, A. D. et al. "Cost of Electrical Interruptions in Commercial Buildings," *Conference Record of the 1975 IEEE I&CPS Technical Conference*, May 5–8, 1975, pp. 123–129.

[22] Ringlee, R. J. and Goode, S. D. "On Procedures for Reliability Evaluation of Transmission Systems," *IEEE Transactions on Power Apparatus and Systems*, Apr. 1970, pp. 527–537.

[23] Sheliga, D. J. "Calculation of Optimum Preventive Maintenance Intervals for Electrical Equipment," *IEEE Transactions on Industry Applications*, Sep./Oct. 1981, pp. 490–495.

Chapter 3
Summary of Equipment Reliability Data

3.1 Introduction. This chapter summarizes the reliability data collected from electrical equipment reliability surveys over a period of 30 years. The chapter is divided into two parts consisting of recent equipment surveys conducted between 1976 and 1989 (i.e., Part 1) and equipment surveys conducted prior to 1976 (i.e., Part 2). Detailed reports on the surveys are given in the appendixes and references. The results of these surveys are discussed and compared. Detailed information contained in the other chapters of this book and pertinent to equipment reliability data is referenced in this chapter. Detailed lists of references on equipment reliability are presented in the appendixes and at the end of this chapter.

A knowledge of the reliability of electrical equipment is an important consideration in the design and operation of industrial and commercial power distribution systems. The failure characteristics of individual pieces of electrical equipment (i.e., components) can be partially described by the following basic reliability statistics:

(1) Failure rate, often expressed as failures per year per component (failures per unit-year)

(2) Downtime to repair or replace a component after it has failed in service, expressed in hours (or minutes) per failure

(3) In some special cases, probability of starting (or operating) is used

Reliability data on the pertinent factors (e.g., cause and type of failures, maintenance procedures, repair method, etc.) is also required to practically characterize the performance of electrical equipment in service. (Refer to Appendixes A and B.)

The reliability performance of industrial and commercial electrical power distribution systems (e.g., economic operation, frequency and duration of equipment and system outages, etc.) can be estimated from a knowledge of the reliability data of individual electrical parts (i.e., components) that are interconnected to form an operating system. The analytical models required for estimating the reliability of various power system configurations are presented in Chapters 2 and 8. Based on the results of these analytical models, the cost of interruptions can be estimated and used in the reliability cost/reliability worth methodology presented in Chapter 7 and Appendix C. The cost of power interruptions to industrial plants and commercial buildings is summarized in Chapter 2.

Electrical equipment reliability data is normally obtained from field surveys of individual industrial and commercial equipment failure reports. The reason for conducting a survey is to provide answers to critical questions pertaining to the failure characteristics of electrical equipment in industrial and commercial installations.

Each survey has a defined objective of obtaining field data on electrical equipment failure characteristics and this determines the form of the questionnaires that are sent to various respondents.

An analysis of the survey returns may or may not provide answers to all the questions posed in the questionnaire. The significance of the survey data obtained is dependent upon many factors, for example, the number of equipment failures reported, their operating history, the survey questionnaire, etc. There will undoubtedly be new questions raised and also some old questions and controversies left unresolved. Items found to be of little significance will be omitted and the survey form simplified to maximize the response for the next survey. The procedure for conducting the survey is given in Appendix F. Information on the determination and analysis of data for reliability studies is presented in Reference [16].[2]

The IEEE Industry Applications Society (IAS) has a continuing program to conduct surveys on the reliability of electrical equipment in industrial and commercial installations (see References [6], [9] - [13], and [15]). The most significant results from these surveys are then summarized for inclusion in a future revision of this standard.

As in previous survey reports, this chapter maintains the standard for credibility of failure rates by identifying categories that contain an insufficient number of failures. If there were less than eight failures, an * is used to indicate a small sample size. It is believed that a minimum of eight field failures is necessary to have a reasonable chance of estimating the failure rate or the average downtime per failure to within a factor of 2 (see Appendix A, Part 1 for details). Both the average downtime per failure data and median downtime per failure data are given so that the effect of a few very long outages on the average downtime can be indicated by a large difference between the average and median values.

An equipment reliability reference guide is shown in Table 9. For each electrical component presented in this chapter, the tables and appendixes that contain reliability data pertinent to that component are presented. Table 10 contains a summary of the failure rate and average and median downtime per failure data for all electrical equipment surveyed. These values are suggested for use in the absence of better data not being available from the reader's own experience.

---

[2]The numbers in brackets correspond to those in the references at the end of this chapter.

Table 9
Equipment Reliability Reference Guide

Electrical Equipment		Reference Tables in Chapter 3								Appendixes
		Part 1	Part 2							
		Surveys 1976–89	Surveys Prior to 1976							
Motors	> 50 hp	32	—	—	—	—	—	41	—	A, B, H
	> 200 hp	23–31	—	—	—	—	—	41	—	
	> 250 hp	33	—	—	—	—	—	41	—	
Motor Starters		—	—	36	37	38	39	41	42	A, B
Generators		12	—	—	—	—	—	41	—	A, B, L
Transformers	Power	13 15 16 17 / 18 19 20 21	—	—	—	—	—	41	42	A, B, G
	Rectifier	14 16 17 18 / 19 20 22 —	—	—	—	—	—	41	42	
Circuit Breakers		—	35	—	37	38	39	41	42	A, B, J, K
Disconnect Switches		—	—	36	37	38	39	41	42	A, B
Bus Duct		—	—	36	37	38	39	41	—	A, B
Switchgear	Bus Insulated	11	—	—	—	—	—	41	—	A, B, E
	Bus Bare	11	—	—	—	—	—	41	—	
Open Wire		—	—	36	37	38	39	41	42	A, B
Cable		—	—	36	37	38	39	41	42	A, B, I
Cable Joints		—	—	36	37	38	39	41	42	
Cable Terminations		—	—	36	37	38	39	41	42	
Electric Utility Power Supplies		—	—	—	—	—	40	41	—	A, B, D

Table 10
Summary of Optional Failure Rate and Average and Median Downtime
per Failure for All Electrical Equipment Surveyed

Equipment	Equipment Subclass	Failure Rate (Failures per Unit-Year)	Actual Hours of Downtime per Failure	
			Industry Average	Median Plant Average
Transformers	Liquid Filled — All	0.0062	356.1[1]	—
	300 – 10 000 kVA	0.0059	297.4[1]	—
	10 000+ kVA	0.0153	1178.5[1]	—
Rectifier Transformers	Liquid Filled 300 – 10 000 kVA	0.0153	1664.0[1]	—
Motors > 200 hp[6]	Induction			
	0 – 1000 V	0.0824	42.5	15.0
	1001 – 5000 V	0.0714	75.1	12.0
	Synchronous			
	1001 – 5000 V	0.0762	78.9	16.0
Circuit Breakers[3]	Fixed (Including Molded Case) — All	0.0052	5.8	4.0
	0 – 600 V — All Sizes	0.0042	4.7	4.0
	0 – 600 A	0.0035	2.2	1.0
	Above 600 A	0.0096	9.6	8.0
	Above 600 V[3]	0.0176	10.6	3.8
	Metalclad Drawout Type — All	0.0030	129.0	7.6
	0 – 600 V — All Sizes	0.0027	147.0[2]	4.0
	0 – 600 A	0.0023	3.2	1.0
	Above 600 A	0.0030	232.0	5.0
	Above 600 V[3]	0.0036	109.0[2]	168.0
Motor Starters	Contact Type: 0 – 600 V	0.0139	65.1	24.5
	Contact Type: 601 – 15 000 V	0.0153	284.0	16.0
Generators	Continuous Service			
	Steam Turbine Driven	0.1691	32.7	—
	Emergency and Standby Units			
	Reciprocating Engine Driven			
	Rate per Hour in Use (0.00536)		478.0	—
	Failures per Start Attempt (0.0135)			
Disconnect Switches	Enclosed	0.006100	1.6	2.8
Switchgear Bus — Indoor and Outdoor[4]	Insulated: 601 – 15 000 V	0.001129	261.0	28.0
	Bare: 0 – 600 V	0.000802	550.0	27.0
	Bare: Above 600 V	0.001917	17.3	36.0
Bus Duct — Indoor and Outdoor (Unit = 1 Circuit Ft)	All Voltages	0.000125	128.0	9.5
Open Wire (Unit = 1000 Circuit Ft)	0 – 15 000 V	0.01890	42.5	4.0
	Above 15 000 V	0.00750	17.5	12.0

Table 10 *(continued)*

Equipment	Equipment Subclass	Failure Rate (Failures per Unit-Year)	Actual Hours of Downtime per Failure	
			Industry Average	Median Plant Average
Cable — All Types of Insulation (Unit = 1000 Circuit Ft)[5]	Above Ground and Aerial			
	0–600 V	0.00141	457.0	10.5
	601–15 000 V — All	0.01410	40.4[2]	6.9
	In Trays Above Ground	0.00923	8.9	8.0
	In Conduit Above Ground	0.04918	140.0	47.5
	Aerial Cable	0.01437	31.6	5.3
	Below Ground and Direct Burial			
	0–600 V	0.00388	15.0	24.0
	601–15 000 V — All	0.00617	95.5[1]	35.0
	In Duct or Conduit	0.00613	96.8	35.0
	Above 15 000 V	0.00336	16.0	16.0
Cable (Unit = 1000 Circuit Ft)	601–15 000 V			
	Thermoplastic	0.00387	44.5	10.0
	Thermosetting	0.00889	168.0	26.0
	Paper Insulated Lead Covered	0.00912	48.9	26.8
	Other	0.01832	16.1	28.5
Cable Joints — All Types of Insulation	601–15 000 V			
	In Duct or Conduit Below Ground	0.000864	36.1	31.2
Cable Joints[5]	601–15 000 V			
	Thermoplastic	0.000754	15.8	8.0
	Paper Insulated Lead Covered	0.001037	31.4	28.0
Cable Terminations[5] — All Types of Insulation	Above Ground and Aerial			
	0–600 V	0.000127	3.8	4.0
	601–15 000 V — All	0.000879	198.0	11.1
	Aerial Cable	0.001848	48.5	11.3
	In Trays Above Ground	0.000333	8.0	9.0
	In Duct or Conduit Below Ground			
	601–15 000 V	0.000303	25.0	23.4
Cable Terminations	601–15 000 V			
	Thermoplastic	0.004192	10.6	11.5
	Thermosetting	0.000307	451.0	11.3
	Paper Insulated Lead Covered	0.000781	68.8	29.2
Miscellaneous	Inverters	1.254000	107.0	185.0
	Rectifiers	0.038000	39.0	52.2

NOTES: (1) See Tables 13 and 14 in this chapter for data comparing replacement time with average repair time of transformers.

(2) See Tables 50, 51, 55, and 56 in Appendix B for results on a special study on the effects of failure repair method and failure repair urgency on the average hours downtime per failure.

(3) See Appendix J for circuit breakers above 63 kV from a CIGRE 13-06 worldwide survey. See Appendix K for a later small IEEE survey.

(4) Units = the number of connected circuit breakers and connected switches.

(5) See Appendix I for utility industry data on underground cable, terminations, and splices.

(6) See Table 32 for motors > 50 hp.

3.2 Part 1: Recent Equipment Reliability Surveys (1976–89)

3.2.1 1979 Switchgear Bus Reliability Data. The reliability of switchgear bus in industrial and commercial applications was investigated in a 1979 survey [11] (see Appendix E), and the summarized failure rate and median outage duration time for the various subcategories of equipment are shown in Table 11. In this survey, the term "units" for a bus is defined as the total number of connected circuit breakers and connected switches. In the previous survey in 1974, the term "units" included the total number of connected circuit breakers or instrument transformer compartments. The total number of plants in the 1979 survey response was considerably greater than the 1974 survey; however, the unit-year sample size was slightly less.

Table 11
Switchgear Bus—Indoor and Outdoor
1979 Survey Data

Industry	Equipment Subclass	Failure Rate (Failures per Unit-Year)	Median Hours Downtime per Failure
All	All	0.001050	28
All	Insulated, Above 600 V	0.001129 (0.001700)	28 (26.8)
All	Bare, All Voltages	0.000977	28
All	Bare, 0–600 V	0.000802 (0.000340)	27 (24.0)
All	Bare, Above 600 V	0.001917 (0.000630)	36 (13.0)
Petroleum/Chemical	Insulated, Above 600 V	0.002020	40
Petroleum/Chemical	Bare, All Voltages	0.002570	28
Petroleum/Chemical	Bare, 0–600 V	0.002761	22
Petroleum/Chemical	Bare, Above 600 V	*	48

NOTES: *Small sample size, less than eight failures.
  (1) Number in parentheses = The result from the 1974 survey.

The 1974 survey generated some controversy concerning bare and insulated buses. As can be seen from Table 11, insulated bus equipment showed a significantly higher failure rate than bare bus above 600 V. An analysis of the 1974 data base revealed that the majority of the data collected came from the petroleum/chemical industry. In the 1979 survey, the petroleum/chemical industry data was separated from the remaining industrial data base and indicated that the number of reported failures in each category was dominated by the petroleum/chemical industry. The bare bus failure was significantly higher and the insulated bus failure rate lower in the 1979 survey than in the 1974 survey.

A comparison of the median downtime per failure in both surveys revealed no significant differences. It is important to emphasize that the duration of an outage is dependent on many factors, and, without supplementary information on the operating procedures, maintenance type, spare parts inventory, etc., the data in these surveys should be viewed as general information.

Some important additional observations based on the 1979 survey are:
(1) Newer bus appears to experience a higher failure rate than older bus. This may be partly explained by improper installation, type of construction of new switchgear, etc., but is not completely consistent with the observation that failure rates are highly dependent on maintenance.
(2) Outdoor bus shows a higher failure rate than indoor bus.
(3) Primary and contributing causes of failures were investigated. Inadequate maintenance was one of the leading "suspected primary causes of failure" and exposure to contaminants (including dust, moisture, and chemicals) was the leading "contributing cause to failure." This tends to support the data showing outdoor bus with a relatively high failure rate.
(4) The survey results on type of failures show a surprisingly high percentage of line-to-line failures, rather than line-to-ground.

3.2.2 1980 Generator Survey Data. The results of the 1980 generator survey data (see Reference [10]) are summarized in Table 12. A "unit" in this survey was defined to include the generator's driver and its ancillary equipment, including the device from which the generator's output is made available to the "outside" world. The term "unit-year" was defined as the summation of the running times reported for each generator.

Table 12
1980 Generator Survey Data

Equipment Subclass	Average Hours Downtime per Failure	Failure Rate
Continuous Service Steam Turbine Driven	32.7	0.16900 failures per unit year
Emergency and Standby Units Reciprocating Engine Driven	478.0	0.00536 failures per hour in use
Reciprocating Engine Driven	*	0.01350 failures per start attempt

NOTES: *Small sample size; less than eight failures.
(1) Appendix L contains data from a recent survey of diesel and gas turbine generators, 600–1800 kW.

Two major categories (i.e., continuously applied units and emergency or standby applied units) emerged from an evaluation of the responses. All of the continuous units were steam turbine driven, and all of the emergency or standby units were reciprocating engine driven. An important point to note on the data for emergency and standby units: Failure to start for automatically started units was counted as a failure; whereas failure to start for manually started units was not counted as a failure.

3.2.3 1979 Survey of the Reliability of Transformers. A survey published in 1973–74 raised some interesting questions and created some controversy (see References [8] and [23]). The most controversial items in this survey concerned: (1) the average outage duration time after a transformer failure, in relation to the failure repair method, and (2) the comparatively high failure rate for rectifier transformers.

The 1979 survey form (see Reference [12]) was improved considerably, taking lessons learned from the 1973–74 version. Items felt to be of little significance in the past have been omitted, and the form was simplified to maximize the response. Data relating specifically to transformer reliability, such as rating, voltage, age, and maintenance, were included in the new form. The most significant categories in the failed unit data are: the causes of the failure, the failure repair method, failure repair urgency and the duration of the failure, and the age at time of the failure. The survey form of the 1979 survey (published in 1983) is shown in Appendix G.

3.2.3.1 Failure Rate and Failure Repair Method for Power and Rectifier Transformers Survey Results. The survey response for power transformers is summarized in Table 13 and for rectifier transformers is summarized in Table 14.

Table 13
Power Transformers (1979 Survey)

Equipment Subclass	Failure Rate (Failures per Unit-Year)	Average Repair Time (Hours per Failure)	Average Replacement Time (Hours per Failure)
All Liquid Filled	0.0062	356.1	85.1
Liquid Filled 300–10 000 kVA	0.0059	297.4	79.3
Liquid Filled > 10 000 kVA	0.0153	1178.5*	192.0*
Dry 300–10 000 kVA	*	*	*

*Small sample size; less than eight failures.

Table 14
Rectifier Transformers (1979 Survey)

Equipment Subclass	Failure Rate (Failures per Unit-Year)	Average Repair Time (Hours per Failure)	Average Replacement Time (Hours per Failure)
All Liquid Filled	0.0190	2316.0	41.4
Liquid Filled 300–10 000 kVA	0.0153	1664.0*	38.7*
Liquid Filled > 10 000 kVA	*	*	*

*Small sample size; less than eight failures.

The survey results for the liquid-filled power transformers compared favorably between the 1973–74 and 1979 surveys: 0.0041 and 0.0062 failures per unit-year, respectively. The 1979 survey also confirmed the fact that the failure rate for rectifier transformers (i.e., 0.0190) is much higher than those for the other transformer categories (i.e., 0.0062). This may be due to the severe duties to which they were subjected and/or the harsh environments in which they are housed.

Tables 13 and 14 include data on average outage duration time versus the failure repair method. The data clearly indicates that the restoration of a unit to service by repair rather than replacement results in a much longer average outage duration time. This is consistent with previous survey results. Despite this fact, in most categories a larger number of units were restored to service by repair. These results show the obvious benefits in having spares at the site or readily available. The data also provides some of the information necessary in the preparation of an economic justification for spares. The averages shown represent only those cases where restoration work was begun *immediately*. Those instances in which the repair or replacement was deferred were excluded to avoid distorting the average outage duration time data.

3.2.3.2 Failure Rate versus Age of Power Transformers. The survey response for power transformer failures as a function of the transformer's age is summarized in Table 15.

Table 15
Failure Rate versus Age of Power Transformers (1979 Survey)

Equipment Subclass	Age[1] (Years)	Number of Units	Sample Size (Unit-Years)	Number of Failures[2]	Failure Rate (Failures per Unit-Year)
Liquid Filled 300–10 000 kVA	1–10	638	2625.5	19	0.0072
Liquid Filled 300–10 000 kVA	11–25	715	8846.5	47	0.0053
Liquid Filled 300–10 000 kVA	> 25	397	5938.0	36	0.0060
Liquid Filled > 10 000 kVA	1–10	27	144.0	0*	—
Liquid Filled > 10 000 kVA	11–25	28	283.5	7*	0.0246*
Liquid Filled > 10 000 kVA	> 25	9	158.0	2*	0.0126*

NOTES: *Small sample size; less than eight failures.
(1) Age was the age of the transformer at the end of the reporting period.
(2) Relay or tap changer faults were not considered in calculation of failure rates or repair and replacement times.

An examination of Table 15 reveals that the failure rates for power transformers was approximately equal in all three age groups. It can be seen that slightly higher failure rates for transformer units aged 1 to 10 years and for units greater than 25 years may be attributable to "infant mortality" and to units approaching the end of their life, respectively.

3.2.3.3 Failure-Initiating Cause. Table 16 summarizes the *failure-initiating cause* data for power and rectifier transformers. This table reveals that a large percentage of transformer failures were initiated by some type of insulation breakdown or transient overvoltages.

Table 16
Failure-Initiating Cause for Power and Rectifier Transformers (1979 Survey)

Failure-Initiating Cause	All Power Transformers		All Rectifier Transformers	
	Number of Failures[1]	Percentage	Number of Failures	Percentage
Transient Overvoltage Disturbance (Switching Surges, Arcing Ground Fault, etc.)	18	16.4%	2	13.3%
Overheating	3	2.7	1	6.7
Winding Insulation Breakdown	32	29.1	2	13.3
Insulation Bushing Breakdown	15	13.6	1	6.7
Other Insulation Breakdown	6	5.5	3	20.0
Mechanical Breaking, Cracking, Loosening, Abrading, or Deforming of Static or Structural Parts	8	7.3	3	20.0
Mechanical Burnout, Friction, or Seizing of Moving Parts	3	2.7	2	13.3
Mechanically Caused Damage from Foreign Source (Digging, Vehicular Accident, etc.)	3	2.7	0	0.0
Shorting by Tools or Other Metal Objects	1	0.9	0	0.0
Shorting by Birds, Snakes, Rodents, etc.	3	2.7	0	0.0
Malfunction of Protective Relay Control Device or Auxiliary Device	5	4.6	0	0.0
Improper Operating Procedure	4	3.6	0	0.0
Loose Connection or Termination	8	7.3	1	6.7
Others	1	0.9	0	0.0
Continuous Overvoltage	0	0.0	0	0.0
Low Voltage	0	0.0	0	0.0
Low Frequency	0	0.0	0	0.0
	110	100.0%	15	100.0%

NOTE: (1) Failure-initiating cause not specified for two failures.

3.2.3.4 Failure-Contributing Cause. Table 17 summarizes the *failure-contributing cause* for power and rectifier transformers. Normal deterioration from age and cooling medium deficiencies were reported to have contributed to a large number of both power and rectifier transformer failures.

Table 17
Failure-Contributing Cause for Power and Rectifier Transformers
(1979 Survey)

Failure-Contributing Cause	All Power Transformers		All Rectifier Transformers	
	Number of Failures[1]	Percentage	Number of Failures[2]	Percentage
Persistent Overloading	1	1.1%	0	0.0%
Abnormal Temperature	5	5.6	1	7.1
Exposure to Aggressive Chemicals, Solvents, Dusts, Moisture, or Other Contaminants	13	14.4	1	7.1
Normal Deterioration from Age	12	13.3	4	28.6
Severe Wind, Rain, Snow, Sleet, or Other Weather Conditions	4	4.5	0	0.0
Lack of Protective Device	2	2.2	0	0.0
Malfunction of Protective Device	7	7.8	0	0.0
Loss, Deficiency, Contamination, or Degradation of Oil or Other Cooling Medium	9	10.0	3	21.5
Improper Operating Procedure or Testing Error	3	3.3	0	0.0
Inadequate Maintenance	7	7.8	3	21.5
Others	27	30.0	1	7.1
Exposure to Nonelectrical Fire or Burning	0	0.0	0	0.0
Obstruction of Ventilation by Foreign Object or Material	0	0.0	0	0.0
Improper Setting of Protective Device	0	0.0	0	0.0
Inadequate Protective Device	0	0.0	1	7.1
	90	100.0%	14	100.0%

NOTES: (1) Failure-contributing cause not specified for 22 failures.
(2) Failure-contributing cause not specified for two failures.

3.2.3.5 Suspected Failure Responsibility. Table 18 summarizes the *suspected failure responsibility* for power and rectifier transformer failures. The respondents believed that manufacturer defects and inadequate maintenance were responsible for the majority of power transformer failures (i.e., 59.3%). Table 18 shows that inadequate operating procedures were a more significant cause of rectifier transformer failures (i.e., 31.2%) than inadequate maintenance.

Table 18
Suspected Failure Responsibility for Power and Rectifier Transformers
(1979 Survey)

Suspected Failure Responsibility	All Power Transformers		All Rectifier Transformers	
	Number of Failures[1]	Percentage	Number of Failures	Percentage
Manufacturer-Defective Component or Improper Assembly	32	33.3%	5	31.2%
Transportation to Site, Improper Handling	1	1.0	0	0.0
Application Engineering, Improper Application	3	3.1	2	12.5
Inadequate Installation and Testing Prior to Start Up	6	6.3	0	0.0
Inadequate Maintenance	25	26.0	2	12.5
Inadequate Operating Procedures	4	4.2	5	31.3
Outside Agency—Personnel	3	3.1	0	0.0
Outside Agency—Others	6	6.3	0	0.0
Others	16	16.7	2	12.5
	96	100.0%	16	100.0%

NOTE: (1) Suspected failure responsibility was not specified for 16 failures.

3.2.3.6 Maintenance Cycle and Extent of Maintenance. The 1973–74 survey asked the respondent to give an opinion of the maintenance quality as excellent, fair, poor, or none. It is very difficult to be completely objective in responding to this type of question. The new survey, therefore, asked for a brief description of the extent of maintenance performed, the idea being to enable the reader to judge the benefits derived from a particular maintenance procedure. The large percentage of failures that resulted from inadequate maintenance shows the importance of a comprehensive preventive maintenance program and compilation of accurate data on the extent and frequency of the maintenance performed. Unfortunately, the response did not lend itself to reporting in tabular form. Maintenance information continues to be the most difficult to obtain and report for all equipment categories.

3.2.3.7 Type of Failure. The 1979 survey limited the choices of failure type to "Winding" and "Other" as shown in Table 19 for power and rectifier transformers. Clearly, the most significant failure type occurred in power transformer windings.

**Table 19
Type of Failure for Power and Rectifier Transformers
(1979 Survey)**

Type of Failure	All Power Transformers		All Rectifier Transformers	
	Number of Failures	Percentage	Number of Failures	Percentage
Winding	59	53%	8	50%
Other	53	47	8	50

3.2.3.8 Failure Characteristics. The failure characteristics of power and rectifier transformers are shown in Table 20. As would be expected, the survey results show that about 75% of transformer failures resulted in their removal from service by automatic protective devices; however, the percentage requiring manual removal was significant. Increasing use of transformer oil or gas analysis could be a factor here, enabling detection of incipient faults in their early stages, and thus permitting manual removal before a major failure occurs.

**Table 20
Failure Characteristics for Power and Rectifier Transformers
(1979 Survey)**

Failure Characteristics	All Power Transformers		All Rectifier Transformers	
	Number of Failures	Percentage	Number of Failures	Percentage
Automatic Removal by Protective Device	83	75%	11	69%
Partial Failure, Reducing Capacity	5	5	0	0
Manual Removal	23	20	5	31

3.2.3.9 Voltage Rating. The failure rates for liquid-filled power transformers and rectifier transformers classified by their voltage ratings are shown in Tables 21 and 22, respectively. An examination of Table 21 reveals the failure rate for the 600–15 000 V transformers (i.e., 0.0052 failures per unit-year) is less than that for the higher voltage units. The lack of data (i.e., small sample sizes) reported for rectifier transformers above 15 kV makes it impossible to draw any definite conclusions as to the effect of voltage or size on their failure rates.

63

Table 21
Failure Rate versus Voltage Rating and Size for Power Transformers
(1979 Survey)

Equipment Subclass	Voltage (kV)	Number of Units	Sample Size (Unit-Years)	Number of Failures	Failure Rate (Failures per Unit-Year)
Liquid Filled 300–10 000 kVA	0.6–15	1626	15 775	82	0.0052
Liquid Filled 300–10 000 kVA	> 15	124	1637	18	0.0110
Liquid Filled > 10 000 kVA	> 15	52	490	9	0.0184

Table 22
Failure Rate versus Voltage Rating for Rectifier Transformers
(1979 Survey)

Equipment Subclass	Voltage (kV)	Number of Units	Sample Size (Unit-Years)	Number of Failures	Failure Rate (Failures per Unit-Year)
All Liquid Filled	0.6–15	65	745	15	0.0201

3.2.4 1983 IEEE Survey on the Reliability of Large Motors. A decision was made by the IEEE Motor Reliability Working Group to focus on motors that were of a critical nature in industrial and commercial installations and, thus, only motors larger than 200 hp were selected to be included in the survey (see Reference [15] and Appendix H). Another decision was made to limit the survey to only include motors that were 15 years old or less in order to focus on motors that were similar to those presently being manufactured and used today.

Failure rates are given for induction, synchronous, wound-rotor, and direct-current motors. Pertinent factors that effect the failure rates of these motors are identified. Data is presented on key variables, such as downtime per failure, failed component, causes of failure, and the time of failure discovery. The results of this recent survey are compared with four other surveys on the reliability of motors (see References [4], [6], [13], and [24]). Details of the survey report are shown in Appendix H. The results of the survey are summarized in this section. The term "large motor" is defined in this section to be any motor whose horsepower rating exceeds 200 hp.

3.2.4.1 Overall Summary of the Failure Rate for Large Motors. The 1983 survey included data reported for 360 failures on 1141 motors with a total service of 5085 unit-years. The overall summary of the survey results for induction, synchronous, wound-rotor, and direct-current motors is shown in Table 23. Calendar time was used in the calculation of the unit-years of service (rather than the running time) to simplify the data collection procedure.

Table 23
Overall Summary for Large Motors Above 200 hp
(IEEE Survey of Industrial and Commercial Installations, 1983–85 [15])

Number of Plants in Sample Size	Sample Size (Unit-Years)	Number of Failures Reported	Equipment Subclass	Failure Rate (Failures per Unit-Year)	Average Hours Downtime per Failure	Median Hours Downtime per Failure
75	5085.0	360	All	0.0708	69.3	16.0
			Induction			
33	1080.3	89	0–1000 V	0.0824	42.5	15.0
52	2844.4	203	1001–5000 V	0.0714	75.1	12.0
5	78.1	2*	5001–15 000 V	*	*	*
1	13.5	—	Not Specified	—	—	—
			Synchronous			
19	459.3	35	1001–5000 V	0.0762	78.9	16.0
2	29.5	3*	5001–15 000 V	*	*	*
			Wound Rotor			
5	137.0	10	0–1000 V	0.0730	*	*
9	251.1	8	1001–5000 V	0.0319	*	*
2	39.0	4*	5001–15 000 V	*	*	*
			Direct Current			
5	122.7	6*	0–1000 V	*	*	*
1	30.0	—	1001–5000 V	—	—	—

*Small sample size; less than eight failures.

A summary of the important conclusions derived from the 1983 survey on the failure rates of large motors are:

(1) Induction and synchronous motors had approximately the same failure rate of 0.07 to 0.08 failures per unit-year.

(2) Induction motors rated 0 to 1000 V and those rated 1001–5000 V had approximately the same failure rates. The response on motors operating above 5000 V was too small to draw any meaningful conclusions.

(3) Wound-rotor motors rated 0 to 1000 V had a failure rate that was about the same as induction motors of the same rating.

(4) The sample size for direct-current motors was too small to draw any meaningful conclusions.

(5) Motors with intermittent duty operation had a failure rate that was about half as great as those with continuous duty.

(6) Motors with less than one start per day had approximately the same failure rate as those motors with between one to ten starts per day, which would indicate that up to ten starts per day do not have a major effect on the motor failure rates.

3.2.4.2 Downtime per Failure versus Repair/Replacement and Urgency for Repair for Large Motors. The comparison of the downtime per motor failure data for "repair" versus "replace with spare" is considered important when deciding whether a spare motor should be purchased when designing a new plant. The downtime per failure survey characteristics for all types of motors grouped together as a category is shown in Table 24.

Table 24
Downtime per Failure versus Repair or Replace with Spare
and Urgency for Repair—All Types of Motors Above 200 hp
(IEEE Survey of Industrial and Commercial Installations 1983–85 [15])

	Number of Failures	Average Hours (Downtime per Failure)	Median Hours (Downtime per Failure)
Repair—Normal Working Hours[1]	87	97.7	24.0
Repair—Round the Clock	45	81.4	72.0
Replace with Spare[2]	111	18.2	8.0
Low Priority	4*	370.0*	400.0*
Not Specified	6*	288.0*	240.0*
Total	251	69.3	14.0

NOTES: *Small sample size; less than eight failures.
    (1) 6570 hours for one failure omitted.
    (2) 960 hours for one failure omitted.

An examination of Table 24 shows the effect on the "repair" time that the "urgency for repair" has had. There were 45 cases of motor failures where the "repair" activities were carried out on a "round-the-clock, all-out" effort. There were four cases of motor failures where "low-priority" urgency resulted in a very long downtime; it is important to exclude these cases when making decisions on the design of industrial and/or commercial power systems. In general, the "average downtime per failure" is about five times larger for "repair" versus "replace with spare."

3.2.4.3 Failed Component—Large Motors. The identified motor component that failed is shown in Table 25 for induction, synchronous, wound-rotor, direct-current, and "all" motors.

Table 25
Failed Component—Large Motors (Above 200 hp)
(IEEE Survey of Industrial and Commercial Installations 1983–85 [15])
(Number of Failures)

Failed Component[1]	Induction Motors	Synchronous Motors	Wound-Rotor Motors	Direct-Current Motors	Total (All Types)
Bearings	152	2	10	2	166
Windings	75	16	6	—	97
Rotor	8	1	4	—	13
Shaft or Coupling	19	—	—	—	19
Brushes or Slip Ring	—	6	8	2	16
External Devices	10	7	1	—	18
Not Specified	40	9	—	2	51
Total	304	41	29	6	380

NOTE: (1) Some respondents reported more than one failed component per motor failure.

It can be seen that the two largest categories reported are motor *bearing* and *winding* failures with 166 and 97 failures, respectively, out of a total of 380 failures. Bearings and windings represent 44% and 26%, respectively, of the total number of motor failures.

3.2.4.4 Failed Component versus Time of Discovery—Large Motors. Data on the failed component versus the time the failure was discovered is shown in Table 26. It can be seen that 60.5% of the failures found during "maintenance or test" are bearings. Many users consider that it is very important to find as many failures as possible during "maintenance or test" rather than "normal operation." Bearings and windings represent 36.6% and 33.1%, respectively, of the failures discovered during "normal operation."

Table 26
Failed Component versus Time of Discovery
(All Types of Motors Above 200 hp)
(IEEE Survey of Industrial and Commercial Installations 1983–85 [15])
(Percentage of Failures)

Failed Component	Time of Discovery		
	Normal Operation	Maintenance or Test	Other
Bearings	36.6%	60.5%	50.0%
Windings	33.1	8.3	28.6
Rotor	5.1	1.8	0.0
Shaft or Coupling	5.8	8.3	14.3
Brushes or Slip Rings	3.1	7.3	0.0
External Devices	5.0	3.7	0.0
Not Specified	11.3	10.1	7.1
Total Percentage of Failures	100.0%	100.0%	100.0%
Total Number of Failures	257	109	14

3.2.4.5 Causes of Large Motor Bearing and Winding Failures. The causes of motor failures categorized according to the failure initiator, the failure contributor, and the failure's underlying cause are shown in Table 27 for induction, synchronous, and "all" motors.

"Mechanical breakage" is the largest failure initiator for *induction* motors. "Normal Deterioration from Age," "High Vibration," and "Poor Lubrication" are the major failure contributors to induction motor failures. "Inadequate Maintenance" and "Defective Component" are the largest underlying causes of induction motor failures.

"Electrical Fault or Malfunction" and "Other Insulation Breakdown" are the major failure initiators for *synchronous* motors. "Normal Deterioration from Age" is the major fault contributor of synchronous motors. "Defective Component" is the largest underlying cause of synchronous motor failures.

Table 27 shows a correlation between *bearing failures* and the *causes of failure*: 50.3% of bearing failures were initiated by "Mechanical Breakage;" 31.3% and 21.8%, respectively, had "Poor Lubrication" and "High Vibration" as failure contributors; and 27.6% blamed "Inadequate Maintenance" as the underlying cause.

Table 27
Causes of Failure versus Motor Type and versus
Bearing and Winding Failures—Motors Above 200 hp
(IEEE Survey of Industrial and Commercial Installations 1983–85 [15])
(Percentage of Failures)

All Motor Types—Failed Component		All Types of Motors %	Induction Motors %	Synchronous Motors %	Causes of Failures
Bearings %	Windings %				
					Failure Initiator
0.0%	4.1%	1.5%	1.4%	0.0%	Transient Overvoltage
12.4	21.4	13.2	14.7	0.0	Overheating
1.9	36.7	12.3	11.9	21.1	Other Insulation Breakdown
50.3	10.2	33.1	37.4	5.2	Mechanical Breakage
3.7	11.2	7.6	5.8	23.7	Electrical Fault or Malfunction
0.0	2.1	0.9	0.7	2.6	Stalled Motor
31.7	14.3	31.4	28.1	47.4	Other
100.0%	100.0%	100.0%	100.0%	100.0%	Total Percentage of Failures
161	98	341	278	38	Total Number of Failures
					Failure Contributor
1.4%	6.5%	4.2%	4.9%	2.7%	Persistent Overloading
0.7	7.6	3.0	3.4	0.0	High Ambient Temperature
2.7	18.5	5.8	6.7	2.7	Abnormal Moisture
0.0	5.4	1.5	1.5	2.7	Abnormal Voltage
0.0	1.1	0.6	0.7	0.0	Abnormal Frequency
21.8	8.7	15.5	17.6	5.4	High Vibration
5.4	6.5	4.2	4.5	2.7	Aggressive Chemicals
31.3	5.4	15.2	16.9	8.1	Poor Lubrication
0.0	7.6	3.9	2.2	2.7	Poor Ventilation or Cooling
20.4	18.5	26.4	24.0	51.4	Normal Deterioration from Age
16.3	14.2	19.7	17.6	21.6	Other
100.0%	100.0%	100.0%	100.0%	100.0%	Total Percentage of Failures
147	92	330	267	37	Total Number of Failures
					Failure Underlying Cause
17.8%	10.9%	20.1%	20.3%	22.2%	Defective Component
14.5	10.9	12.9	15.9	0.0	Poor Installation/Testing
27.6	19.6	21.4	22.8	11.1	Inadequate Maintenance
2.0	6.5	3.6	3.3	2.8	Improper Operation
0.7	0.0	0.6	0.8	0.0	Improper Handling/Shipping
7.9	7.6	6.1	6.5	2.8	Inadequate Physical Protection
2.6	15.2	5.8	5.3	11.1	Inadequate Electrical Protection
7.2	5.4	6.8	5.7	5.6	Personnel Error
2.0	3.3	3.9	2.8	13.9	Outside Agency—Not Personnel
5.9	4.3	4.9	4.9	0.0	Motor-Driven Equipment Mismatch
11.8	16.3	13.9	11.7	30.5	Other
100.0%	100.0%	100.0%	100.0%	100.0%	Total Percentage of Failures
152	92	309	246	36	Total Number of Failures

Table 27 also shows a correlation between *winding failures* and the *causes of failure*: 36.7% of the winding failures had "Other Insulation Breakdown" as the initiator; 18.5% and 18.5%, respectively, had "Normal Deterioration from Age" and "Abnormal Moisture" as failure contributors; and 19.6% had "Inadequate Maintenance" and 15.2% had "Inadequate Electrical Protection" as the underlying cause.

It is of interest to note that "Inadequate Maintenance" was the largest underlying cause of both bearing and winding failures. A special study of 71 failures attributed to "Inadequate Maintenance" is shown in Table 28. It can be clearly seen that 59.1% of the motor components that failed were bearings, that 52.1% of the failures were *initiated* by "Mechanical Breakage," and 43.7% of the failures had "Poor Lubrication" as a *failure contributor*.

Table 28
Failures Caused by "Inadequate Maintenance" versus "Failed Component," "Failure Initiator," and "Failure Contributor" (All Types of Motors Above 200 hp)
(IEEE Survey of Industrial and Commercial Installations 1983–85 [15])
(Number of Failures in Percent)

%	Failed Component
59.1%	Bearing
25.4	Winding
1.4	Rotor
0.0	Shaft or Coupling
8.5	Brushes or Slip Rings
1.4	External Device
4.2	Other
100.0%	Total Percentage (Number of Failures = 71)
%	**Failure Initiator**
0.0%	Transient Overvoltage
4.2	Overheating
14.1	Other Insulation Breakdown
52.1	Mechanical Breakage
2.8	Electrical Fault or Malfunction
0.0	Stalled Motor
26.8	Other
100.0%	Total Percentage (Number of Failures = 71)
%	**Failure Contributor**
0.0%	Persistent Overloading
4.2	High Ambient Temperature
7.0	Abnormal Moisture
0.0	Abnormal Voltage
0.0	Abnormal Frequency
4.2	High Vibration
9.9	Aggressive Chemical
43.7	Poor Lubrication
1.4	Poor Ventilation/Cooling
18.3	Normal Deterioration from Age
11.3	Other
100.0%	Total Percentage (Number of Failures = 71)

3.2.4.6 Other Significant Results. Several additional parameters were reported in Reference [15] in terms of their effect on the failure rate of motors above 200 hp. These included the effect of horsepower, speed, enclosure, environment, duty cycle, service factor, average number of starts per day, grounding practice, maintenance quality, maintenance cycle, type of maintenance performed, and months since last maintenance prior to the failure. Some combinations of these parameters, two at a time, have also been studied and reported (see Reference [14]).

3.2.4.6.1 Open versus Enclosed Motors. The following significant conclusions were reached:

(1) Open motors had a higher failure rate than weather-protected or enclosed motors.

(2) Indoor motors had a higher failure rate for open motors than for weather-protected or enclosed motors.

(3) Outdoor motors had a lower failure rate than indoor motors because most outdoor motors were weather protected or enclosed, and most indoor motors were open.

3.2.4.6.2 Service Factor. The 1.15 service factor (S.F.) induction motors had a higher reported failure rate than 1.0 S.F. induction motors; but the opposite was true for synchronous motors.

3.2.4.6.3 Speed and Horsepower. The failure rate for induction motors did not vary significantly among the three speed categories (i.e., 0–720 rev/min, 721–1800 rev/min, and 3600 rev/min). The highest failure rate was in the middle speed category, while the lowest failure rate was in the 3600 rev/min category. The 201–500 hp induction motors had approximately the same failure rate as 501–5000 hp induction motors in each of the three speed ranges studied.

Synchronous motors in the speed category 0–720 rev/min had a higher failure rate than synchronous motors in the 721–1800 rev/min category. There were no respondents for the 3600 rev/min category.

3.2.4.7 Data Supports Chemical Industry Motor Standard. Reliability data for induction motors from both the 1983 IEEE survey and the 1973–74 IEEE survey (see Appendixes A and B) supported the need for several of the features incorporated into IEEE Std 841-1986, IEEE Recommended Practice for Chemical Industry Severe Duty Squirrel-Cage Induction Motors—600 V and Below (ANSI) [2].[3] The IEEE surveys show the need for improved reliability of bearings and windings and, in some cases, the need for better physical protection against aggressive chemicals and moisture. Some of the more significant recommendations for an IEEE Std 841-1986 (ANSI) [2] motor include:

(1) TEFC enclosure

(2) Maximum 80 °C rise at 1.0 service factor

(3) Contamination protection for bearings and grease reservoirs

(4) Three-year continuous L-10 bearing life

(5) Maximum bearing temperature of 45 °C rise (50 °C rise on two-pole motors)

---

[3]IEEE publications are available from the Institute of Electrical and Electronics Engineers, IEEE Service Center, 445 Hoes Lane, Piscataway, NJ 08855-1331.

(6) Cast iron frame construction

(7) Nonsparking fan

(8) Single connection point per phase in terminal box

(9) Maximum sound power level of 90 dBA

(10) Corrosion-resistant paint, internal joints and surfaces, and hardware

IEEE Std 841-1986 (ANSI) [2] was tailored for the petroleum/chemical industry; however, it can be beneficial for other industries with similar requirements.

3.2.4.8 Comparison of 1983 Motor Survey with Other Motor Surveys. One of the primary purposes of comparing results of the 1983 motor survey with previous surveys and other surveys (see References [3], [4], and [24]) is to attempt to identify trends in the failure characteristics of motors (i.e., changing failure rates with time, varying causes of motor failures, assessing the impact of maintenance practices, etc.).

3.2.4.8.1 1983 EPRI and 1983–85 IEEE Surveys. The size and scope of the IEEE Working Group and EPRI motor surveys is shown in Table 29. The motor failure rate of 0.035 failures per unit-year in the EPRI sponsored study of the electric utility industry is about *half* the IEEE failure rate of 0.0708 failures per year.

The percentage of motor failures classified by component in the two surveys is shown in Table 30. Similar results were obtained in these two studies on the failed component, with bearing-, winding-, and rotor-related percentages that were each about the same.

Table 29
Size and Scope Comparison of the IEEE 1983–85 Motor Survey [15]
with the EPRI Sponsored Motor Survey in Electric Utility Power Plants [3]

Parameter	IEEE Working Group	EPRI Phase I
Horsepower	> 200	100 and up
Number of Companies/Utilities	33	56
Number of Plants or Units	75	132
Number of Motors	1141	4797
Total Population (Unit-Years)	5085	24 914[1]
Total Failures	360	871[1]
Failure Rate (All Motors)	0.0708	0.035[1]

NOTE: (1) To first failure.

Table 30
Failure by Component Comparison — IEEE 1983–85 Motor Survey [15]
and EPRI Sponsored Survey [3]
(Percentage of Failures)

IEEE Working Group	EPRI Phase I
44% Bearings	41% Bearing Related
26% Windings	37% Stator Related
8% Rotors/Shafts/Couplings	10% Rotor Related

Table 31 shows some differences between the two studies on the causes of failures. The IEEE survey found "Inadequate Maintenance," "Poor Installation/Testing," and "Misapplication" to be a significant percentage of the causes of motor failures; while the EPRI study attributed a larger percentage to the manufacturer. In addition, the EPRI study had a much larger percentage of failures attributed to "Other or Not Specified." Additional results from the EPRI sponsored study were also given in a later paper (see Reference [4]).

Table 31
Cause of Failure Comparison — IEEE 1983–85 Motor Survey [15] and EPRI Sponsored Motor Survey [3]

Falure Cause	EPRI Phase I		IEEE Working Group		Failure Cause
	Number	Percent	Number	Percent	
Manufacturer Design Workmanship	401	32.8%	62	17.2%	Defective Component
Misoperation	124	10.2	32	8.9	Improper Operation/Personnel Error
Misapplication	83	6.8	52	14.5	Misapplication Motor-Driven Equipment Mismatch Inadequate Electrical Protection Inadequate Physical Protection
—			66	18.3	Inadequate Maintenance
—			40	11.1	Poor Installation/Testing
—			12	3.3	Outside Agency Other than Personnel
—			2	0.6	Improper Handling/Shipping
Other or Not Specified	613	50.2	94	26.1	Other or Not Specified
Total	1221	100.0%	360	100.0%	

3.2.4.8.2 1982 Doble Data and 1983–85 IEEE Surveys. A 1982 Doble survey (see Reference [24]) in the electric utility industry (for motors 1000 hp and up and not over 15 years of age) reported 68 insulation-related failures in 2078 unit-years of service during the year 1981. This gives an insulation-related failure rate of 0.033 failures per unit-year. This can be compared with a winding failure rate of 26% times 0.0708, which equals 0.018 failures per unit-year that can then be calculated from the 1983–85 IEEE survey of motors above 200 hp and not older than 15 years data in Tables 29 and 30.

3.2.4.8.3 IEEE Surveys 1973–74 and 1983–85. Table 32 shows the results from the 1973–74 IEEE motor reliability survey [13] of industrial plants. This survey covered motors 50 hp and larger, and had no limit on the age of the motor. These results can be compared to Table 23 for the 1983–85 IEEE survey of motors above 200 hp and not older than 15 years. The 1983–85 failure rates of induction motors and synchronous motors were about double those from the 1973–74 survey for motors 601–15 000 V.

Table 32
1973-74 IEEE Overall Summary for
Motors 50 hp and Larger [13]

Number of Plants in Sample Size	Sample Size (Unit-Years)	Number of Failures Reported	Equipment Subclass	Failure Rate (Failures per Unit-Year)	Average Hours Downtime per Failure	Median Hours Downtime per Failure
—	42 463	561	All	0.0132	111.6	—
			Induction			
17	19 610	213	0-600 V	0.0109	114.0	18.3
17	4229	172	601-15 000 V	0.0404	76.0	91.5
			Synchronous			
2	13 790	10	0-600 V	0.0007	35.3	35.3
11	4276	136	601-15 000 V	0.0318	175.0	153.0
6	558	31	Direct Current	0.0556	37.5	16.2

3.2.4.8.4 AIEE 1962 and 1983-85 IEEE Surveys. Table 33 shows the results from the 1962 AIEE motor reliability survey [6] of industrial plants. This survey covered motors 250 hp and larger and had no limit on the age of the motor. The failure rates for both induction motors and synchronous motors from the 1962 AIEE survey are within 30% of those shown in Table 23 for the 1983-85 IEEE survey of motors above 200 hp and not older than 15 years. The two surveys conducted 21 years apart show remarkably similar results.

Table 33
1962 AIEE Overall Summary for
Motors 250 hp and Larger, U.S. and Canada [6]

Number of Plants in Sample Size	Sample Size (Unit-Years)	Number of Failures Reported	Equipment Subclass	Failure Rate (Failures per Unit-Year)	Average Hours Downtime per Failure	Median Hours Downtime per Failure
46	1420	140	Induction	0.0986	78.0	70.0
53	600	39	Synchronous	0.0650	149.0	68.0

3.3 Part 2: Equipment Reliability Surveys Conducted Prior to 1976

3.3.1 Introduction. From 1973 through 1975, the Power Systems Reliability Subcommittee of the IEEE Industrial Power Systems Department conducted and published surveys of electrical equipment reliability in industrial plants (see References [9] and [13]). Those reliability surveys of electrical equipment and electric utility power supplies were extensive, and summaries of the following pertinent reliability data are given in this section:

(1) Failure rate and outage duration time for electrical equipment and electric utility power supplies

(2) Failure characteristic or failure modes of electrical equipment; that is, the effect of the failure on the system

(3) Causes and types of failures of electrical equipment

(4) Failure repair method and failure repair urgency

(5) Method of service restoration after a failure

(6) Loss of motor load versus time of power outage

In addition, reference is made to summaries of pertinent reliability data and information that are contained in other chapters, including

(7) Maximum length of time of an interruption of electrical service that will not stop plant production

(8) Plant restart time after service is restored, following a failure that caused a complete plant shutdown

(9) Cost of power interruptions to industrial plants and commercial buildings

(10) An example showing that the two power sources in a double-circuit utility supply may not be completely independent

(11) Equipment failure rate multipliers versus maintenance quality

(12) Percentage of failures caused by inadequate maintenance versus month since maintained

All of the reliability data summarized in the above twelve items was taken from the IEEE surveys of industrial plants (see References [9] and [13]) and commercial buildings (see Reference [7]). The detailed reports are given in Appendixes A, B, C, and D. More recent surveys on "switchgear bus," "transformers," "large motors," and "cable, terminations, and splices" are included in Appendixes E, G, H, and I, respectively. Recent surveys on circuit breakers are shown in Appendixes J and K. A 1989 survey on diesel and gas turbine generating units is included in Appendix L.

3.3.2 Reliability of Electrical Equipment (1974 Survey). The term "electrical equipment" in this section includes all the electrical equipment listed in Table 34.

Table 34
In-Plant Electrical Equipment List

Electrical Equipment	
Circuit breakers (some)	Open wire
Motor starters	Cable
Disconnect switches — enclosed	Cable joints (some)
Bus duct	Cable terminations

In compiling the data for the 1974 survey, a failure was defined as any trouble with a power system component that causes any of the following effects:

(1) Partial or complete plant shutdown, or below-standard plant operation

(2) Unacceptable performance of user's equipment

(3) Operation of the electrical protective relaying or emergency operation of the plant electric system

(4) De-energization of any electric circuit or equipment

A failure on a public utility supply system may cause the user to have either of the following:

(1) A power interruption or loss of service

(2) A deviation from normal voltage or frequency outside the normal utility profile

A failure on an in-plant component causes a forced outage of the component, that is, the component is unable to perform its intended function until it is repaired or replaced. The terms "failure" and "forced outage" are often used synonymously.

All of the electrical equipment categories listed in this section have eight or more failures. This is considered an adequate sample size (see Reference [16]) in order to have a reasonable chance of determining a failure rate within a factor of 2. Failure rate and average downtime per failure data for an additional six categories of equipment are contained in Reference [13] (see Appendix A).

The additional categories of equipment that have between four and seven failures and thus might be considered by some as too small a sample size include

(1) Circuit breakers used as motor starters

(2) Disconnect switches — open

(3) Cable joints, 601-15 000 V, above ground and aerial

(4) Cable joints, 601-15 000 V, thermosetting

(5) Fuses

(6) Protective relays

3.3.2.1 Failure Modes of Circuit Breakers. The failure modes of "metalclad drawout" and "fixed-type" circuit breakers are shown in Table 35. Of primary concern to industrial plants is the large percentage of circuit breaker failures (i.e., 42%) which "opened when it should not." This type of circuit breaker failure can significantly affect plant processes and may result in a total plant shutdown. Also, a large percentage (i.e., 32%) of the circuit breakers "failed while in service (not while opening or closing)." Appendixes J and K and Reference [23] contain additional detailed information on circuit breaker reliability.

3.3.2.2 Failed Characteristics of Other Electrical Equipment. The failure characteristics of electrical equipment (excluding transformers and circuit breakers) are shown in Table 36. The dominant failure characteristic for this equipment is that it "Failed in service." A large percentage of the damage to "Motor Starters" (i.e., 36%), "Disconnect Switches" (i.e., 18%), and "Cable Terminations" (i.e., 12%) was discovered during testing or maintenance; however, the remaining electrical equipment did not significantly exhibit this failure characteristic.

3.3.2.3 Causes and Types of Failures of Electrical Equipment. The following data is presented in Tables 37 and 38:

(1) Failures, damaged part

(2) Failure type

(3) Suspected failure responsibility

(4) Failure-initiating cause

(5) Failure-contributing cause

The data presented in Table 38 indicate that the respondents suspected "Inadequate maintenance" and "Manufacturer-defective component" were responsible for a significant percentage of the failures for several categories of electrical equipment.

Table 35
Failure Modes of Circuit Breakers[1] (1974 Survey)
(Percentage of Total Failures in Each Failure Mode)

| All Circuit Breakers % | Metalclad Drawout | | | Failed Type[2] | | Failure Characteristics |
	All %	601–15 000 V %	0–600 V All Sizes %	0–600 V All Sizes %	All %	
5%	5%	2%	7%	8%	6%	Failed to close when it should
9	12	21	0	0	2	Failed while opening
42	58	49	71	5	4	Opened when it should not
7	6	4	9	5	4	Damaged while successfully opening
2	1	0	0	0	4	Damaged while closing
32	16	24	10	77	73	Failed while in service (not while opening or closing)
1	0	0	0	0	2	Failed during testing or maintenance
1	2	0	3	0	0	Damage discovered during testing or maintenance
1	0	0	0	5	5	Other
100%	100%	100%	100%	100%	100%	Total percentage
165	117	53	59	39	48	Number of failures in total percentage
8	7	0	7	1	1	Number not reported
173	124	53	66	40	49	Total failures

NOTES: (1) Appendix K contains some limited data from a later IEEE survey. Appendix J contains data for circuit breakers above 63 kV from a CIGRE 13-06 worldwide survey with a very large population.
(2) Includes molded case.

Table 36
Failure Characteristics of Other Electrical Equipment (1974 Survey)

Motor Starters %	Disconnect Switches %	Bus Duct %	Open Wire %	Cable %	Cable Joints %	Cable Terminations %	Failure Characteristics
37%	72%	90%	68%	92%	96%	80%	Failed in service
5	3	5	2	2	4	2	Failed during testing or maintenance
36	18	0	1	2	0	12	Damage discovered during testing or maintenance
20	6	5	6	3	0	6	Partial failure
2	1	0	23	1	0	0	Other

Table 37
Failure, Damaged Part, and Failure Type (1974 Survey)

Circuit Breakers %	Motor Starters %	Dis-connect Switches %	Bus Duct %	Open Wire %	Cable %	Cable Joints %	Cable Termi-nations %	Failure, Damaged Part
0%	5%	0%	15%	0%	5%	0%	0%	(1) Insulation — winding
2	0	1	10	1	0	0	12	(2) Insulation — bushing
19	10	14	65	6	83	91	74	(3) Insulation — other
1	0	0	0	0	3	0	0	(4) Mechanical — bearings
11	16	9	0	0	0	0	0	(5) Mechanical — other moving parts
6	2	30	0	4	1	0	4	(6) Mechanical — other
6	13	8	0	3	1	0	0	(7) Other electrical auxiliary device
28	2	1	0	3	1	0	0	(8) Other electrical protective device
1	0	0	0	0	0	0	0	(9) Tap changer — no-load type
0	0	0	0	0	0	0	0	(10) Tap changer — load type
26	52	37	10	83	6	9	10	(99) Other
								Failure Type
33%	14%	15%	70%	34%	73%	71%	55%	(1) Flashover or arcing involving ground
10	20	4	30	23	1	9	4	(2) All other flashover or arcing
19	55	47	0	25	7	20	37	(3) Other electrical defect
11	11	14	0	6	5	0	4	(4) Mechanical defect
27	0	20	0	12	14	0	0	(99) Other

Table 38
Suspected Failure Responsibility, Failure-Initiating Cause, and Failure-Contributing Cause (1974 Survey)

Circuit Breakers %	Motor Starters %	Dis-connect Switches %	Bus Duct %	Open Wire %	Cable %	Cable Joints %	Cable Termi-nations %	Suspected Failure Responsibility
23%	18%	29%	26%	0%	16%	0%	0%	(1) Manufacturer — defective component
0	0	0	0	0	0	0	0	(2) Transportation to site — defective handling
4	51	6	16	2	8	0	18	(3) Application engineering — improper application
3	0	4	5	9	14	50	38	(4) Inadequate installation and testing prior to start up
23	8	13	16	30	10	18	22	(5) Inadequate maintenance
6	3	39	0	2	3	0	0	(6) Inadequate operating procedures
5	0	1	5	5	4	5	0	(7) Outside agency — personnel
1	0	0	0	21	6	2	8	(8) Outside agency — other
35	20	8	32	31	39	25	14	(9) Other
								Failure-Initiating Cause
13%	1%	4%	0%	26%	26%	11%	12%	(1) Transient overvoltage disturbances (lightning, switching surges, arcing ground fault in ungrounded system)
0	0	0	0	0	0	0	0	(2) Overvoltage
3	1	4	30	21	1	0	2	(3) Overheating
18	8	5	20	8	29	40	50	(4) Other insulation breakdown
13	8	17	45	7	24	31	24	(21) Mechanical breaking, cracking, loosening, abrading, or deforming of static or structural parts
5	6	2	0	0	0	0	0	(22) Mechanical burnout, friction, or seizing of moving parts
1	0	20	0	10	7	0	4	(23) Mechanically caused damage from foreign source (digging, vehicular accident, etc.)
2	5	0	5	14	2	0	2	(41) Shorting by tools or metal objects
1	1	0	0	3	0	0	2	(42) Shorting by birds, snakes, rodents, etc.

Table 38 *(continued)*

Circuit Breakers %	Motor Starters %	Dis- connect Switches %	Bus Duct %	Open Wire %	Cable %	Cable Joints %	Cable Termi- nations %	Failure-Initiating Cause
1	0	0	0	0	0	0	0	(51) Loss of control power
11	63	0	0	0	0	0	0	(52) Malfunction of protective relay control device, or auxiliary device
0	0	3	0	0	0	0	0	(61) Low voltage
0	0	0	0	0	0	0	0	(62) Low frequency
33	7	45	0	11	10	18	4	(99) Other
								Failure-Contributing Cause
4%	0%	8%	6%	0%	2%	0%	0%	(1) Persistent overloading
1	0	3	0	0	0	2	0	(2) Above normal temperatures
0	0	1	0	0	0	0	0	(3) Below-normal temperatures
2	0	0	0	28	14	13	10	(4) Exposure to aggressive chemicals or solvents
3	0	4	17	1	8	22	12	(5) Exposure to abnormal moisture or water
0	0	0	0	3	2	0	0	(6) Exposure to nonelectrical fire or burning
0	0	0	0	0	1	0	0	(8) Obstruction of ventilation by foreign objects or material
17	40	5	49	3	30	29	24	(9) Normal deterioration from age
1	0	0	11	30	16	2	16	(10) Severe wind, rain, snow, sleet, or other weather conditions
2	0	0	0	1	0	0	0	(11) Protective relay improperly set
1	2	0	0	0	0	0	0	(12) Loss or deficiency of lubricant
0	0	0	0	0	0	0	0	(13) Loss or deficiency of oil or cooling medium
10	3	0	6	2	3	0	8	(14) Misoperation or testing error
3	1	26	0	2	1	0	0	(15) Exposure to dust or other contaminants
56	54	53	11	30	24	32	30	(99) Other

Table 39
Failure Repair Method and Failure Repair Urgency (1974 Survey)

Circuit Breakers %	Motor Starters %	Dis-connect Switches %	Bus Duct %	Open Wire %	Cable %	Cable Joints %	Cable Termi-nations %	Failure Repair Method
51%	33%	30%	65%	70%	47%	87%	60%	(1) Repair of failed component in place or sent out for repair
49	67	70	35	9	53	13	34	(2) Repair by replacement of failed component with spare
0	0	0	0	21	0	0	6	(99) Other
								Failure Repair Urgency
73%	66%	20%	80%	55%	66%	56%	53%	(1) Requiring round-the-clock all-out efforts
22	34	80	15	26	28	22	31	(2) Requiring repair work only during regular workday, perhaps with some overtime
5	0	0	5	0	6	22	16	(3) Requiring repair work on a nonpriority basis
0	0	0	0	19	0	0	0	(99) Other

3.3.2.4 Failure Repair Method and Failure Repair Urgency. The "Failure Repair Method" and the "Failure Repair Urgency" had a significant effect on the "average downtime per failure." Table 39 shows the percentages of these two parameters for eight classes of electrical equipment. A special study on this subject is reported in Tables 50, 51, 55, and 56 of Reference [16] (see Appendix B) for circuit breakers and cables (see Note 2 in Table 10 of this chapter).

3.3.2.5 Reliability of Electric Utility Power Supplies to Industrial Plants. The "failure rate" and the "average downtime per failure" of electric utility supplies to industrial plants are given in Table 40. Additional details from this survey published in 1975 are given in Reference [9] (see Appendix D). A total of 87 plants participated in the IEEE survey covering the period from January 1, 1968 through October 1974.

The survey results shown in Table 40 have distinguished between power failures that were terminated by a switching operation versus those requiring repair or replacement of equipment. The latter have a much longer outage duration time. Some of the conclusions that can be drawn from the IEEE data are:

(1) The failure rate for single-circuit supplies is about 6 times that of multiple-circuit supplies that operate with all circuit breakers closed; the average duration of each outage is about 2.5 times as long.

Table 40
IEEE Survey of Reliability of Electric Utility Supplies to Industrial Plants [9]
(1975 Survey)
(See Tables II, III, IV, and V in Appendix D for additional details.)

	Failures per Unit-Year [1]			Average Duration (Minutes per Failure) [1]		
	λ_S	λ_R	λ	r_S	r_R	r
Single-Circuit Utility Supplies						
Voltage Level						
V ≤ 15 kV	0.905	2.715	3.621	3.5	165	125
15 kV < V ≤ 35 kV	—	1.657	1.657	—	57	57
V > 35 kV	0.527	0.843	1.370	1.5	59	37
All	0.556	1.400	1.956	2.3	110	79
Multiple-Circuit Utility Supplies (All Voltage Levels)						
Switching Scheme						
All Breakers Closed	0.255	0.057	0.312	8.5	130	31
Manual Throwover	0.732	0.118*	0.850	8.1	84*	19
Automatic Throwover	1.025	0.171	1.196	0.6	96	14
All	0.453	0.085	0.538	5.2	110	22
Multiple-Circuit Utility Supplies (All Switching Schemes)						
Voltage Level						
V ≤ 15 kV	0.640	0.148	0.788	4.7	149	32
15 kV < V ≤ 35 kV	0.500	0.064*	0.564	4.0	115*	17
V > 35 kV	0.357	0.067	0.424	6.1	184	34
Multiple-Circuit Utility Supplies (All Circuit Breakers Closed)						
Voltage Level						
V ≤ 15 kV	0.175	0.088*	0.263	0.7	335*	112
15 kV < V ≤ 35 kV	0.342	0.019*	0.361	7.0	120*	13
V > 35 kV	0.250	0.061	0.311	11.0	203	49

NOTES: *Small sample size; less than eight failures.

(1) Failure rates λ_S and λ_R and average durations r_S and r_R are, respectively, rates and durations of failures terminated by switching and by repair or replacement. Unsubscripted rates and durations are overall values.

(2) Failure rates for multiple-circuit supplies that operate with either a manual or an automatic throwover scheme are comparable to those for single-circuit supplies; but throwover schemes have a smaller average failure duration than single-circuit supplies.

(3) Failure rates are highest for utility supply circuits operated at distribution voltages and lowest for circuits operated at transmission voltages (greater than 35 kV).

It is important to note that the data in Table 40 show that the two power sources of a double-circuit utility supply are not completely independent. This is analyzed in an example in 7.1.15, where (for the one case analyzed) the actual failure rate of a double-circuit utility supply is more than 200 times larger than the calculated value for two completely independent utility power sources.

Utility supply failure rates vary widely in various locations. One of the significant factors in this difference is believed to be different exposures to lightning storms. Thus, average values for the utility supply failure rate may not be appropriate for use at any one location. Local values should be obtained, if possible, from the utility involved, and these values should be used in reliability and availability studies.

An earlier IEEE reliability survey of electric power supplies to industrial plants was published in 1973 and is reported in Table 3 of Reference [13] (see Appendix A). The earlier survey had a smaller data base and is not believed to be as accurate as the one summarized in Table 40. The earlier survey of electric utility power supplies had lower failure rates.

3.3.2.6 Method of Electrical Service Restoration to Plant. The 1973–75 IEEE data on "method of electrical service restoration to plant" is shown in Table 41. A percentage breakdown of the method of restoration to plant is ranked as follows:

(1) Replacement of failed component with spare   22%
(2) Repair of failed component   22%
(3) Other   22%
(4) Utility service restored   12%
(5) Secondary selection — manual   11%
(6) Primary selection — manual   7%
(7) Primary selection — automatic   2%
(8) Secondary selection — automatic   2%
(9) Network protector operation — automatic   0%

The most common methods of service restoration to plant are replacement of the failed component with a spare or the repair of the failed component. The primary selection or secondary selection is used only 22% of the time. This would indicate that most power distribution systems in this IEEE survey were radial.

3.3.2.7 Equipment Failure Rate Multiplier versus Maintenance Quality. The relationship between maintenance practice and equipment failures is discussed in detail in Chapter 5. Equipment failure rate multipliers versus maintenance quality are given in Chapter 5 for transformers, circuit breakers, and motors. These multipliers were determined in a special study (Part 6 of Reference [16]) (see Appendix B). The failure rate of motors is very sensitive to the quality of maintenance.

The percentage of failures due to "inadequate maintenance" versus the "time since maintained" is given in Chapter 5 for circuit breakers, motors, open wire, transformers, and all electrical equipment classes combined. A high percentage of electrical equipment failures was blamed on "inadequate maintenance" if there had been no maintenance for more than 2 years prior to the failure.

3.3.2.8 Reliability Improvement of Electrical Equipment in Industrial Plants Between 1962 and 1973. The failure rates for electrical equipment (except for motor starters) in industrial plants appeared to have improved considerably during the 11 year interval between the 1962 AIEE reliability survey (see Reference [6]) and the 1973–74 IEEE reliability survey (see Reference [13]). Table 42 shows

Table 41
Method of Service Restoration (1974 Survey)

Method	Total %	Electric Utility Power Supplies %	Transformers %	Circuit Breakers %	Motor Starters %	Motors %	Generators %	Disconnect Switches %	Switchgear Bus—Insulated %	Switchgear Bus—Bare %	Bus Duct %	Open Wire %	Cable %	Cable Joints %	Cable Terminations %
(1) Primary selection—manual	7%	1%	3%	6%	0%	5%	20%	0%	58%	25%	20%	13%	14%	28%	19%
(2) Primary selection—automatic	2	8	0	1	0	0	0	0	0	5	0	4	5	8	0
(3) Secondary selection—manual	11	1	25	6	0	14	33	0	17	10	10	2	20	32	23
(4) Secondary selection—automatic	2	1	3	8	0	0	0	0	0	0	0	1	0	8	4
(5) Network protector operation—automatic	0+	0	0	0	0	0	0	0	0	5	0	0	0	0	0
(6) Repair of failed component	22	5	25	11	12	30	20	3	17	20	35	31	42	24	27
(7) Replacement of failed component with spare	22	2	39	38	10	29	14	77	0	10	35	6	2	0	12
(8) Utilty service restored	12	81	0	1	0	0	13	0	0	0	0	1	1	0	0
(9) Other	22	1	5	29	78	22	0	20	8	25	0	42	16	0	15
Total percentage	100%	100%	100%	100%	100%	100%	100%	100%	100%	100%	100%	100%	100%	100%	100%
Total number reported	1204	171	75	160	68	318	15	69	12	20	20	103	122	25	26

Table 42
Failure Rate Improvement Factor of
Electrical Equipment in Industrial Plants
During 11 Year Interval Between 1962
AIEE Survey and 1973 IEEE Survey

Equipment Category	Failure Rate Ratio AIEE (1962) / IEEE (1973)
Cable	
Nonleaded in Underground Conduit	9.7
Nonleaded, Aerial	5.8
Lead Covered in Underground Conduit	3.4
Nonleaded in Above-Ground Conduit	1.6
Cable Joints and Terminations	
Nonleaded	5.3
Leaded	2.0
Circuit Breakers	
Metalclad Drawout, 0-600 V	6.0
Metalclad Drawout, Above 600 V	2.9
Fixed 2.4-15 kV	2.5
Disconnect Switches	
Open, Above 600 V	3.4
Enclosed, Above 600 V	1.6
Open Wire	3.4
Transformers	
Below 15 kV, 0-500 kVA[1]	2.0
Below 15 kV, Above 500 kVA	2.0
Above 15 kV	1.6
Motor Starters, Contactor Type	
0-600 V	1.3
Above 600 V	1.3

NOTE: (1) 300-750 kVA for 1973.

how much the failure rates have improved for several equipment categories. These data are calculated from a 1974 report (see Reference [13]). In 1962, circuit breakers had failure rates that were 2.5 to 6 times higher than those reported in 1973. The largest improvements in equipment failure rates have occurred on cables and circuit breakers. The authors discussed some of the reasons for the failure rate improvements during the 11 year interval. It would appear that manufacturers, application engineering, installation engineering, and maintenance personnel have all contributed to the overall reliability improvement.

The authors also make a comparison between the surveys of the "actual down-time per failure" for all the equipment categories shown in the table in Reference

[13]. However, in general, the "actual downtime per failure" was greater in 1973 than in 1962.

3.3.2.9 Loss of Motor Load versus Time of Power Outage. A special study was reported in Table 47 of Reference [13] (see Appendix B) on the loss of motor load versus duration of power outages. When the duration of power outages is longer than 10 cycles, most plants lose motor load. However, when the duration of power outages is between 1 and 10 cycles, only about one-third of the plants lose their motor load.

Test results of the effect of fast bus transfers on load continuity are reported in Reference [5]. This includes 4 kV induction and synchronous motors with the following type of loads:

(1) Forced draft fan
(2) Circulating water pump
(3) Boiler feed booster pump
(4) Condensate pump
(5) Gas recirculation fan

A list of prior papers on the effect of fast bus transfer on motors is also contained in Reference [4].

3.3.2.10 Critical Service Loss Duration Time. What is the maximum length of time that an interruption of electrical service will *not* stop plant production? The median value for all plants is 10 seconds. See Table 2 in Chapter 2 for a summary of the IEEE survey of industrial plants.

What is the maximum length of time before an interruption to electrical service is considered critical in commercial buildings? The median value of all commercial buildings is between 5 and 30 minutes. See Table 3 in Chapter 2 for a summary of IEEE survey of commercial buildings.

3.3.2.11 Plant Restart Time. What is the plant restart time after service is restored following a failure that has caused a complete plant shutdown? The median value of the plant restart time for all plants is 4 hours. See Table 4 in Chapter 2 for a summary of the IEEE survey of industrial plants.

3.3.2.12 Other Sources of Reliability Data. The reliability data from industrial plants that are summarized in 3.3.2 are based upon Reference [13], which was published during 1973–75. Reference [6] is an earlier reliability survey of industrial plants that was published in 1962. Portions of that data are tabulated in 3.2.4.8.4.

Many sources of reliability data on similar types of electrical equipment exist in the electric utility industry. The Edison Electric Institute (EEI) has collected and published reliability data on power transformers, power circuit breakers, metalclad switchgear, motors, excitation systems, and generators (see References [17]–[23]). Most EEI reliability activities do not collect outage duration time data. The North American Electric Reliability Council (NERC) collects and publishes reliability and availability data on generation prime mover equipment.

Failure rate data and outage duration time data for power transformers, power circuit breakers, and buses are given in Reference [16]. These data have come from electric utility power systems.

Very little other published data is available on failure modes of power circuit breakers and on the probability of a circuit breaker not operating when called upon to do so. An extensive worldwide reliability survey of the major failure modes of power circuit breakers above 63 kV on utility power systems has been made by CIGRE 13-06 Working Group as shown in Appendix J. Failure rate data and failure per operating cycle data have been determined for each of the major failure modes. Outage duration time data has also been collected. In addition, data has been collected on the costs of scheduled preventive maintenance; this includes the man-hours per circuit breaker per year and the cost of spare parts consumed per circuit breaker per year.

IEEE Std 500-1984 (ANSI) [1] is a reliability data manual for use in the design of nuclear power generating stations. The equipment failure rates therein cover such equipment as annunciator modules, batteries and chargers, blowers, circuit breakers, switches, relays, motors and generators, heaters, transformers, valve operators and actuators, instruments, controls, sensors, cables, raceways, cable joints, and terminations. No information is included on equipment outage duration times.

The Institute of Nuclear Power Operations (INPO) organization operates the Nuclear Plant Reliability Data System (NPRDS), which collects failure data on electrical components in the safety systems of nuclear power plants. Outage duration time data is collected on each failure. The NPRDS data base contains more details than IEEE Std 500-1984 (ANSI) [1]; but INPO has followed a policy of not publishing its data.

3.4 References. The information in this chapter shall be used in conjunction with the following publications:

[1] IEEE Std 500-1984, IEEE Guide to the Collection and Presentation of Electrical, Electronic, Sensing Component, and Mechanical Equipment Reliability Data for Nuclear Power Generating Stations (ANSI).

[2] IEEE Std 841-1986, IEEE Recommended Practice for Chemical Industry Severe Duty Squirrel-Cage Induction Motors — 600 V and Below (ANSI).

[3] Albrecht, P. F., Appiarius, J. C., Cornell, E. P., Houghtaling, D. W., McCoy, R. M., Owen, E. L., and Sharma, D. K. "Assessment of the Reliability of Motors in Utility Applications," *IEEE Transactions on Energy Conversion*, vol. EC-2, Sep. 1987, pp. 396–406.

[4] Albrecht, P. F., Appiarius, J. C., Cornell, E. P., Houghtaling, D. W., McCoy, R. M., Owen, E. L., and Sharma, D. K. "Assessment of the Reliability of Motors in Utility Applications Updated," *IEEE Transactions on Energy Conversion*, vol. EC-1, Mar. 1986, pp. 39–46.

[5] Averill, E. L. "Fast Transfer Test of Power Station Auxiliaries," *IEEE Transactions on Power Apparatus and Systems*, vol. PAS-96, May/Jun. 1977, pp. 1004–1009.

[6] Dickinson, W. H. "Report of Reliability of Electrical Equipment in Industrial Plants," *AIEE Transactions*, part II, Jul. 1962, pp. 132–151.

[7] IEEE Committee Report. "Cost of Electrical Interruptions in Commercial Buildings," *IEEE-ICPS Technical Conference Record*, 75-CHO947-1-1A, Toronto, Canada, May 5–8, 1975, pp. 124–129. (see Appendix C)

[8] IEEE Committee Report. "Reasons for Conducting a New Reliability Survey on Power, Rectifier, and Arc-Furnace Transformers," *IEEE-ICPS Technical Conference Record*, May 1979, pp. 70–75.

[9] IEEE Committee Report. "Reliability of Electric Utility Supplies to Industrial Plants," *IEEE-ICPS Technical Conference Record*, 75-CHO947-1-1A, Toronto, Canada, May 5–8, 1975, pp. 131–133. (see Appendix D)

[10] IEEE Committee Report. "Report of Generator Reliability Survey of Industrial Plants and Commercial Buildings," *IEEE-ICPS Technical Conference Record*, CH1543-8-1A, May 1980, pp. 40–44.

[11] IEEE Committee Report. "Report of Switchgear Bus Reliability Survey of Industrial Plants," *IEEE Transactions on Industry Applications*, Mar./Apr. 1979, pp. 141–147. (see Appendix E)

[12] IEEE Committee Report. "Report of Transformer Reliability Survey of Industrial and Commercial Buildings," *IEEE Transactions on Industry Applications*, 1983, pp. 858–866. (see Appendix G)

[13] IEEE Committee Report. "Report on Reliability Survey of Industrial Plants, Parts 1–6," *IEEE Transactions on Industry Applications*, Mar./Apr. 1974, pp. 213–252; Jul./Aug. 1974, pp. 456–476; Sep./Oct. 1974, p. 681. (see Appendixes A and B)

[14] McWilliams, D. W., Patton, A. D., and Heising, C. R. "Reliability of Electrical Equipment in Industrial Plants—Comparison of Results from 1959 Survey and 1971 Survey," *IEEE-ICPS Technical Conference Record*, 74CHO855-71A, Denver, CO, Jun. 2–6, 1974, pp. 105–112.

[15] O'Donnel, P. "Report of Large Motor Reliability Survey of Industrial Plants & Commercial Buildings," *IEEE Transactions on Industry Applications*, Parts 1 and 2, Jul./Aug. 1985, pp. 853–872; Part 3, Jan./Feb. 1987, pp. 153–158. (see Appendix H)

[16] Patton, A. D. "Determination and Analysis of Data for Reliability Studies," *IEEE Transactions on Power Apparatus and Systems*, vol. PAS-87, Jan. 1968, pp. 84–100.

[17] "Report on Equipment Availability for 10 Year Period 1965–74," *EEI Publication no. 75-50* (Prime Mover Generation Equipment).

NOTE: These data are now collected and published by the North American Electric Reliability Council (NERC).

[18] "Report on Excitation System Troubles—1975," *EEI Publication no. 76-78*, Dec. 1976. (Later data have also been published.)

[19] "Report on Generator Troubles—1975," *EEI Publication no. 76-82*, Dec. 1976. (Later data have also been published.)

[20] "Report on Metalclad Switchgear Troubles—1975," *EEI Publication no. 76-82*, Dec. 1976. (Later data have also been published.)

[21] "Report on Motor Troubles—1975," *EEI Publication no. 76-79*, Dec. 1976. (Later data have also been published.)

[22] "Report on Power Circuit Breaker Troubles — 1975," *EEI Publication no. 76-81*, Dec. 1976. (Later data have also been published.)

[23] "Report on Power Transformer Troubles — 1975," *EEI Publication no. 76-80*, Dec. 1976. (Later data have also been published.)

[24] "Summary of Replies to the 1982 Technical Questionnaire on Rotating Machinery (Motors 1000 hp and Up)," *Unpublished Report at Doble Conference*, Apr. 1982, Boston, MA.

Chapter 4
Evaluating and Improving the
Reliability of an Existing Plant

4.1 Introduction. The 1974 survey of electrical equipment reliability in industrial plants (see Reference [4][4]) and subsequent investigations showed the utility supply as being the largest single component affecting the reliability of an industrial plant. (See Table 3 in Appendix A and Table 40 in Chapter 3.) Industrial users may or may not be in a position to improve the utility supply reliability and, as a result, must also focus their attention on critical areas within their own plants. A logical approach to the analysis of options available in the industrial plant (in terms of both utility supply and plant distribution) will lead to the greatest reliability improvement for the least cost. In many instances, reliability improvements can be obtained without any cost by making the proper inquiries.

Most industrial users simply "hook up" to the utility system and do not fully recognize that their requirements can have an impact on how the utility supplies them. A utility is somewhat bound by the system available at the plant site and the investment that can be made per revenue dollar. However, most utilities are willing to discuss the various supply systems that are available to their customers. Many times, an option is available (sometimes with financial sharing between the user and the utility) that will meet the exact reliability needs of an industrial plant.

A thorough and properly integrated investigation of the entire electric system (plant and supply) will pinpoint the components or subsystems having unacceptable reliability. Some important general inquiries are listed below. Many of these questions apply to both the utility and the plant distribution systems.

(1) How is the system supposed to operate?
(2) What is the physical condition of the electric system?
(3) What will happen if faults occur at different points?
(4) What is the probability of a failure and its duration?
(5) What is the critical duration of a power interruption that will cause significant financial loss? (That is, will a 1 minute interruption cost production dollars or merely be an inconvenience?)
(6) Is there any fire or health hazard that will be precipitated by an electrical fault or a power loss?
(7) Is any equipment vulnerable to voltage dips or surges?

---

[4]The numbers in brackets correspond to those in the references at the end of this chapter.

The answers to these and similar questions, if properly asked, can and will result in savings to the industrial user if they produce *action*.

The question at this point should be "How do I get started?" However, another question could be, "Why bother?" The answer to the former question is covered in this chapter, and the answer to the latter question is based on the following analogy. When preparing for a long trip, a motorist will make sure that his car is in good working condition before he leaves. He will check the brakes, engine, transmission, tires, exhaust system, etc., to see that they are in good condition and make the required repairs. For the motorist knows that "on-the-road" breakdowns and failures are expensive, time consuming, and can be hazardous. In an industrial plant, an unplanned electrical failure will consume valuable production time as well as dollars and may cause injury to personnel. Circuit breakers, relays, meters, transformers, wireways, etc., need periodic checks and preventive maintenance (see Chapter 5) to improve the likelihood of trouble-free performance. Some plants have been shut down completely by events such as a ballast failure. These "shutdowns" are commonly caused by improper settings in protective devices, circuit breaker contacts that were welded shut, or relays that were not set (or did not react) properly. This chapter shows the plant engineer how to minimize downtime by analyzing the system.

4.2 Utility Support Availability. Loss of incoming power will cause an interruption to critical areas unless alternate power sources are available. Therefore, the reliability of the incoming power is of paramount importance to the plant engineer. It can be stated that different plants and even circuits within a plant vary in their response to loss of power. In some cases, production will not be significantly affected by a 10 minute power interruption. In other cases, a 10 millisecond interruption or disturbance will cause significant loss. The plant engineer should assess the plant's vulnerability and convey his or her requirements to the local utility (as well as his or her own management). (See 2.2 in Chapter 2 for information on economic loss versus unavailability of incoming power.)

The local utility should be able to supply a listing of the number, type, and duration of power interruptions over the preceding 3 to 5 year period. The utility should also be able to predict the future average performance based on its historical data and planned construction projects. In addition, the utility may be able to supply the "feeder" performance of other circuits near the facility under investigation. A second alternative would be to obtain a diagram of the utility feed and evaluate its availability using Chapter 2 methods. As a last resort, the average numbers in this book will provide a good base (see Table 40 in Chapter 3).

The utility's history of interruptions can be compared with recorded plant dollar loss in verifying process vulnerability. By assigning a dollar loss to each interruption, it will be possible to determine a relationship between the duration of a power loss and a monetary loss for a particular industrial plant. When the actual outage cost is higher or lower than would be predicted, the cause of the deviation should be determined (that is, a 15 minute power loss at a shift change will be less costly than one during peak production). With a refined cost formula in hand, the cost of available options versus projected losses can be evaluated.

Occasionally, a plant will experience problems at times other than during a recorded outage. These problems may be caused by voltage dips (or, more rarely, voltage spikes) which are difficult to trace. With problems such as these, it is necessary to begin recording the exact date and time of these occurrences and ask the utility to search for faults or other system disturbances at (or near) the specific times that they have been recorded. It would be wise to convey the fault times to the utility reasonably soon after the fault (that is, call them the following day). It must be emphasized that, unless these problems are significant in terms of dollars lost, safety, or frequency (that is, every other day), it is not reasonable to pursue the cause of voltage dips since they are a natural phenomenon in the expansive system operated by a utility. Frequent dips can be caused by large motor starts, welder inrush, or intermittent faults in the plant's distribution system (or even by a neighbor's system).

It is also reasonable to cover "what if" questions with the utility and to weigh their answers in any supply decision. A list of questions include

(1) How long will the plant be without power if:
    (a) The main transformer fails?
    (b) The feed to the main transformer fails?
    (c) The pole supporting the plant feed is struck by a vehicle and downed?
    (d) The utility main line fuse or protector interrupts?
    (e) The utility main feed breaker opens for a fault?
    (f) The utility substation transformer fails?
    (g) The utilty substation feeds are interrupted?

(2) What kind of response time can be expected from the utility for loss of power:
    (a) During a lightning storm?
    (b) During a low trouble period (that is, "normal" conditions)?
    (c) During a snow or ice storm?
    (d) During a heat storm (that is, long periods of high temperatures)?

(3) What should be done when the plant experiences an interruption?
    (a) Who should be called? A name and number should be made available to *all* responsible personnel. Alternates and their numbers should also be included.
    (b) What information should be given to those called?
    (c) How should plant personnel be trained to respond?
    (d) Can plant personnel restore power by switching utility lines, and who should be contacted to obtain permission to switch?

(4) Are there any more reliable feeds near the plant, and what is the cost of extending them to the plant? (That is, is a spare feed available, and what is the cost to make it available?)
    (a) Is this additional feed from the same station or from another station?
    (b) What is the probability (frequency and duration) of both the main and the spare feeds being interrupted simultaneously?
    (c) What is the reliability improvement obtained from the additional (or alternate) feed?

(5) Will the utility's protective equipment coordinate with the plant's service circuit breaker? If not, what can be done to coordinate these series protective devices?

(6) What is the available short-circuit current, and are there plans to change the system so as to affect the short-circuit current?

All of the above questions may not apply to all plants, but should be matched with specific plant requirements.

There is an important fact to consider when a multiple-ended feed is being considered. While service is maintained during the loss of one of the feeds, a voltage depression will be seen until the fault is cleared by proper relay action. Therefore, the plant will see a voltage dip for *any* faults on *all* incoming feeds. If the plant is affected with equal severity by either a voltage dip or a short-duration (several seconds) interruption, a multiple-ended supply (with secondary tie) may actually worsen plant reliability. This is just one example of the need to carefully evaluate the current supply situation in conjunction with the net improvement of various proposals.

4.3 Where to Begin — The Plant Single-Line Diagram. The "blueprint" for electrical analysis is the "single-line diagram." The existence of a single-line diagram is essential for any plant electrical engineer, manager, or operator. It is the "road map" to any part of the electric system. In fact, a single-line diagram should be prepared even if the ensuing analysis is not done.

The single-line diagram should begin at the incoming power supply. Standard IEEE symbols should be used in representing electrical components (see IEEE Std 315-1975 (ANSI) [3]). It is usually impractical to show all circuits in a plant on a single schematic; so the initial single-line diagram should show only major components, circuits, and panels. More detailed analysis may be required in critical areas (described later), and additional single-line diagrams should be prepared for these areas as required.

Since an analysis is being made from the single-line diagram, the type, size, and rating of each device as well as its unavailability should be shown on the diagram. The diagram should include at least the following information:

(1) Incoming lines (voltage and size — capacity and rating)
(2) Generators (in plant)
(3) Incoming main fuses, potheads, cutouts, switches, and main and tie breakers
(4) Power transformers (rating, winding connection, and grounding means)
(5) Feeder breakers and fused switches
(6) Relays (function, use, and type)
(7) Potential transformers (size, type, and ratio)
(8) Current transformers (size, type, and ratio)
(9) Control transformers
(10) All main cable and wire runs with their associated isolating switches and potheads (size and length of run)
(11) All substations, including integral relays and main panels and the exact nature of the load in each feeder and on each substation

The single-line diagram may show planned, as well as actual, feeder, circuit breaker, and substation loads (actual measurements should be taken). In most industrial plants, load is added (or deleted) in small increments, and the net effect is not always seen until some part of the system becomes overloaded (or underloaded). Many times, circuits are added without appropriate modification of the standard settings on the associated upstream circuit breakers. In addition, original designs may not have included special attention to the critical areas of production. With these thoughts in mind, the following information should be added to the single-line diagram:

(1) The original system should be identified. The exact nature of the new loads and their approximate locations should be noted.
(2) Critical areas of the system should be highlighted.
(3) The component reliability numbers from Chapter 3 should be inserted so that the reliability performance of the plant can be analyzed on an "if new" basis. (It is preferable to use numbers indigenous to a particular plant whenever this information is available.)

The above information may be too voluminous for clear representation on a single drawing. It may, therefore, be advantageous to include the incoming supply and main feeder circuit breakers (at least) and even major equipment (very large motors or groups requiring the entire capacity of a main feeder position) on one diagram. The load end of the feeders can be detailed on one or more subsequent drawings. After completion of the single-line diagrams, a comprehensive analysis can begin. However, the general inspection covered in 4.4 can, and should, be made concurrent with the preparation of a plant single-line diagram.

The single-line diagram is a picture of an ever-changing electric system. The efforts in preparing the diagram and analyzing the system should, therefore, be augmented by a means to capture new pictures of the system (or of proposed systems) as changes are made (or proposed). Therefore, a procedure should be formalized to ensure that all proposals undergo reliability scrutiny (as well as single-line diagram update), and that their effect on the total system is analyzed before the proposal is approved. This process not only maintains the system's integrity; but it also minimizes expense by more effective utilization of existing electrical facilities.

4.4 Plant Reliability Analysis. An inspection analysis of the physical condition of a plant's distribution system can be utilized (hopefully on a continual basis) to improve plant reliability. The following inspection requires little, if any, capital investment while providing a favorable increase in reliability:

(1) Equipment should be periodically checked for proper condition, and programs should be initiated for preventive maintenance procedures as required. (See Chapter 5 for further information.)
    (a) Oil in transformers and circuit breakers should be periodically checked for mineral, carbon, and water content as well as level and temperature.
    (b) Molded case circuit breakers should be exercised periodically (that is, operated "on" to "off" to "on").

    (c) Terminals should be tightened. Each terminal should be inspected for discoloration (overheating) which is generally caused by either a bad connection or equipment overload. Cabinets, etc., should be checked for excessive warmth. Remember that circuit breakers and fuses interrupt as a result of heat in the overload mode.

    (d) Surge arresters should be checked for their readiness to operate.

(2) Distribution centers should be checked to see that spare fuses are available. Spare circuit breakers may also be necessary for odd sizes or special applications.

(3) Switches, disconnect switches, bus work, and grounds should be checked for corrosion, and unintentional entry of water or corrosive foreign material. It may be wise to operate suspected switches to see that their mechanisms are free, so that faults can be properly isolated and switches safely re-fused.

(4) The mechanical part of the electric system should be checked.

    (a) The conduit, duct, cable tray, and busway systems should be well supported mechanically, and the grounding system should be electrically continuous. Employees can be shocked or injured if a circuit faults to ground without a solid continuous return path to the source interrupter. Supports, such as wood poles, should be checked for excessive rusting or rotting, which would significantly reduce their mechanical strength.

    (b) Open wire circuits should be checked for insulator and surge arrester failure and contamination.

    (c) The system's key locations (open area distribution centers and lines) should be checked for foreign growth, such as trees, weeds, shrubs, etc., as well as for general accessibility. The distribution centers should be free from storage of trash, flammables, or even general plant inventory.

    (d) Permanent and portable wiring should be checked for fraying or other loss of insulating value.

    (e) In general, the system should be checked for any obvious situations where accidents could precipitate an interruption.

(5) The electrical supply room(s) should be thoroughly checked.

    (a) The relay and control power fuses should be intact (not blown).

    (b) All indicating lights should be operable and clearly visible.

    (c) All targets should be reset so that none show a tripping. Counters (if any) should be checked and the count (number) should be recorded.

    (d) The control power, batteries, emergency lighting, and emergency generation should be tested and checked to see that they are operational. In many cases, plants have been unable to transfer to their spare circuit or start their standby generator because of dead batteries.

(6) Switches, conduits, busways, and duct systems should be checked for overheating. This could be caused by overload equipment, severely unbalanced loads, or poor connections.

4.5 Circuit Analysis and Action. The first subsequent investigation, following completion of the plant single-line diagram is the analysis of the system to pinpoint design problems. Key critical or vulnerable areas, and overdutied or improperly protected equipment can be located by the following procedure:

(1) Assign faults to various points in the system and note their effect on the system. For example, assume that the cable supply to the air conditioning compressor failed. How long could operations continue? Is any production cooling involved? Are any computer rooms cooled by this system? What would happen if a short circuit (or ground fault) occurred on the secondary terminals of a unit substation? Consideration should be given to relay action (including backup protection), service restoration procedures, etc., in this "what if" analysis. This review could be called a failure mode and effects analysis.

(2) Calculate feeder loads to verify that all equipment is operating within its rating (do not forget current transformers and other auxiliary equipment). Graphic or demand ammeters (as required) should be used to gather up-to-date information. Fault duties should also be considered (see Chapter 5 in IEEE Std 141-1986 (ANSI) [1]).

(3) Perform a relay coordination analysis (see IEEE Std 242-1986 (ANSI) [2] or Chapter 4 in IEEE Std 141-1986 (ANSI) [1]).

(a) Are the relays and fuses properly set or rated for the current load levels?

(b) Is there any new load that has reduced critical circuit reliability (or increased vulnerability)?

Obviously, overloaded equipment should be replaced or load transferred so that the equipment can be operated well within its rating. The major protection points — outside the critical areas — should be capable of keeping the system intact by clearing faults and allowing the critical process to continue. The probability of jeopardizing the critical circuits by extraneous electrical faults should be minimized, either by physically isolating the critical circuits or by judicial use and proper maintenance of protective devices to electrically sever and isolate faults from critical circuits.

With isolation criteria secure, the investigation should move to the critical circuits themselves to see that proper backup equipment is available and that restoration procedures are adequate. For example, a conveyor system with large rollers may have one motor for each roller, or several hundred motors. The failure rate is 0.0109 per unit-year for the motors, or two motor failures can be expected annually for a plant with 200 motors. The typical downtime is 65 hours (but could be less for this specific example). In this case, there should be a means of separating the motor from the systems and allowing the conveyor system to continue operation (possibly allowing the roller to idle until the end of a shift), and several spare motors should be available to minimize downtime.

Most plants have a population of motors large enough to expect several failures per year. The large variety usually precludes the maintenance of a spare motor stock (although their availability can be checked with local distributors). Highly critical nonstandard equipment may require spares. However, each component of the electrical system should be viewed in its relationship to the critical process and downtime. (Relay or fuse coordination again plays an important role here.)

The worth of carrying spare parts should be carefully weighed when long process interruptions could result from a single component failure.

4.6 Other Vulnerable Areas. In many plants, the major process is controlled by a small component. This component may be a rectifier system, a computer, or a retrofitted magnetic or punched-tape system. The continuity of the electric feed to this controller is just as important to the process as the main machine itself. By proper application of power sources within the device (usually large banks of capacitors and/or batteries) or external uninterruptible power sources, the control can cause the equipment to go into a "safe-hold" position if the power source is interrupted. This continuity (availability) is important to note when thousands of dollars worth of products are being machined in one operation (such as in the aircraft industry). The accuracy and efficacy of a computer or a computer-based process is directly related to the "quality" of its environment. This quality is determined by more than just the continuity of the electric supply. Voltage dips, line noise, ineffective grounding, extraneous electrical and magnetic fields, temperature changes, and even excessively high humidity can adversely affect the accuracy of a computer (or to a lesser extent, a microprocessor). To minimize the probability of errors, the computer should be properly shielded and grounded. It may even be beneficial to install a continuous, uninterruptible power supply or transient suppressor equipment on computer circuits where the controlled process is critical.

Testing facilities should have a backup power supply where interruptions could abort long-term testing (that is, tests that span large periods of time). It is important to note that only sufficient power needs to be supplied to operate the test itself.

Another area of importance is the lighting required for safe operation of the machines. A failure in a particular lighting circuit may reduce the area lighting to a level below what is necessary to maintain a safe watch over production. Two means of overcoming this vulnerability are (1) emergency task lighting and (2) sufficient lighting such that a single circuit outage does not reduce lighting to an unacceptable level. Another important lighting consideration is the fact that some metal halide lights require as long as 15 minutes to restart after being extinguished. Since even severe voltage dips can extinguish this type of lighting (a dip that may go virtually unnoticed by production equipment), supplementary lighting is necessary when metal halide light is a primary source of illumination. Other new high output lamps will restart in 1 to 6 minutes; but this too can cause production problems.

Air, oil, and water systems are frequently important auxiliary inputs upon which production depends. A compressor outage can, for example, cause significant production loss. While failures in these systems are usually mechanical in nature, electrical failures are not uncommon. Pumps are often integral parts of the cooling system in large transformers or even in rectifier circuits, and loss of coolant circulation could either shut down the equipment or significantly reduce production output. Therefore, pumps should be well maintained (mechanically and electrically) when they comprise a significant part of the system, and spare parts may be a wise investment. Ventilation can also be critical to cooling, and ventilator fans are often neglected—until they fail. Hence, periodic maintenance and/or spare ventilator motors may be a good investment.

Some plants rely on a single cable to supply their entire electrical requirements, and many plants rely on single cables for major blocks of load. In these cases, it may be prudent to take several precautionary steps. One possible step would be the periodic testing of cables (see Reference [5]). Another measure would be the use of spare cables or the storage of a single "portable" cable with permanently made ends (and provisions for installing the portable cable at the various cable terminations in the plant distribution system). Lastly, *advance* (documented) arrangements could be made with a local contractor or the local utility for the use of their portable cables (and/or services) on an emergency basis.

Premature equipment failure can result from electrical potential that is either too high, too low, excessively harmonic laden, or unbalanced (and also a combination of any or all of these). Voltage tolerances are fairly well established by NEMA and ANSI. However, in Reference [6], a means is provided to evaluate a situation where more than one area deviates from rating. It must be noted that some situations are offsetting, such as a high voltage (less than 10% high) and unbalanced voltage.

It is important to record and log voltage levels (of all three phases) at various strategic points on a periodic basis (that is, annually) and to occasionally determine the harmonic content in the plant's distribution system. The widespread use of solid-state switching devices has caused an increase in harmonic content in the plant power; but it has been unofficially reported that such devices must approach 50% of the plant load before significantly detrimental effects occur. However, the engineer must look at harmonic content in conjunction with other criteria to determine whether there is cause for a significant loss of life in his or her equipment. Filter circuits are generally used to remove harmful harmonics, and their nature is beyond the scope of this publication. Fluorescent lighting also produces harmonics; but these harmonics are "blocked" by the use of delta–wye transformers.

4.7 Conclusion. The plant engineer should analyze his or her system electrically and physically and inquire about the utility's system. In this analysis, the engineer should:
  (1) See that faults are properly isolated and that critical loads are not vulnerable to interruption or delayed repair.
  (2) Analyze the critical areas and evaluate the need for special restoration equipment, spare parts, or procedures.
  (3) Based on probability and economic analysis, make capital or preventive maintenance investments as indicated by the analysis.
  (4) Make carefully documented contingency (catastrophe) plans.
  (5) Check the quality of the power supply from the utility and throughout the plant to determine if the equipment is vulnerable to premature failure.
  (6) Develop preventive maintenance, checking, and logging procedures to ensure the continuous optimum reliable performance of the plant.

4.8 References. The information in this chapter shall be used in conjunction with the following publications:

[1] IEEE Std 141-1986, IEEE Recommended Practice for Electric Power Distribution for Industrial Plants (ANSI).

[2] IEEE Std 242-1986, IEEE Recommended Practice for Protection and Coordination of Industrial and Commercial Buildings (ANSI).

[3] IEEE Std 315-1975 (Reaff. 1989), IEEE Standard Graphic Symbols for Electrical and Electronics Diagrams (ANSI).

[4] IEEE Committee Report. "Report on Reliability Survey of Industrial Plants," Parts 1-6, *IEEE Transactions on Industry Applications*, vol. IA-10, Mar./Apr., Jul./Aug., Sep./Oct. 1974, pp. 213-252, 456-476, 681.

NOTE: Reference [4] is reprinted in Appendixes A and B.

[5] Lee, R. "New Developments in Cable System Testing," *IEEE Transactions on Industry Applications*, vol. IA-13, May/Jun. 1977.

[6] Linders, J. R. "Effects of Power Supply Variations on AC Motor Characteristics," *IEEE Transactions on Industry Applications*, vol. IA-8, Jul./Aug. 1972, pp. 383-400.

Chapter 5
Electrical Preventive Maintenance

5.1 Introduction. The objective of this chapter is to examine the "why" of electrical preventive maintenance and the role it can play in the reliability of distribution systems for industrial plants and commercial buildings. Details of "when" and "how" can be obtained from other sources (see References [1] and [8]-[15].[5])

Of the many factors involved in reliability, electrical preventive maintenance usually receives meager emphasis in the design phase and operation of electric distribution systems when, in fact, it can be a key factor in high reliability. Large expenditures for electric systems are made to provide the desired reliability; however, failure to provide timely and high-quality preventive maintenance leads to system or component malfunction or failure and prevents obtaining the intended design goal.

5.2 Definitions. The following definitions should be used in conjunction with this chapter:

electrical equipment. A general term including materials, fittings, devices, appliances, fixtures, apparatus, machines, etc., used as a part of, or in connection with, an electric installation.

electrical preventive maintenance. A system of planned inspection, testing, cleaning, drying, monitoring, adjusting, corrective modification, and minor repair of electric equipment to minimize or forestall future equipment operating problems or failures; which, depending upon equipment type, may require exercising or proof testing.

5.3 Relationship of Maintenance Practice and Equipment Failure. The Reliability Subcommittee of the IEEE Industrial and Commercial Power Systems Committee published the results of a survey that included the effect of maintenance quality on the reliability of electrical equipment in industrial plants (see Reference [11]). Each participant in the survey was asked to give his or her opinion of the maintenance quality in his or her plant. A major portion of the electrical equipment population covered in the survey had a maintenance quality that was classified as "excellent" or "fair." It is interesting to note that maintenance quality had a significant effect on the percentage of all failures blamed on "inadequate maintenance."

---

[5]The numbers in brackets correspond to those in the references at the end of this chapter.

As shown in Table 43, of the 1469 failures reported from all causes, "inadequate maintenance" was blamed for 240, or 16.4% of all the failures.

The IEEE data also showed that "months since maintenance" is an important parameter when analyzing the failure data of electrical equipment. Table 44 shows the data on failures caused by inadequate maintenance for circuit breakers, motors, open wire, transformers, and all equipment classes combined. The percentage of failures blamed on "inadequate maintenance" shows a close correlation with "failure, months since maintained."

From the IEEE data obtained, it was possible to calculate "failure rate multipliers" for transformers, circuit breakers, and motors based upon "maintenance quality." These "failure rate multipliers" are shown in Table 45 and can be used to adjust the equipment failure rates shown in Chapter 3. "Perfect" maintenance quality has zero failures caused by inadequate maintenance.

5.4 Design for Electrical Preventive Maintenance. Electrical preventive maintenance should be a prime consideration for any new electrical equipment installation. Quality, installation, configuration, and application are fundamental prerequisites in attaining a satisfactory preventive maintenance program. A system that is not adequately engineered, designed, and constructed will not provide reliable service, regardless of how good or how much preventive maintenance is accomplished.

One of the first requirements in establishing a satisfactory and effective preventive maintenance program is to have good quality electrical equipment that is properly installed. Examples of this are as follows:

  (1) Large exterior bolted covers on switchgear or large motor terminal compartments are not conducive to routine electrical preventive maintenance inspections, cleaning, and testing. Hinged and gasketed doors with a three-point locking system would be much more satisfactory.
  (2) Space heater installation in switchgear or an electric motor is a vital necessity in high humidity areas. This reduces condensation on critical insulation components. The installation of ammeters in the heater circuit is an added tool for operating or maintenance personnel to monitor their operation.
  (3) Motor insulation temperatures can be monitored by use of resistance temperature detectors, which provide an alarm indication at a selected temperature (depending on the insulation class). Such monitoring indicates that the motor is dirty and/or air passages are plugged.

The distribution system configuration and features should be such that maintenance work is permitted without load interruption or with only minimal loss of availability. Often, electrical equipment preventive maintenance is not done or is deferred because load interruption is required to a critical load or to a portion of the distribution system. This may require the installation of alternate electrical equipment and circuits to permit routine or emergency maintenance on one circuit while the other one supplies the critical load that cannot be shut down.

Electrical equipment that is improperly applied will not give reliable service regardless of how good or how much preventive maintenance is accomplished. The most reasonably accepted measure is to make a corrective modification.

Table 43
Number of Failures versus Maintenance Quality
for All Equipment Classes Combined

Maintenance Quality	Number of Failures		Percent of Failures Due to Inadequate Maintenance
	All Causes	Inadequate Maintenance	
Excellent	311	36	11.6%
Fair	853	154	18.1%
Poor	67	22	32.8%
None	238	28	11.8%
Total	1469	240	16.4%

Table 44
Percentage of Failure Caused from
Inadequate Maintenance versus Month Since Maintained

Failure, (months since maintained)	All Electrical Equipment Classes Combined	Circuit Breakers	Motors	Open Wire	Trans-formers
Less than 12 months ago	7.4%	*12.5%	8.8%	*0	*2.9%
12–24 months ago	11.2%	19.2%	8.8%	*22.2%	*2.6%
More than 24 months ago	36.7%	77.8%	44.4%	38.2%	36.4%
Total	16.4%	20.8%	15.8%	30.6%	11.1%

*Small sample size; less than seven failures caused by inadequate maintenance.

Table 45
Equipment Failure Rate
Multipliers versus Maintenance Quality

Maintenance Quality	Trans-formers	Circuit Breakers	Motors
Excellent	0.95	0.91	0.89
Fair	1.05	1.06	1.07
Poor	1.51	1.28	1.97
All	1.0	1.0	1.0
Perfect maintenance	0.89	0.79	0.84

5.5 Electrical Equipment Preventive Maintenance. Electrical equipment deterioration is normal. But, left unchecked, the deterioration can progress and cause a malfunction or an electrical failure. Electrical equipment preventive maintenance procedures should be developed to accomplish four basic functions, that is, keep it clean, dry, sealed tight, and minimize the friction. Water, dust, high or low ambient temperature, high humidity, vibration, component quality, and countless other conditions can affect proper operation of electrical equipment. Without an effective electrical preventive maintenance program, the risk of a serious electrical failure increases.

A common cause of electrical failure is dust and dirt accumulation and the presence of moisture. This can be in the form of lint, chemical dust, day-to-day accumulation of oil mist and dirt particles, etc. These deposits on the insulation, combined with oil and moisture, become conductors and are responsible for tracking and flashovers. Deposits of dirt can cause excessive heating and wear, and decrease apparatus life. Electrical apparatus should be operated in a dry atmosphere for best results; but this is often impossible, so precautions should be established to minimize the entrance of moisture. Moisture condensation in electrical apparatus can cause copper or aluminum oxidation and connection failure.

Loose connections are another cause of electrical failures. Electrical connections should be kept tight and dry. Creep or cold flow is a major cause of joint failure. Mounting hardware and other bolted parts should be checked during routine electrical equipment servicing.

Friction can affect the freedom of movement of electrical devices and can result in serious failure or difficulty. Dirt on moving parts can cause sluggishness and improper electrical equipment operations, such as arcing and burning. Checking the mechanical operation of devices and manually or electrically operating any device that seldom operates should be standard practice.

Procedures and practices should be initiated to substantiate that electrical equipment is kept clean, dry, sealed tight, and with minimal friction by visual inspection, exercising, and proof testing. Electrical preventive maintenance should be accomplished on a regularly scheduled basis as determined by inspection experience and analysis of any failures that occur.

An electrical preventive maintenance program certainly will not eliminate all failures; but it will minimize their occurrence. Some of the key elements in establishing a program are as follows:

(1) Establish an "equipment service library" consisting of bulletins, manuals, schematics, parts lists, failure analysis reports, etc. The bulletins and manuals are normally provided by the electrical equipment manufacturer. Often they are not taken very seriously after equipment installation and are lost, misplaced, or discarded; but this documentation is vital to the development of electrical preventive maintenance procedures and to aid in training.

(2) In addition to the above documentation, each in-service failure should be thoroughly investigated and the cause determined and documented. Generally, it will be found that timely and adequate electrical preventive maintenance could have prevented the failure. If correctable by electrical preventive maintenance, the corrective action should be included on the

work list. If the failure was caused by a weak component, then all identical equipment should be modified as soon as possible. "Failure analysis" plays a major part in an electrical preventive maintenance program.

(3) Provide the training necessary to accomplish the program that has been established. The techniques utilized in the performance of an electrical preventive maintenance program are extremely important. The success or failure of it relies on the qualifications and know-how of the personnel performing the work; therefore, training in electrical preventive mainte-nance techniques is a major objective. Servicing of electrical equipment requires better-than-average skills and special training. Properly trained and adequately equipped maintenance personnel must have a very thorough knowledge of the equipment operation. They must be able to make a thorough inspection and also accomplish repairs. For example, special train-ing in the use of dc high-potential dielectric tests or megger tests as well as the interpretation of the results may be required.

(4) A good record system should be developed that will show the repairs required by equipment over a long period of time. On each regular inspec-tion, variations from normal conditions should be noted. The frequency and magnitude of the work should then be increased or decreased according to an analysis of the data. Avoid performing too much maintenance work as this can contribute to failures. The records should reflect the availability of spare parts, service attitude of equipment manufacturers, major equipment failures to date, and time required for repairs, etc. These records are not only useful in planning and scheduling electrical preventive maintenance work; they are also useful in evaluating equipment performance for future purchases.

5.6 References. The information in this chapter shall be used in conjunction with the following publications:

[1] ANSI/NFPA 70-1990, National Electrical Code.[6]

[2] IEEE C57.106-1977, IEEE Guide for Acceptance and Maintenance of Insulating Oil in Equipment (ANSI).[7]

[3] IEEE Std 43-1974 (Reaff. 1984), IEEE Recommended Practice for Testing Insu-lation Resistance of Rotating Machinery (ANSI).

[4] IEEE Std 56-1977 (Reaff. 1982), IEEE Guide for Insulation Maintenance of Large AC Rotating Machinery (10 000 kVA and Larger) (ANSI).

[5] IEEE Std 62-1978, IEEE Guide for Field Testing Power Apparatus Insulation.

---

[6] ANSI publications are available from the Sales Department of the American National Standards Institute, 1430 Broadway, New York, NY 10018. NFPA publications are available from Publication Sales, National Fire Protection Association, 1 Batterymarch Park, P.O. Box 9101, Quincy, MA 02269-9101.

[7] IEEE publications are available from the Institute of Electrical and Electronics Engineers, IEEE Service Center, 445 Hoes Lane, Piscataway, NJ 08855-1331.

[6] IEEE Std 95-1977 (Reaff. 1982), IEEE Recommended Practice for Insulation Testing of Large AC Rotating Machinery with High Direct Voltage (ANSI).

[7] IEEE Std 450-1987, IEEE Recommended Practice for Maintenance, Testing, and Replacement of Large Lead Storage Batteries for Generating Stations and Substations (ANSI).

[8] Curdts, E. B. *Insulation Testing by D-C Methods*, Technical Publication 22T1-1971, James G. Biddle Company, Plymouth Meeting, PA, p. 2.

[9] *Factory Mutual Systems Transformer Bulletin 14-8*, Oct. 1976, Public Information Division, 1151 Boston-Providence Turnpike, Norwood, MA.

[10] Hubert, C. I. *Preventive Maintenance of Electrical Equipment*, McGraw-Hill Book Company.

[11] IEEE Committee Report. Report on Reliability Survey of Industrial Plants, part 6, *IEEE Transactions on Industry Applications*, vol. IA-10, Jul./Aug. and Sep./Oct. 1974, pp. 456–476, 681.

NOTE: Reference [11] is reprinted in its entirety in Appendix B.

[12] *Maintenance Hints*, Westinghouse Electric Corporation, Pittsburgh, PA.

[13] Miller, H. N. *DC Hypot Testing of Cables, Transformers, and Rotating Machinery*, Manual P-16086, Associated Research Inc., Chicago, IL.

[14] Shaw, E. T. *Inspection and Test of Electrical Equipment*, Westinghouse Electric Corporation, Pittsburgh, PA.

[15] Smeaton, R. W. *Motor Application and Maintenance Handbook*, McGraw-Hill Book Company, New York, NY 1969.

Chapter 6
Emergency and Standby Power

6.1 Introduction. When a reliability analysis has been completed, the rate at which power failures occur and the expected duration of those power failures can be predicted at most points to utilization equipment. This knowledge can be used to determine whether there is a need to increase the reliability of delivered power supplied to particular utilization points. Emergency or standby power can be used readily to improve availability and reduce the interruption frequency of delivered power. A cost/reliability trade-off decision must be made to improve reliability of power only to those areas that can justify the cost for such improvement. Various types of emergency or standby systems are ideally suited to providing large improvements to relatively small sections of a power system.

6.2 Interruption Frequency and Duration. An evaluation of each piece of utilization equipment must be made to determine actual needs. The difference between interruption frequency and duration of supplied power must be clearly understood. Interruption frequency is the "expected (average) number of power interruptions to a load per unit time, usually expressed as interruptions per year." Expected interruption duration is the "average duration of a single load interruption event." Interruption frequency and duration requirements for control power to a boiler would certainly be greater than those for a room air conditioner.

Many power-consuming operations require a very low interruption frequency with much less concern for interruption duration. A power failure during the vulcanizing cycle of a rubber manufacturing process will cause loss of steam and errors in the time/temperature control for proper curing. This results in the product being scrapped. The difference in loss between a power failure of 1 minute duration and one of 30 minutes duration is minimal. Thus, a power system that experiences 2 power failures of 30 minutes each is more desirable than a system that experiences 6 power failures of 1 minute each.

Other power utilization equipment demands short interruption duration with much less concern for interruption frequency. A power failure to a process that stamps out metal parts will cause little loss due to the power failure itself; but there will be production loss directly related to the length of the power failure. Thus, a power system that experiences 6 power failures of 1 minute each is more suitable than one that experiences 2 power failures of 30 minutes each.

6.3 System Selection. The type of emergency or standby power system to use depends on what the system is expected to accomplish. Can the equipment or process tolerate a power failure of 1 millisecond, 10 seconds, or 1 minute? For how long a period of time does the emergency or standby power system have to perform its intended function? For hours, minutes, or seconds?

An offline system is one that is dormant until it is called upon to operate, such as a diesel generator that is started up when a power failure occurs. An online system is one that is operating at all times, such as an inverter supplied by dc power via the primary power source through a battery charger. The above system utilizes batteries on float charge to supply the inverter if a power failure strikes the primary power source.

The selection of an offline engine-driven generator for the rubber manufacturing process mentioned above would be a misapplication. The offline system can reduce the interruption duration; but it cannot reduce the interruption frequency. The transfer device or devices have a failure rate of their own and, thus, actually reduce the reliability of delivered power. When the primary power source is being utilized, a failure of the transfer device may cause loss of power, which would not have occurred if the offline system had not been installed. The selection of an offline system for the metal stamping process is a proper application.

A study must be undertaken to determine the systems capable of performing the desired function. Systems are available to provide reliable power to overcome the problems encountered due to power failures ranging from milliseconds to many hours. More than one type of system may be suitable for a particular application. Selection of the proper system will then depend on first cost, operating and owning cost (such as maintenance and fuel requirements), system reliability, output power quality, expansion capacity, and environmental considerations.

6.4 Descriptions and Applications of Available Systems

6.4.1 General. The following information contains data on some commonly used systems. While power sources, such as solar and chemical, may become viable in the future, they are not in common use and will not be discussed here.

6.4.2 Engine-Driven Generators. These units are available in sizes from 1 kW to several thousand kW. Fuels commonly used are diesel, gasoline, and natural or liquefied petroleum (LP) gas. If kept warm, they will come dependably on line in 8 to 15 seconds. Diesel units are generally heavier duty, have less costly fuel, and fire danger is lower than for gasoline units. Gasoline-driven units range up to 100 kW and have a lower initial cost than diesel sets. Natural and LP gas engines provide quick starting after long shutdown periods because of the inherently fresh fuel. One of the drawbacks may be the lack of assurance of fuel supplies when the system is needed. Engine-driven generators are generally applied as offline units for reducing downtime or in combination with a mechanical-stored energy system or a small uninterruptible power supply to improve both reliability and availability of delivered power.

6.4.3 Turbine-Driven Generators. Two types of turbines can be used for prime movers of either steam or gas. Since steam is generally not available when a power failure has occurred, only the gas prime mover will be discussed.

Gas turbines can utilize various grades of oil as well as natural and propane gas. Sizes generally range from 100 kW to several thousand kW. Gas turbine generators can be placed on line in 20 seconds for smaller units and in up to several minutes for larger units. They can more easily be rooftop mounted since their physical size and weight per kW are less than for engine-driven units. Turbine-driven generators are almost exclusively applied as offline systems.

6.4.4 Mechanical-Stored Energy Systems. This type of system is comprised of a rotating flywheel that converts its rotating kinetic energy into electric power. It is generally applied as an online system. Depending on the frequency requirements of the load, a typical mechanical-stored energy system can ride through a power failure for up to 2 seconds. Thus, its main use is as a buffer to mechanically filter out transients. (See Fig 7.) A supply time of 15 seconds can be attained by using an eddy current clutch and driving the flywheel at a higher speed than the generator it operates. This type of system may allow an engine-driven prime mover to come up to speed, either to drive a separate generator or to maintain the speed of the flywheel and its associated generator. (See Fig 8.) Such systems have a history of maintenance problems. There are several other hybrid systems that utilize dc drive motors, batteries, engines, and turbines.

6.4.5 Inverter/Battery Systems. A simple offline inverter system is shown in Fig 9. The system is not an uninterruptible power supply. The transfer time for a mechanical transfer will cause a power interruption of 60 to 190 milliseconds. A static transfer switch is more costly; but it will result in a much shorter interruption. The contactor closes upon loss of primary power and is in the circuit to prevent continual energization of the static inverter whose efficiency is approximately 70% and, thus, wastes energy while energized.

Figure 10 shows the most widely used system for supplying uninterruptible power. The load is basically free of power interruptions, transient disturbances, and voltage and frequency variations. A failure of the inverter will cause a loss of power until the inverter is repaired or until prime power can be connected directly to the load.

Figure 11 shows a redundant uninterruptible power supply with static switches to clear a faulted inverter. The batteries for this system are required to supply power only until the diesel generators can be placed on line. The system in Fig 11 is much more reliable than that shown in Fig 10, but is more expensive. Installation requirements can be impressive for the battery. A battery sized to provide power for a 250 kW inverter for 1 hour will weigh approximately 25 tons.

**Fig 7
Simple Inertia-Driven "Ride Through" System**

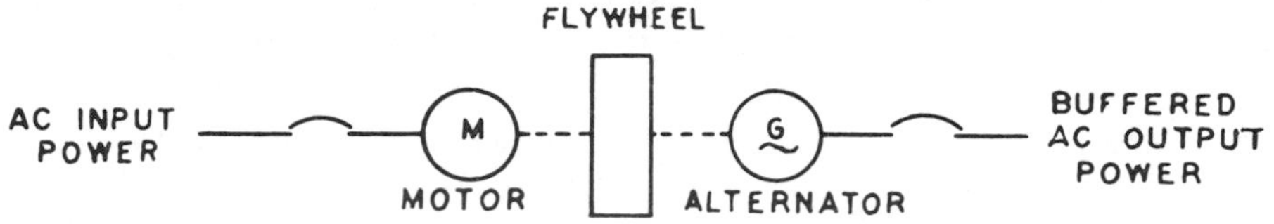

FLYWHEEL

AC INPUT POWER — M (MOTOR) — G (ALTERNATOR) — BUFFERED AC OUTPUT POWER

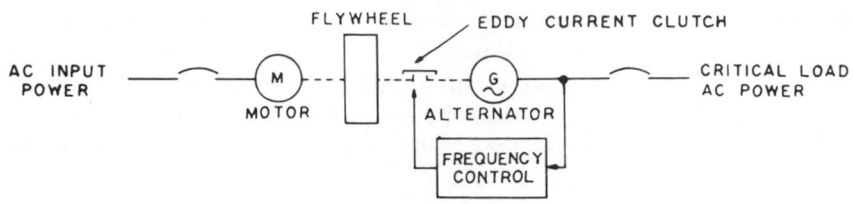

**Fig 8
Constant Frequency Inertia System**

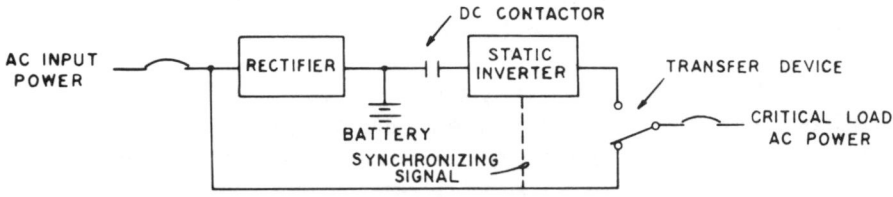

**Fig 9
Short Interruption Static Inverter System**

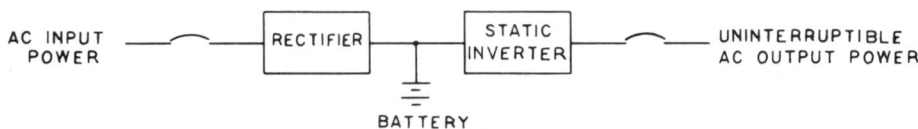

**Fig 10
Nonredundant Uninterruptible Power Supply**

6.4.6 Mechanical Uninterruptible Power Supplies. Figure 12 shows a typical rotating uninterruptible power supply. The ac motor drives the dc generator, which in turn supplies power for the dc motor, which drives the ac generator. The battery will provide power for the dc motor upon loss of primary power. The ac generator provides uninterruptible power to the load. The lack of moving parts in the static inverters and rectifiers has proven to be a strong selling point over the mechanical uninterruptible power supplies.

6.5 Selection and Application Data. The figures and system descriptions presented here are only a few of the many types of systems and hybrid systems available. For comprehensive selection and application data, reference is made to IEEE Std 446-1987, IEEE Recommended Practice for Emergency and Standby Power Systems for Industrial and Commercial Applications (ANSI) [1].[8]

---

[8]The numbers in brackets correspond to those in the references at the end of this chapter. IEEE publications are available from the Institute of Electrical and Electronics Engineers, IEEE Service Center, 445 Hoes Lane, Piscataway, NJ 08855-1331.

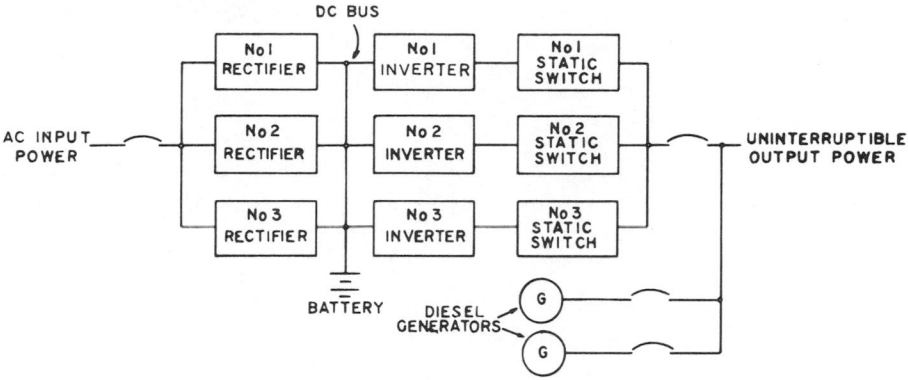

Fig 11
Redundant Uninterruptible Power Supply

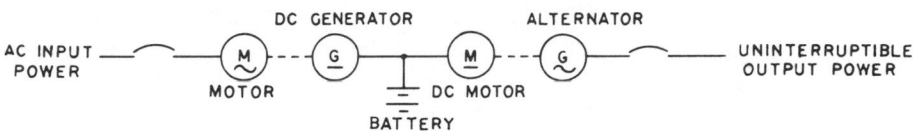

Fig 12
Rotating Uninterruptible Power Supply

6.6 Reference. The information in this chapter shall be used in conjunction with the following publication:

[1] IEEE Std 446-1987, IEEE Recommended Practice for Emergency and Standby Power Systems for Industrial and Commercial Applications (ANSI).

Chapter 7
Examples of Reliability
Analysis and Cost Evaluation

7.1 Examples of Reliability and Availability Analysis of Common Low-Voltage Industrial Power Distribution Systems

7.1.1 Quantitative Reliability and Availability Predictions. In this chapter, a description is given of how to make quantitative reliability and availability predictions for proposed new configurations of industrial power distribution systems. Six examples are worked out, including a simple radial system, a primary-selective system, and a secondary-selective system. A brief tabulation is also given of the pertinent reliability data needed in order to make the reliability and availability predictions. The simple radial system analyzed had an average number of forced hours of downtime per year that was 19 times larger than a secondary-selective system; the failure rate was 6 times larger. The importance of two separate power supply sources from the electric utility has been identified and analyzed. This approach could be used to assist in cost/reliability trade-off decisions in the design of the power distribution system.

7.1.2 Introduction. An industrial power distribution system may receive power at 13.8 kV from an electric utility and then distribute the power throughout a plant for use at the various locations. One of the questions often raised during the design of the power distribution system is whether there is a way of making a quantitative comparison of the failure rate and the forced hours downtime per year of a secondary-selective system with a primary-selective system and a simple radial system. This comparison could be used in cost/reliability and cost/availability trade-off decisions in the design of the power distribution system. The estimated cost of power outages at the various plant locations could be factored into the decision as to which type of power distribution system to use. The decisions could be based upon "total owning cost over the useful life of the equipment" rather than "first cost."

Six examples of common low-voltage industrial power distribution systems are analyzed in this section:

    (1) Example 1 — Simple radial system
    (2) Example 2 — Primary-selective system to 13.8 kV utility supply
    (3) Example 3 — Primary-selective system to load side of 13.8 kV circuit breaker
    (4) Example 4 — Primary-selective system to primary of transformer
    (5) Example 5 — Secondary-selective system
    (6) Example 6 — Simple radial system with spares

Only forced outages of the electrical equipment are considered in the six examples. It is assumed that scheduled maintenance will be performed at times when 480 V power output is not needed. The frequency of scheduled outages and the average duration can be estimated, and, if necessary, these can be added to the forced outages given in the six examples.

When making a reliability study, it is necessary to define what is a failure of the 480 V power. Some of the failure definitions for 480 V power that are often used are as follows:

(1) Complete loss of incoming power for more than 1 cycle
(2) Complete loss of incoming power for more than 10 cycles
(3) Complete loss of incoming power for more than 5 seconds
(4) Complete loss of incoming power for more than 2 minutes

Definition (3) will be used in the six examples given. This definition of failure can have an effect in determining the necessary speed of automatic switchover equipment that is used in primary-selective or secondary-selective systems. In some cases, when making reliability studies, it might be necessary to further define what is "complete loss of incoming power;" for example, "voltage drops below 70%."

One of the main benefits of a reliability and availability analysis is that a disciplined look is taken at the alternative choices in the design of the power distribution system. By using published reliability data collected by a technical society from industrial plants, the best possible attempt is made to use historical experience to aid in the design of the new system.

7.1.3 Definition of Terminology. The definition of terms is given in Chapter 1 and 2.1.3. The units that are being used for "failure rate" and "average downtime per failure" are:

λ = Failure rate (failures per year).
r = Average downtime per failure (hours per failure = Average time to repair or replace a piece of equipment after a failure. In some cases, this is the time to switch to an alternate circuit when one is available.

7.1.4 Procedure for a Reliability and Availability Analysis. The "minimal cut-set" method for system reliability evaluation is described in 2.1.6, 2.1.8, and 2.1.9. The quantitative reliability indexes that are used in the six examples are the failure rate and the forced hours downtime per year. These are calculated at the 480 V point of use in each example. The failure rate λ is a measure of unreliability. The product of λr (failure rate $\times$ average downtime per failure) is equal to the forced hours downtime per year and can be considered a measure of forced unavailability since a scale factor of 8760 converts one quantity into the other. The average downtime per failure r could be called "restorability."

The necessary formulas for calculating the reliability indexes of the minimal cut-set approach are given in Eqs 1 and 2 in 2.1.9 and Eqs 5 and 6 in 2.1.11.1. A sample using these formulas is shown in Fig 13 for two components in series and two components in parallel. In these samples, the scheduled outages are assumed to be 0 and the units for λ and r are, respectively, failures per year and hours downtime per failure. The formula in Fig 13 assumes the following:

(1) The component failure rate is constant with age.

$$\boxed{\lambda_1, r_1} \quad\quad \boxed{\lambda_2, r_2}$$

$$f_s = \lambda_1 + \lambda_2$$

$$f_s r_s = \lambda_1 r_1 + \lambda_2 r_2$$

$$r_s = \frac{\lambda_1 r_1 + \lambda_2 r_2}{\lambda_1 + \lambda_2}$$

(a)

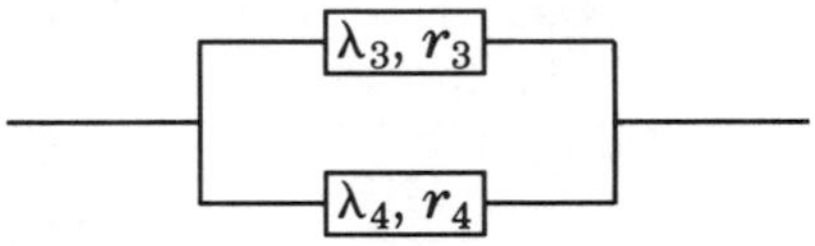

$$\boxed{\lambda_3, r_3} \quad \boxed{\lambda_4, r_4}$$

$$f_p = \frac{\lambda_3 \lambda_4 (r_3 + r_4)}{8760}$$

$$f_p r_p = \frac{\lambda_3 r_3 (\lambda_4 r_4)}{8760}$$

$$r_p = \frac{r_3 r_4}{r_3 + r_4}$$

(b)

Nomenclature:
f = Frequency of failures.
λ = Failures per year.
r = Average hours of downtime per failure.
s = Series.
p = Parallel.

NOTE: These formulas are approximate and should only be used when both $\frac{\lambda_3 r_3}{8760}$ and $\frac{\lambda_4 r_4}{8760}$ are less than 0.01.

Fig 13
Formulas for Reliability Calculations
(a) Reparable Components in Series (Both Must Work for Success)
(b) Reparable Components in Parallel (One or Both Must Work for Success)

(2) The outage time after a failure has an exponential distribution. (Probability of outage time exceeding τ is $\epsilon^{-\tau/r}$).

(3) Each failure event is independent of any other failure event.

(4) The component forced unavailability is small:

$$\frac{\lambda_i r_i}{8760} < 0.01$$

The reliability data to be used for the electrical equipment and the electric utility supply are given in 7.1.5.

Table 46
Reliability Data from 1973–74 IEEE Reliability Survey of Industrial Plants
(See Reference [8])

Equipment Category	λ, Failures per Year	r, Hours of Downtime per Failure	$\lambda \cdot r$, Forced Hours of Downtime per Year	Data Source in IEEE Survey [8], Table No.
Protective relays	0.0002	5.0	0.0010	19
Metalclad drawout circuit breakers				
0–600 V	0.0027	4.0	0.0108	5, 50
Above 600 V	0.0036	83.1*	0.2992	5, 51
Above 600 V	0.0036	2.1†	0.0076	5, 51
Power cables (1000 circuit ft)				
0–600 V, above ground	0.00141	10.5	0.0148	13
601–15 000 V, conduit below ground	0.00613	26.5*	0.1624	13, 56
601–15 000 V, conduit below ground	0.00613	19.0†	0.1165	13, 56
Cable terminations				
0–600 V, above ground	0.0001	3.8	0.0004	17
601–15 000 V, conduit below ground	0.0003	25.0	0.0075	17
Disconnect switches enclosed	0.0061	3.6	0.0220	9
Transformers				
601–15 000 V	0.0030	342.0*	1.0260	4, 48
601–15 000 V	0.0030	130.0†	0.3900	4, 48
Switchgear bus—bare				
0–600 V (connected to 7 breakers)	0.0024	24.0	0.0576	10
0–600 V (connected to 5 breakers)	0.0017	24.0	0.0408	10
Switchgear bus—insulated				
601–15 000 V (connected to 1 breaker)	0.0034	26.8	0.0911	10
601–15 000 V (connected to 2 breakers)	0.0068	26.8	0.1822	10

*Repair failed unit.
†Replace with spare.

7.1.5 Reliability Data from 1973–75 IEEE Surveys. In order to make a reliability and availability analysis of a power distribution system, it is necessary to have data on the reliability of each component of electrical equipment used in the system. Ideally, these reliability data should come from field use of the same type of equipment under similar environmental conditions and stress levels. In addition, there should be a sufficient number of field failures in order to represent an adequate sample size. It is believed that eight field failures are the minimum number necessary in order to have a reasonable chance of determining a failure rate to within a factor of 2. The types of reliability data needed on each component of electrical equipment are:

(1) Failure rate (failures per year)
(2) Average downtime to repair or replace a piece of equipment after a failure (hours per failure)

These reliability data on each component of electrical equipment can then be used to represent historical experience for use in cost/reliability and cost/availability trade-off studies in the design of new power distribution systems.

From 1973–75, the Power Systems Reliability Subcommittee of the Industrial and Commercial Power Systems Committee conducted and published surveys of electrical equipment reliability in industrial plants (see References [7] and [8]).[9] See Appendixes A, B, and D for the data. See Chapter 3 for a summary of these data and data from later surveys. These reliability surveys of electrical equipment and electric utility power supplies were extensive. The pertinent failure rate and average downtime per failure information for the electrical equipment are given in Table 46. In compiling these data, a failure was defined as any trouble with a power system component that causes any of the following effects:

(1) Partial or complete plant shutdown, or below-standard plant operation
(2) Unacceptable performance of user's equipment
(3) Operation of the electrical protective relaying or emergency operation of the plant electric system
(4) De-energization of any electric circuit or equipment

A failure on a public utility supply system may cause the user to have either of the following:

(1) A power interruption or loss of service
(2) A deviation from normal voltage or frequency outside the normal utility profile

A failure of an in-plant component causes a forced outage of the component, that is, the component is unable to perform its intended function until it is repaired or replaced. The terms "failure" and "forced outage" are often used synonymously.

In addition to the reliability data for electrical equipment shown in Table 46, there are some "failure modes" of circuit breakers that require backup protective equipment to operate, for example, "failed to trip" or "failed to interrupt." Both of these failure modes would require that a circuit breaker farther up the line be opened, and this would result in a larger part of the power distribution system being

---

[9]The numbers in brackets correspond to those in the references at the end of this chapter. IEEE publications are available from the Institute of Electrical and Electronics Engineers, IEEE Service Center, 445 Hoes Lane, Piscataway, NJ 08855-1331.

disconnected. Reliability data on the "failure modes of circuit breakers" are shown in Table 47. These data are used for the 480 V circuit breakers in all six examples discussed in this section. It will be assumed that the "flashed over while open" failure mode for circuit breakers and disconnect switches has a failure rate of 0.

Table 47
Failure Modes of Circuit Breakers
Percentage of Total Failures in Each Failure Mode (See Table 35)

Percent of Total Failures (All Voltages)	Failure Characteristic
	Backup protective equipment required
9	Failed while opening
	Other circuit breaker failures
7	Damaged while successfully opening
32	Failed while in service (not while opening or closing)
5	Failed to close when it should
2	Damaged while closing
42	Opened when it shouldn't
1	Failed during testing or maintenance
1	Damage discovered during testing or maintenance
1	Other
100	Total Percentage

The failure rate and average downtime per failure data for the electric utility power supplies are given in Table 48. These include both single-circuit and double-circuit reliability data. The two power sources in a double-circuit utility supply are not completely independent, and the reliability and availability analysis must take this into consideration. This subject is discussed further in 7.1.15.

Table 48
IEEE Survey of Reliability of Electric Utility Power Supplies to Industrial Plants
(1975 Survey)
(See Table 40)

Number of Circuits (All Voltages)	λ, Failures per Year	r, Hours of Downtime per Failure	$\lambda \cdot r$, Forced Hours of Downtime per Year
Single circuit	1.956	1.32	2.582
Double circuit Loss of both circuits*	0.312	0.52	0.1622
Calculated value for loss of Source 1 (while Source 2 is OK)	1.644	0.15[†]	0.2466

*Manual switchover time of 9 min to source 2.
[†]Data for double circuits had all circuit breakers closed.

7.1.6 Example 1—Reliability and Availability Analysis of a Simple Radial System

7.1.6.1 Description of Simple Radial System. A simple radial system is shown in Fig 14. Power is received at 13.8 kV from the electric utility. Then it goes through a 13.8 kV circuit breaker inside the industrial plant, 600 feet of cable in underground conduit, an enclosed disconnect switch, to a transformer that reduces

Fig 14
Simple Radial System—Example 1

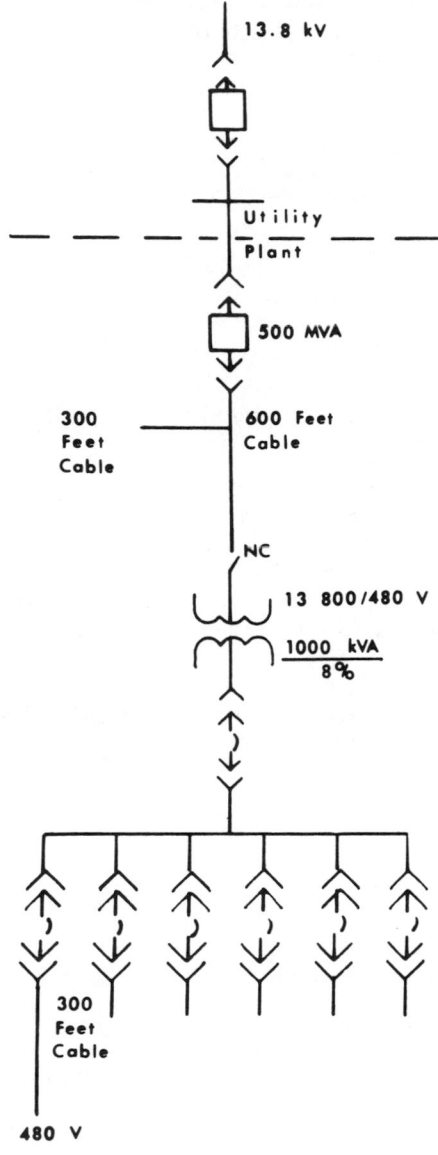

13.8 kV

Utility
Plant

500 MVA

300
Feet
Cable

600 Feet
Cable

NC

13 800/480 V

1000 kVA
8%

300
Feet
Cable

480 V

the voltage to 480 V, then through a 480 V main circuit breaker, a second 480 V circuit breaker, 300 feet of cable in above ground conduit, to the point where the point is used in the industrial plant.

7.1.6.2 Results—Simple Radial System. The results from the reliability and availability calculations are given in Table 49. The failure rate and the forced hours downtime per year are calculated at the 480 V point of use.

The relative ranking of how each component contributes to the failure rate is of considerable interest. This is tabulated in Table 50.

Table 49
Simple Radial System —
Reliability and Availability of Power at 480 V — Example 1

Component	λ, Failures per Year	$\lambda \cdot r$, Forced Hours of Downtime per Year
13.8 kV power source from electric utility	1.956	2.582
Protective relays (3)	0.0006	0.0030
13.8 kV metalclad circuit breaker	0.0036	0.2992*
Switchgear bus—insulated (connected to 1 breaker)	0.0034	0.0911
Cable (13.8 kV); 900 ft, conduit below ground	0.0055	0.1458*
Cable terminations (6) at 13.8 kV	0.0018	0.0450
Disconnect switch (enclosed)	0.0061	0.0220
Transformer	0.0030	1.0260*
480 V metalclad circuit breaker	0.0027	0.0108
Switchgear bus—bare (connected to 7 breakers)	0.0024	0.0576
480 V metalclad circuit breaker	0.0027	0.0108
480 V metalclad circuit breakers (5) (failed while opening)	0.0012	0.0048
Cable (480 V); 300 ft conduit above ground	0.0004	0.0044
Cable terminations (2) at 480 V	0.0002	0.0008
Total at 480 V output	1.9896	4.3033

*Data for hours of downtime per failure are based upon *repair failed unit*.

Table 50
Simple Radial System — Relative Ranking of Failure Rates

	λ, Failures per Year
1. Electric utility	1.956
2. 13.8 kV cable and terminations	0.0073
3. Disconnect switch	0.0061
4. 13.8 kV circuit breaker	0.0036
5. Switchgear bus—insulated	0.0034
6. Transformer	0.0030
7. 480 V circuit breaker	0.0027
8. 480 V circuit breaker (main)	0.0027
9. Switchgear bus—bare	0.0024
10. 480 V circuit breakers (5) (failed while opening)	0.0012
11. 480 V cable and terminations	0.0006
12. Protective relays (3)	0.0006
Total	1.9896

The relative ranking of how each component contributes to the forced hours downtime per year is also of considerable interest. This is given in Table 51.

It might be expected that the power distribution system would be shut down once every 2 years for scheduled maintenance for a period of 24 hours. These shutdowns would be in addition to the outage data given in Tables 49–51.

Table 51
Simple Radial System — Relative Ranking
of Forced Hours of Downtime per Year

	$\lambda \cdot r$, Forced Hours of Downtime per Year
1. Electric utility	2.582
2. Transformer	1.0260*
3. 13.8 kV circuit breaker	0.2992*
4. 13.8 kV cable and terminations	0.1908*
5. Switchgear bus — insulated	0.0911
6. Switchgear bus — bare	0.0576
7. Disconnect switch	0.0220
8. 480 V circuit breaker	0.0108
9. 480 V circuit breaker (main)	0.0108
10. 480 V cable and terminations	0.0052
11. 480 V circuit breakers (5) (failed while opening)	0.0048
12. Protective relays (3)	0.0030
Total	4.3033

*Data for hours of downtime per failure are based upon *repair failed unit.*

7.1.6.3 Conclusions — Simple Radial System. The electric utility supply is the largest contributor to both the failure rate and the forced hours downtime per year at the 480 V point of use. A significant improvement can be made in both the failure rate and the forced hours downtime per year by having two sources of power at 13.8 kV from the electric utility. The improvements that can be obtained are shown in Examples 2, 3, and 4, using "primary-selective system" and in Example 5, using "secondary-selective system."

The transformer is the second largest contributor to forced hours downtime per year. The transformer has a very low failure rate; but the long outage time of 342 hours after a failure results in a large forced hours downtime per year. The 13.8 kV circuit breaker is the third largest contributor to forced hours downtime per year, and the fourth largest are the 13.8 kV cables and terminations. This is a result of the average outage time after a failure of 83.1 hours for the 13.8 kV circuit breaker and 26.5 hours for the 13.8 kV cable.

The long outage time after a failure for the transformer, 13.8 kV circuit breaker, and the 13.8 kV cable are all based upon "repair failed unit." These outage times after a failure can be reduced significantly if the "replace with spare" times shown in Table 46 are used instead of "repair failed unit." This is done in Example 6, using a simple radial system with spares.

7.1.7 Example 2—Reliability and Availability Analysis of Primary-Selective System to 13.8 kV Utility Supply

7.1.7.1 Description—Primary-Selective System to 13.8 kV Utility Supply. The primary-selective system to the 13.8 kV utility supply is shown in Fig 15. It is a simple radial system with the addition of a second 13.8 kV power source from the electric utility; the second power source is normally disconnected. In the event that there is a failure in the first 13.8 kV utility power source, then the second 13.8 kV utility power source is switched on to replace the failed power source. Assume that the two utility power sources are synchronized.

Example 2a—Assume a 9 minute "manual switchover time" to utility power source no. 2 after a failure of source no. 1.

Example 2b—Assume an "automatic switchover time" of less than 5 seconds after a failure is assumed. (Loss of 480 V power for less than 5 seconds is not counted as a failure.)

7.1.7.2 Results—Primary-Selective System to 13.8 kV Utility Supply

Example 2a—If the time to switch to a second utility power source takes 9 minutes after a failure of the first source, then there would be a power supply failure of 9 minutes in duration. Using the data from Table 48 for double-circuit utility supplies, this would occur 1.644 times per year (1.956−0.312). This in addition to losing both power sources simultaneously 0.312 times per year for an average outage time of 0.52 hour. If these utility supply data are added together and substituted into Table 49 on the simple radial system, it would result in reducing the forced hours downtime per year at the 480 V point of use from 4.3033 to 2.1291. The failure rate would stay the same at 1.9896 failures per year. These results are given in Table 52.

Table 52
Simple Radial System and Primary-Selective System
to 13.8 kV Utility Supply—Reliabilty and Availability
Comparison of Power at 480 V Point of Use

Distribution System	λ, Failures per Year	$\lambda \cdot r$, Forced Hours Downtime per Year
Example 1 Simple radial system	1.9896	4.3033
Example 2a Primary-selective system to 13.8 kV utility supply (with 9 min switchover after a supply failure)	1.9896	2.1291
Example 2b Primary-selective system to 13.8 kV utility supply (with switchover in less than 5 s after a supply failure)*	0.3456	1.8835

*Loss of 480 V power for less than 5 s is not counted as a failure.

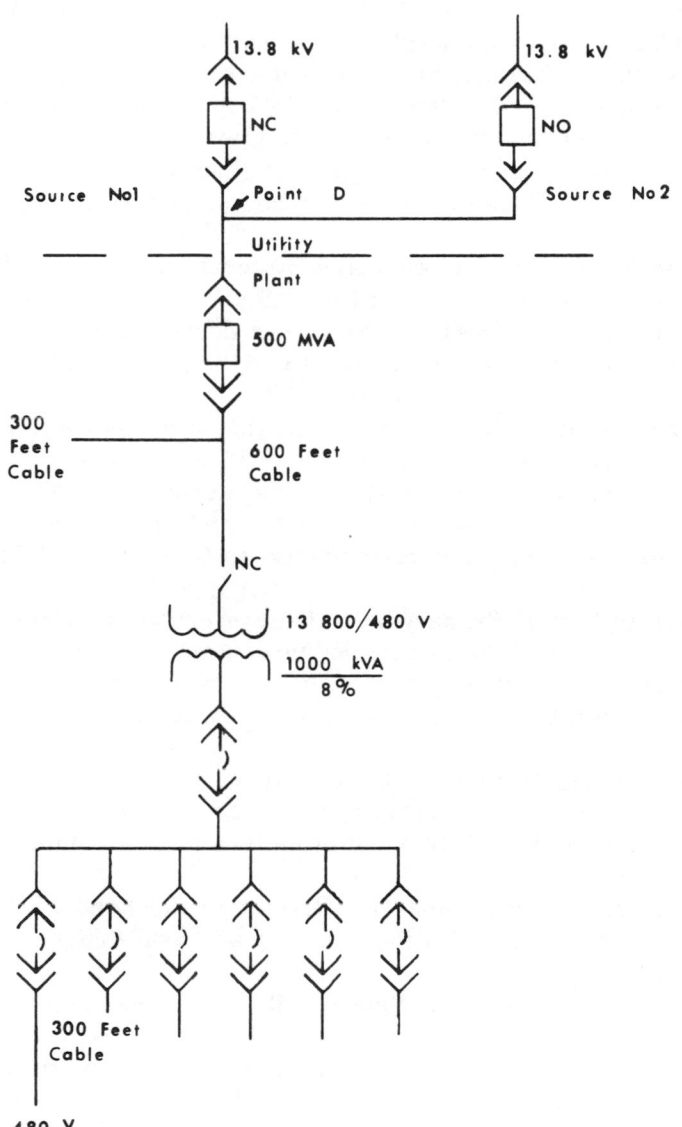

Fig 15
Primary-Selective System to 13.8 kV Utility Supply — Example 2

Example 2b — If the time to switch to a second utility power source takes less than 5 seconds after a failure of the first source, then there would be no failure of the electric utility power supply. The only time a failure of the utility power source would occur is when both sources fail simultaneously. It will be assumed that the data shown in Table 48 are applicable for the loss of both power supply circuits simultaneously. This is 0.312 failures per year with an average outage time of 0.52 hour. If these values of utility supply data are substituted into Table 49, it would result in reducing the forced hours downtime per year from 4.3033 to 1.8835 hours per year at the 480 V point of use; the failure rate would be reduced from 1.9896 to 0.3456 failures per year. These results are also given in Table 52.

7.1.7.3 Conclusion — Primary-Selective System to the 13.8 kV Utility Supply. The use of primary-selective system to the 13.8 kV utility supply with 9 minute manual switchover time reduces the forced hours downtime per year at the 480 V point of use by about 50%; but the failure rate is the same as for a simple radial system.

The use of automatic throwover equipment that could sense a failure of one 13.8 kV utility supply and switchover to the second supply in less than 5 seconds would give a 6 to 1 improvement in the failure rate at the 480 V point of use (a loss of 480 V power for less than 5 seconds is not counted as a failure).

7.1.8 Example 3 — Primary-Selective System to Load Side of 13.8 kV Circuit Breaker

7.1.8.1 Description of Primary-Selective System to Load Side of 13.8 kV Circuit Breaker. Figure 16 shows a single-line diagram of the power distribution system for the primary-selective system to the load side of 13.8 kV circuit breaker. What are the failure rate and the forced hours downtime per year at the 480 V point of use?

Example 3a — Assume a 9 minute manual switchover time.

Example 3b — Assume an automatic switchover can be accomplished in less than 5 seconds after a failure (loss of 480 V power for less than 5 seconds is not counted as a failure).

7.1.8.2 Results — Primary-Selective System to the Load Side of 13.8 kV Circuit Breaker. The results from the reliability and availability calculations are given in Table 53.

7.1.8.3 Conclusions — Primary-Selective System to the Load Side of 13.8 kV Circuit Breaker. The forced hours downtime per year at the 480 V point of use in Example 3 (primary-selective system to the load side of 13.8 kV circuit breaker) is about 10% lower than in Example 2 (primary-selective system to 13.8 kV utility supply). The failure rate is about the same.

Table 53
Primary-Selective System to Load Side of 13.8 kV Circuit Breaker— Reliability and Availability Comparison of Power at 480 V Point of Use

Component	Example 3a (9 min switchover time)		Example 3b (switchover in less than 5 seconds[†])	
	λ, Failures per Year	$\lambda \cdot r$, Forced Hours of Downtime per Year	λ, Failures per Year	$\lambda \cdot r$, Forced Hours of Downtime per Year
13.8 kV power source (loss of only source 1)	1.644			
Protective relays (3)	0.0006			
13.8 kV metalclad circuit breaker	0.0036			
Total through 13.8 kV circuit breaker with 9 min switchover after a failure of source 1 (and source 2 is OK)	1.6482	0.2472		
Loss of both 13.8 kV power sources simultaneously	0.312	0.1622	0.312	0.1622
Switchgear bus—insulated (connected to 2 breakers)	0.0068	0.1822	0.0068	0.1822
Total to point E	1.9670	0.5916	0.3188	0.3444
Cable (13.8 kV); 900 ft, conduit below ground	0.0055	0.1458*	0.0055	0.1458*
Cable terminations (6) at 13.8 kV	0.0018	0.0450	0.0018	0.0450
Disconnect switch (enclosed)	0.0061	0.0220	0.0061	0.0220
Transformer	0.0030	1.0260*	0.0030	1.0260*
480 V metalclad circuit breaker	0.0027	0.0108	0.0027	0.0108
Switchgear bus—bare (connected to 7 breakers)	0.0024	0.0576	0.0024	0.0576
480 V metalclad circuit breaker	0.0027	0.0108	0.0027	0.0108
480 V metalclad circuit breakers (5) (failed while opening)	0.0012	0.0048	0.0012	0.0048
Cable (480 V); 300 ft, conduit above ground	0.0004	0.0044	0.0004	0.0044
Cable terminations (2) at 480 V	0.0002	0.0008	0.0002	0.0008
Total at 480 V output	1.9930	1.9196	0.3448	1.6724

*Data for hours of downtime per failure are based upon *repair failed unit*.
†Loss of 480 V power for less than 5 s is not counted as a failure.

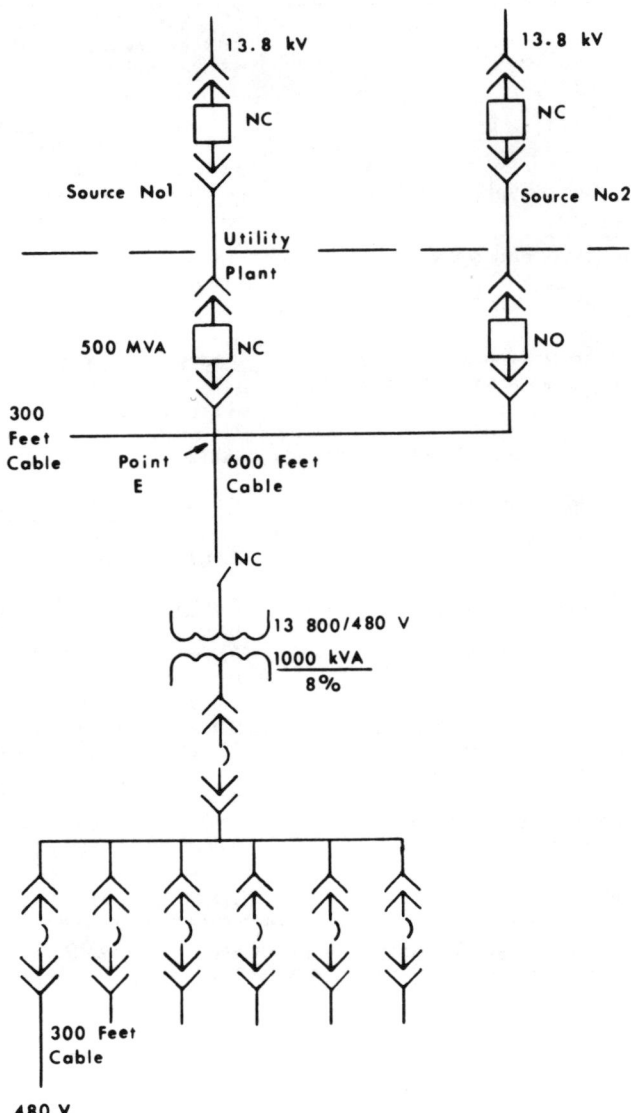

**Fig 16
Primary-Selective System to the Load Side
of 13.8 kV Circuit Breaker—Example 3**

7.1.9 Example 4—Primary-Selective System to Primary of Transformer

7.1.9.1 Description of Primary-Selective System to Primary of Transformer.
Figure 17 shows a single-line diagram of the power distribution system for the primary-selective system to primary of transformer. What are the failure rate and the forced hours downtime per year at the 480 V point of use? Assume 1 hour switchover time.

7.1.9.2 Results—Primary-Selective System to the Primary of Transformer.
The results from the reliability and availability calculations are given in Table 54.

7.1.9.3 Conclusions—Primary-Selective System to the Primary of Transformer.
The forced hours downtime per year at the 480 V point of use in Example 4 (primary-selective system to primary of transformer) is about 32% lower than for the simple radial system shown in Example 1. The failure rate is the same in Examples 1 and 4.

Table 54
Primary-Selective System to Primary of Transformer—Reliability and Availability Comparison of Power at 480 V Point of Use

Component	Example 4 (switchover time 1 h)	
	λ, Failures per Year	$\lambda \cdot r$, Forced Hours of Downtime per Year
13.8 kV power source from electric utility (loss of source 1)	1.644	
Protective relays (3)	0.0006	
13.8 kV metalclad circuit breaker	0.0036	
Switchgear bus—insulated (connected to 1 breaker)	0.0034	
Cable (13.8 kV); 900 ft, conduit below ground	0.0055	
Cable terminations (6) at 13.8 kV	0.0018	
Disconnect switch (enclosed)	0.0061	
Total through disconnect switch with 1 h switchover after a failure of source 1 (and source 2 is OK)	1.6650	1.6650
Loss of both 13.8 kV power sources simultaneously	0.312	0.1622
Total to point F	1.9770	1.8272
Transformer	0.0030	1.0260[*]
480 V metalclad circuit breaker	0.0027	0.0108
Switchgear bus—bare (connected to 7 breakers)	0.0024	0.0576
480 V metalclad circuit breaker	0.0027	0.0108
480 V metalclad circuit breakers (5) (failed while opening)	0.0012	0.0048
Cable (480 V); 300 ft conduit above ground	0.0004	0.0044
Cable terminations (2) at 480 V	0.0002	0.0008
Total at 480 V output	1.9896	2.9424

[*]Data for hours of downtime per failure are based upon *repair failed unit*.

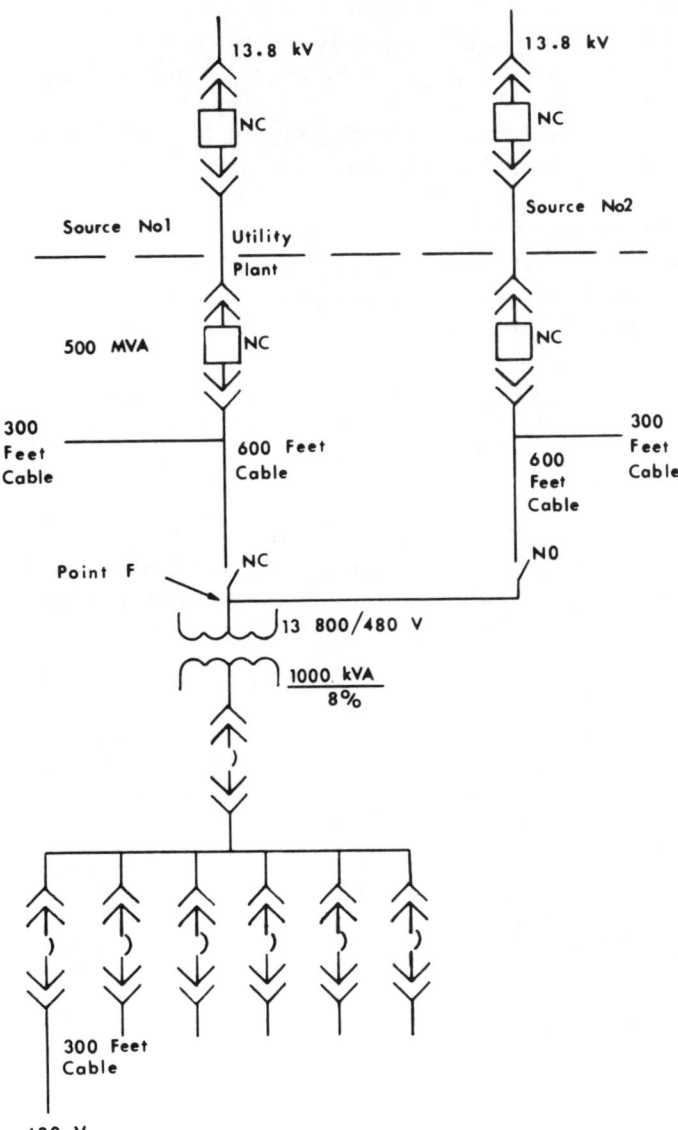

Fig 17
Primary-Selective System to
Primary of Transformer — Example 4

7.1.10 Example 5—Secondary-Selective System

7.1.10.1 Description of Secondary-Selective System. Figure 18 shows a single-line diagram of the power distribution system for a secondary-selective system. What are the failure rate and forced hours of downtime per year at the 480 V point of use?

Example 5a—Assume a 9 minute manual switchover time.

Example 5b—Assume an automatic switchover can be accomplished in less than 5 seconds after a failure (loss of 480 V power for less than 5 seconds is not counted as a failure).

Fig 18
Secondary-Selective System—Example 5

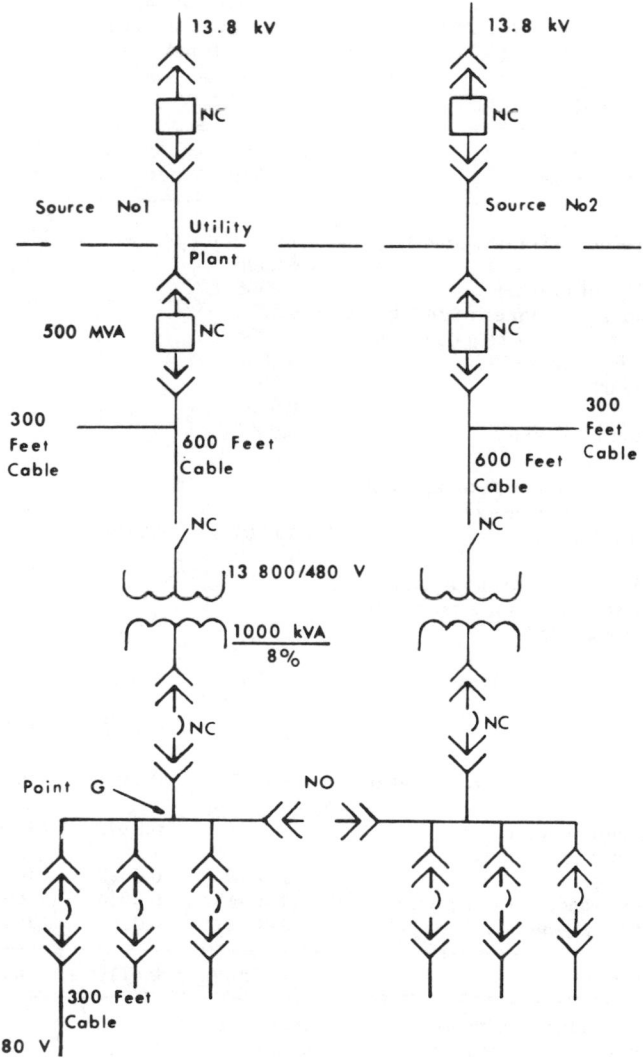

7.1.10.2 Results — Secondary-Selective System. The results from the reliability and availability calculations are given in Table 55.

7.1.10.3 Conclusions — Secondary-Selective System. The simple radial system in Example 1 had an average forced hours downtime per year that was 19 times larger than the secondary-selective system in Example 5b with automatic switchover in less than 5 seconds. The failure rate of the simple radial system was six times larger than the secondary-selective system in Example 5b with automatic switchover in less than 5 seconds.

Table 55
Secondary-Selective System — Reliability and Availability
Comparison of Power at 480 V Point of Use

Component	Example 5a (9 min. switchover time)		Example 5b (switchover in less than 5 s[†])	
	λ, Failures per Year	$\lambda \cdot r$, Forced Hours of Downtime per Year	λ, Failures per Year	$\lambda \cdot r$, Forced Hours of Downtime per Year
13.8 kV power source (loss of only source 1)	1.644			
Protective relays (3)	0.0006			
13.8 kV metalclad circuit breaker	0.0036			
Switchgear bus — insulated (connected to 1 breaker)	0.0034			
Cable (13.8 kV); 900 ft, conduit below ground	0.0055			
Cable terminations (6) at 13.8 kV	0.0018			
Disconnect Switch (enclosed)	0.0061			
Transformer	0.0030			
480 V metalclad circuit breaker	0.0027			
Total through 480 V main circuit breaker with 9 min switchover after a failure of source 1 (and source 2 is OK)	1.6707	0.2506		
Total through 480 V main circuit breaker with switchover in less than 5 s after a failure of source 1 (and source 2 is OK)			0	0
Loss of both power sources simultaneously	0.312	0.1622	0.312	0.1622
Total to point G	1.9827	0.4128	0.312	0.1622
Switchgear bus — bare (connected to 5 breakers)	0.0017	0.0408	0.0017	0.0408
480 V metalclad circuit breaker	0.0027	0.0108	0.0027	0.0108
480 V metalclad circuit breakers (2) (failed while opening)	0.0005	0.0020	0.0005	0.0020
Cable (480 V); 300 ft, conduit above ground	0.0004	0.0044	0.0004	0.0044
Cable terminations (2) at 480 V	0.0002	0.0008	0.0002	0.0008
Total at 480 V output	1.9882	0.4716	0.3175	0.2210

*Data for hours downtime per failure are based upon *repair failed unit.*
†Loss of 480 V power for less than 5 s is not counted as a failure.

7.1.11 Example 6—Simple Radial System with Spares

7.1.11.1 Description of Simple Radial System with Spares. Figure 14 shows a single-line diagram of the power distribution system for a simple radial system. What are the failure rate and forced hours of downtime per year of the 480 V point of use if all of the following spare parts are available and can be installed as a replacement in these average times?:

(1) 13.8 kV circuit breaker (inside plant only)—2.1 hours
(2) 900 feet of cable (13.8 kV)—19 hours
(3) 1000 kVA transformer—130 hours

The above three "replace with spare" times were obtained from Table 46 and are the actual values obtained from the IEEE Committee Report on the Reliability Survey of Industrial Plants [8]. The times are much lower than the "repair failed unit" times that were used in Examples 1 through 5.

7.1.11.2 Results—Simple Radial System with Spares. The results of the reliability and availability calculations are given in Table 56. They are compared with those of the simple radial system in Example 1 using average outage times based upon "repair failed unit."

7.1.11.3 Conclusions—Simple Radial System with Spares. The simple radial system with spares in Example 6 had a forced hours downtime per year that was 22% lower than the simple radial system in Example 1.

7.1.12 Overall Results from Six Examples. The results for the six examples are compared in Table 57, which shows the failure rates and the forced hours downtime per year at the 480 V point of use.

These data do not include outages for scheduled maintenance of the electrical equipment. It is assumed that scheduled maintenance will be performed at times when 480 V power output is not needed. If this is not possible, then outages for scheduled maintenance would have to be added to the numbers shown in Table 57. This would affect a simple radial system much more than a secondary-selective system because of the redundancy of electrical equipment in the latter.

7.1.13 Discussion—Cost of Power Outages. The forced hours of downtime per year is a measure of forced unavailability and is equal to the product of (failures per year) × (average hours) downtime per failure. The average downtime per failure could be called "restorability" and is a very important parameter when the forced hours of downtime per year are determined. The cost of power outages in an industrial plant is usually dependent upon both the failure rate and the restorability of the power system. In addition, the cost of power outages is also dependent on the "plant restart time" after power has been restored (see Reference [3]). The "plant restart time" would have to be added to the "average downtime per failure" r in Table 57 when cost versus reliability and availability studies are made in the design of the power distribution system.

The IEEE Committee Report on the Reliability Survey of Industrial Plants [8] found that the average "plant restart time" after a failure that caused complete plant shutdown was 17.4 hours. The median value was 4.0 hours.

7.1.14 Discussion—Definition of Power Failure. A failure of 480 V power was defined in these six examples as a complete loss of incoming power for more than 5 seconds. This is consistent with the results obtained from the IEEE Committee

Table 56

Simple Radial System with Spares — Reliability and Availability
Comparison of Power at 480 V Point of Use

Component	Example 1 Simple Radial			Example 6 Simple Radial With Spares		
	λ, Failures per Year	r, Hours of Downtime per Failure	$\lambda \cdot r$, Forced Hours of Downtime per Year	λ, Failures per Year	r, Hours of Downtime per Failure	$\lambda \cdot r$, Forced Hours of Downtime per Year
13.8 kV power source from electric utility	1.956		2.582	1.956		2.582
Protective relays (3)	0.0006		0.0030	0.0006		0.0030
13.8 kV metalclad circuit breaker	0.0036	83.1*	0.2292*	0.0036	2.1[†]	0.0076[†]
Switchgear bus—insulated (connected to 1 breaker)	0.0034		0.0911	0.0034		0.0911
Cable (13.8 kV); 900 ft, conduit below ground	0.0055	26.5*	0.1458*	0.0055	19.0[†]	0.1045[†]
Cable terminations (6) at 13.8 kV	0.0018		0.0450	0.0018		0.0450
Disconnect switch (enclosed)	0.0061		0.0020	0.0061		0.0220
Transformer	0.0030	342.*	1.0260*	0.0030	130[†]	0.3900[†]
480 V metalclad circuit breaker	0.0027		0.0108	0.0027		0.0108
Switchgear bus—bare (connected to 7 breakers)	0.0024		0.0576	0.0024		0.0576
480 V metalclad circuit breaker	0.0027		0.0108	0.0027		0.0108
480 V metalclad circuit breakers (5) (failed while opening)	0.0012		0.0048	0.0012		0.0048
Cable (480 V); 300 ft, conduit above ground	0.0004		0.0044	0.0004		0.0044
Cable terminations (2) at 480 V	0.0002		0.0008	0.0002		0.0008
Total at 480 V output	1.9896		4.3033	1.9896		3.3344

*Data for hours of downtime per failure are based upon *repair failed unit.*
[†]Data for hours of downtime per failure are based upon *replace with spare.*

Table 57
Reliability and Availability Comparison at
480 V Point of Use for Several Power Distribution Systems

Distribution System	Example	Switchover in Less than 5 s‡		Switchover Time 9 min			
		λ, Failures per Year	$\lambda \cdot r$, Forced Hours of Downtime per Year	λ, Failures per Year	$\lambda \cdot r$, Forced Hours of Downtime per Year	λ, Failures per Year	$\lambda \cdot r$, Forced Hours of Downtime per Year
Simple radial	1					1.9896	4.3033*
Simple radial with spares	6					1.9896	3.3344†
Primary-selective to 13.8 kV utility supply	2	0.3456	1.8835*	1.9896	2.1291*		
Primary-selective to load side of 13.8 kV circuit breaker	3	0.3448	1.6724*	1.9930	1.9196*		
Primary-selective to primary of transformer (1 h switchover)	4					1.9896	2.9424*
Secondary-selective	5	0.3175	0.2210*	1.9882	0.4716*		

*Data for hours of downtime per failure are based upon *repair failed unit* for 13.8 kV circuit breaker, 13.8 kV cable, and transformer.
†Data for hours of downtime per failure are based upon *replace with spare* for 13.8 kV circuit breaker, 13.8 kV cable, and transformer.
‡Loss of 480 V power for less than 5 s is not counted as a failure.

Report on the Reliability Survey of Industrial Plants [8], which found a median value of 10 seconds for the "maximum length of power failure that will not stop plant production."

7.1.15 Discussion—Electric Utility Power Supply. Previous reliability studies (see References [1], [5], and [6]) have drawn conclusions similar to those made in this chapter. All of these previous studies have identified the importance of two separate power supply sources from the electric utility. The Power System Reliability Subcommittee made a special effort to collect reliability data on double-circuit utility power supplies in an IEEE survey (see Reference [7]). These data are summarized in Table 48 and were used in Examples 2 through 5. The two power sources in a double-circuit utility supply are not completely independent, and the reliability and availability analysis must take this into consideration. The importance of this point is shown in Table 58, where a reliability and availability comparison is made between the actual double-circuit utility power supply and the calculated value from two completely independent utility power sources.

Table 58
Comparison of Actual and Calculated Reliability
and Availability of Double-Circuit Utility Power Supply
(Failure Defined as Loss of Both Power Sources)

	λ, Failures per Year	$\lambda \cdot r$, Forced Hours of Downtime per Year
Actual single-circuit utility power supply from IEEE survey [7]	1.956*	2.582*
Actual double-circuit utility power supply from IEEE survey [7]	0.312*	0.1622*
Calculated two utility power sources at 13.8 kV that are completely independent	0.0012†	0.0008†

*Taken from Table 48.
†Calculated using single-circuit utility power supply data and the formula for parallel reliability shown in Fig 13.

The actual double-circuit utility power supply has a failure rate more than 200 times larger than two completely independent utility power sources. This is due to the fact that the two sources are usually not completely independent, and due to the fact that utility outages are often related to severe weather. The calculated value assumes utility outage probability is uniform for the entire 8760 hours per year. This is not true, for example, if most utility outages were associated with just thunderstorms. The actual double-circuit utility power supply data came from an IEEE survey (see Reference [7]) and are based upon 77 outages in 246 unit-years of service at 45 plants with "all circuit breakers closed." This is a broad composite from many industrial plants in different parts of the country.

It is believed that utility supply failure rates vary widely in various locations. One significant factor in this difference is believed to be different exposures to lightning storms. Thus, average values for the utility supply failure rate may not be valid for any one location. Local values should be obtained, if possible, from the utility involved, and these values should be used in reliability and availability studies.

No examples are included here on the reliability and availability improvement that could be obtained by using local generation rather than purchased power from an electric utility. However, it is of interest to note the very high reliability of location generation equipment that is found in the IEEE Committee Report on the Reliability Survey of Industrial Plants (see Reference [8] and Table 12 in Chapter 3).

7.1.16 Other Discussion. The reliability and availability analysis in the six examples was done for 480 V low-voltage power distribution systems. It is believed that 600 V systems would have similar reliability and availability.

One of the assumptions made in the reliability and availability analysis is that the failure rate of the electrical equipment remains constant with age. It is believed that this assumption does not introduce significant errors in the conclusions. However, it is suspected that the failure rate of cables may change somewhat with age. In addition, data collected by the Edison Electric Institute on failures of power transformers above 2500 kVA show that the failure rate is higher during the first few years of service. See Table 15 in Chapter 3 for the results of an IEEE transformer reliability survey of industrial plants. The reliability data collected in other IEEE surveys (see Reference [8]) did not attempt to determine how the failure rate varied with age for any electrical equipment studied.

A logical question to ask is, "How accurate are reliability and availability predictions?" It is believed that the predicted failure rates and forced outage hours per year are at best only accurate to within a factor of 2 to what might be achieved in the field. However, the relative reliability and availability comparison of the alternative power distribution systems studied should be more accurate than 2 to 1.

The Rome Air Development Center of the U.S. Air Force has had considerable experience comparing the predicted reliability of electronic systems with the actual reliability results achieved in the field. These results (see Reference [2]) show that there is approximately a 12% chance that the field failure rate will be more than 2 to 1 worse than the reliability prediction made using a reliability handbook for electronic equipment (see Reference [10]). It might be expected that the prediction of the reliability of industrial power systems would have an accuracy similar to that obtained by the U.S. Air Force with electronic systems.

Some of the errors introduced when making reliability and availability predictions using published industry failure rates for electrical equipment are:

(1) All details that could contribute to unreliability are not included in the study.
(2) Some of the contributions from human error may not be properly included.
(3) Equipment failure rates can be influenced by the adequacy of the preventive maintenance program used (see References [8] and [11]). Contamination from the environment can also have an influence on equipment failure rates.
(4) Correct conclusions can be made from statistical analysis on the average. But some plants will never experience these "average" problems. For example, several plants will never have a transformer failure.

In spite of these limitations, it is believed that reliability and availability analysis can be very useful in cost/reliability and cost/availability trade-off studies during the design phase of the power distribution system.

7.1.17 Spot Network. A spot network would have a calculated reliability and availability approximately the same as the automatic switchover secondary-selective system (see References [5] and [6]). In addition, it would have the benefit of no momentary outage in the event of a failure of any of the 13.8 kV cables or equipment since bus voltage is not lost on a spot network.

7.1.18 Protective Devices Other than Drawout Circuit Breakers. The six examples in this chapter used drawout circuit breakers as protective devices. Other types of protective devices are also available for use on power systems. The examples in this chapter attempted to show how to make reliability and availability calculations. No attempt was made to study the effect on the reliability and availability of different types of protective devices nor to draw conclusions that any particular type of protective device was more cost effective than another.

7.2 Cost Data Applied to Examples of Reliability and Availability Analysis of Common Low-Voltage Industrial Power Distribution Systems

7.2.1 Cost Evaluation of Reliability and Availability Predictions. Cost evaluations are made in this section on the reliability and availability predictions of four power distribution system examples from 7.1. The "revenue requirements method" described in 2.2.3.1 is utilized in order to determine the most cost-effective system.

7.2.2 Description of Cost Evaluation Problem. Management insists that the engineer utilize an economic evaluation in any capital improvement program. The elements to be included and a method of mathematically equating the cost impact to be expected from electrical interruptions and downtimes against the cost of a new system were presented in 2.2. It was pointed out that there are several acceptable ways of accomplishing the detailed economic analysis for evaluation of systems with varying degrees of reliability. One of those considered acceptable, the revenue requirements (RR) method, was presented in detail, and this method will be used in the analysis of the four examples.

The four examples included are:

Example 1 — Simple radial system — Single 13.8 kV utility supply

Example 2b — Primary-selective system to 13.8 kV utility supply (dual) — Switchover time less than 5 seconds

Example 4 — Primary-selective system to primary of transformer — 13.8 kV utility supply (dual) — Manual switchover in 1 hour

Example 5b — Secondary-selective system with switchover time less than 5 seconds

Table 57 lists the expected failures per year and the average downtime per year for each of the examples. These data will be used to show which of the examples has the minimum revenue requirement and will make allowances for:

(1) Plant start-up time
(2) Revenues lost
(3) Variable expenses saved
(4) Variable expenses incurred

(5) Investment

(6) Fixed investment charges

One of the benefits of such a rigidly structured analysis is that the presentation is made in a sequential manner utilizing cost/failure data prepared with the assistance of management. With this arrangement, the results of the evaluation are less likely to be questioned than if a less sophisticated method were used.

7.2.3 Procedures for Cost Analyses. Utilizing the single-line diagrams for the four examples, a component quantity take-off of each system was made, and present-day installed unit costs were assigned for each component. In the case of the dual 13.8 kV utility company's supply, the basic cost of the second supply was estimated on the basis of a hypothetical case, assuming that only a one-time cost would be incurred. The extension of the costs results in the overall installed cost for each of the four examples. A summary of the installed costs is presented in Table 59. The total installed costs for each example are listed again after item (12) in Table 60.

The RR method is used to calculate the total cost in dollars per year of both the "installed cost" and the "cost of unreliability" for the four examples. The methods for making these calculations are tabulated in Table 60. The reliability data and the assumed cost values used are described in the next two sections.

7.2.4 Reliability Data for Examples. Table 57 can be used to determine the failures per year, λ, and the "average hours downtime per failure," r, for each of the examples. The value of r is determined from the division of $\lambda \cdot r$ by λ. The values of r and λ for the four examples are shown after (1) and (10), respectively, in Table 60.

7.2.5 Assumed Cost Values. The following common cost factors were assumed in 1976 and updated in 1987 for use in all four of the examples:

10 hours/failure — Plant start-up time after a failure, s.

$16 000/hour — Revenues lost per hour of plant downtime, g_p.

$12 000/hour — Variable expenses saved per hour of plant downtime, x_p.

$40 000/failure — Variable expenses incurred per failure, X_i.

0.4 per year — Fixed investment charge factor, F.

These values are shown in Table 60 after (2), (4), (5), (8), and (13), respectively.

7.2.6 Results and Conclusions. The minimum revenue requirements for each of the four examples are shown in item (15) at the bottom of Table 60. Some of the conclusions that can be made are tabulated below:

Example 1—Simple Radial System. This system requires the least initial investment ($123 400); but its minimum revenue requirements of $225 754 per year is second highest of the four examples analyzed.

Example 2b—Primary-Selective System to 13.8 kV Utility Supply (Dual) with Switchover Time Less than 5 Seconds. This system requires an initial investment of $283 400 or 2.3 times that of the simple radial system; however, the minimum revenue requirement is $148 990 per year, which is the least of the four examples.

Based on the data presented, Example 2b would be selected since it has the lowest minimum revenue requirement.

Table 59
Installed Costs*

Item	Unit Cost	Example 1 Simple Radial System Single 13.8 kV Utility Supply		Example 2b Primary-Selective System† to 13.8 kV		Example 4 Primary-Selective System† to Primary of Transformer Utility Supply		Example 5b Secondary-Selective System†	
		Quantity	Total Cost	Quantity	Total Cost	Quantity	Total Cost	Quantity	Total Cost
Utility service standby charge		—	—	LS	$160 000	LS	$160 000	LS	$160 000
Basic equipment									
High-voltage circuit breaker, each	$40 000	1	$40 000	1	40 000	2	80 000	2	80 000
High-voltage circuit cable, linear feet	24	600	14 400	600	14 400	1200	28 800	1200	28 800
1000 kVA transformer with 2-position switch, each	40 000	1	40 000	1	40 000	—	—	2	80 000
1000 kVA transformer with 3-position switch, each	46 000	—	—	—	—	1	46 000	—	—
1600 A low-voltage circuit breaker, each	12 000	1	12 000	1	12 000	1	12 000	3	36 000
600 A MCCB, each	5000	1	5000	1	5000	1	5000	1	5000
Low-voltage cable, linear feet	40	300	12 000	300	12 000	300	12 000	300	12 000
Subtotal — Basic equipment cost			123 400		123 400		183 800		241 800
Total cost			123 400		283 400		343 800		401 800

*All cost estimates were made in 1976 and updated to July 1987.
†Estimates based on the assumption that the utility company's alternate primary service will required 4 miles of 13.8 kV pole line and a 4000 kVA reserve capacity in the utility company's substation.

Table 60
Sample Reliability Economics Problem*

	Example 1	Example 2b	Example 4	Example 5b
	Simple Radial System Single 13.8 kV Utility Supply	Primary-Selective System to 13.8 kV Utility Supply[†]	Primary-Selective System[†] Primary of Transformer	Secondary-Selective System[‡]
(1) r = Component repair time or transfer time to restore service, whichever is less, hours per failure	2.16	5.45	1.48	0.69
(2) $s^{§}$ = Plant start-up time, hours per failure	10	10	10	10
(3) $r + s$ = [Items (1) + (2)]	12.16	15.45	11.48	10.69
(4) $g_p{}^{§}$ = Revenues lost per hour of plant downtime, \$/h	\$16 000	\$16 000	\$16 000	\$16 000
(5) $x_p{}^{§}$ = Variable expenses saved, \$/h	\$12 000	\$12 000	\$12 000	\$12 000
(6) $g_p - x_p$ = [Items (4) - (5)], Value of lost production, \$/h	\$4000	\$4000	\$4000	\$4000
(7) $(g_p - x_p)(r + s)$ = [Items (6) × (3)], \$/failure	\$48 640	\$61 800	\$45 920	\$42 760
(8) $X_i{}^{§}$ = Variable expenses incurred per failure, \$/failure	\$40 000	\$40 000	\$40 000	\$40 000
(9) Items (7) + (8)	\$88 640	\$101 800	\$85 920	\$82 760
(10) λ = Failure rate per year	1.99	0.35	1.99	0.32
(11) Items (9) × (10) = X, \$/year	\$176 394	\$35 630	\$170 980	\$26 483
(12) $C^{§}$ = Investment, \$	\$123 400	\$283 400	\$343 800	\$401 800
(13) $F^{§}$ = Fixed investment charge factor, per year	0.4	0.4	0.4	0.4
(14) CF = Fixed investment charges, \$/year	\$49 360	\$113 360	\$137 520	\$160 720
(15) $G = X + CF$ [Items (11) + (14)], Minimum revenue requirement, \$/year	\$225 754	\$148 990	\$308 500	\$187 203
Economic choice		Example 2b		

*All cost estimates were made in 1976 and updated to July 1987.
†Manual switchover time 1 h.
‡Switchover time less than 5 s.
§Assumed values in this sample problem.

Example 4—Primary-Selective System to Primary of Transformer, 13.8 kV Utility Supply (Dual)—Manual Switchover Time of 1 Hour. This system shows the next to highest initial cost of $343 800 and the highest minimum revenue requirement of $308 500 per year. A major contributor to the high minimum revenue requirement is that, while a dual system has been provided, the utility supplies 1 hour manual switchover requirement increases the failure rate and downtime to account for its high minimum revenue requirement. If an automatic switchover were utilized, the example would be competitive with Example 2b.

Example 5b—Secondary-Selective System with Switchover Time Less than 5 Seconds. This system requires the highest initial investment ($401 800) and produces the next to the lowest minimum revenue requirements of $187 203 per year.

7.3 References. The information in this chapter shall be used in conjunction with the following publications:

[1] Dickinson, W. H., Gannon, P. E., Heising, C. R., Patton, A. D., and McWilliams, D. W. "Fundamentals of Reliability Techniques as Applied to Industrial Power Systems," Conference Record 1971, *IEEE Industrial Commercial Power Systems Technical Conference*, 71C18-IGA, pp. 10–31.

[2] Feduccia, A. J. and Klion, J. *How Accurate Are Reliability Predictions?* Rome Air Development Center, 1968 Annual Symposium on Reliability, IEEE catalog no. 68C33-R, pp. 280–287.

[3] Gannon, P. E. "Cost of Interruptions: Econonomic Evaluation of Reliability," *IEEE Industrial and Commercial Power Systems Technical Conference*, Los Angeles, CA, May 10–13, 1976.

[4] Garver, D. P., Montmeat, F. E., and Patton, A. D. "Power System Reliability I—Measures of Reliability and Methods of Calculation," *IEEE Transactions on Power Apparatus and Systems*, Jul. 1964, pp. 727–737.

[5] Heising, C. R. "Reliability and Availability Comparison of Common Low-Voltage Industrial Power Distribution Systems," *IEEE Transactions on Industrial Generator Applications*, vol. IGA-6, Sep./Oct. 1970, pp. 416–424.

[6] Heising, C. R. and Dunkijacobs, J. R. "Application of Reliability Concepts to Industrial Power Systems," Conference Record 1972, *IEEE Industry Applications Society Seventh Annual Meeting*, 72-CHO685-8-1A, pp. 287–296.

[7] IEEE Committee Report. "Reliability of Electric Utility Supplies to Industrial Plants," *IEEE Technical Conference*, 75-CHO947-1-1A, pp. 131–133.

[8] IEEE Committee Report. "Reliability Survey of Industrial Plants," *IEEE Transactions on Industry Applications*, Mar./Apr., Jul./Aug., Sep./Oct. 1974, pp. 213–252, 456–476, and 681.

NOTE: References [7] and [8] are reprinted in Appendixes A, B, and D.

[9] Patton, A. D. "Fundamentals of Power System Reliability Evaluation," *IEEE Industrial and Commercial Power Systems Technical Conference*, Los Angeles, CA, May 10–13, 1976.

[10] *Reliability Stress and Failure Rate Data for Electronic Equipment*, MIL-HDBK-217A, Department of Defense, Dec. 1, 1965.

[11] Wells, S. J. "Electrical Preventive Maintenance," *IEEE Industrial and Commercial Power Systems Technical Conference*, Los Angeles, CA, May 10–13, 1976.

Chapter 8
Basic Concepts of Reliability Analysis
by Probability Methods

8.1 Introduction. This chapter provides the theoretical background for the reliability analysis used in other chapters, Chapter 2 in particular. Some basic concepts of probability theory are discussed as these are essential to the understanding and development of quantitative reliability analysis methods. Definitions of terms commonly used in system reliability analysis are also included. The three methods discussed are the cut-set, the state-space, and the network reduction methods.

8.2 Definitions. Some commonly used terms in system reliability analysis are defined here. These terms are also used in the wider context of system reliability activities. Additional definitions more specifically related to power distribution systems are given in 2.1.3.

component. A piece of electrical or mechanical equipment, a line or circuit, or a section of a line or circuit, or a group of items that is viewed as an entity for the purposes of reliability evaluation.

failure. The termination of the ability of an item to perform a required function. (See 2.1.3 for a more detailed definition that is applicable to industrial and commercial power distribution systems.)

failure rate (forced outage rate). The mean number of failures of a component per unit exposure time. Usually exposure time is expressed in years and failure rate is expressed in failures per year.

mean time between failures (MTBF). The mean exposure time between consecutive failures of a component. It can be estimated by dividing the exposure time by the number of failures in that period, provided that a sufficient number of failures has occurred in that period.

mean time to repair (MTTR). The mean time to repair a failed component. It can be estimated by dividing the summation of repair times by the number of repairs, and, therefore, it is practically the average repair time.

system. A group of components connected or associated in a fixed configuration to perform a specified function of distributing power.

8.3 Basic Probability Theory. This section discusses some of the basic concepts of probability theory. An appreciation of these ideas is essential to the understanding and development of reliability analysis methods.

8.3.1 Sample Space. Sample space is the set of all possible outcomes of a phenomenon. For example, consider a system of three distribution links. Assuming that each link exists either in the operating or "up" state or in the failed or "down" state, the sample space is

S = (1U, 2U, 3U), (1D, 2U, 3U), (1U, 2D, 3U), (1U, 2U, 3D), (1D, 2D, 3U),
    (1D, 2U, 3D), (1U, 2D, 3D), (1D, 2D, 3D)

Here iU, iD denote that the component i is up or down, respectively. The possible outcomes of a system are also called "system states," and the set of all possible system states is called "system-state space."

8.3.2 Event. In the example of three distribution links, the descriptions (1D, 2D, 3U), (1D, 2U, 3D), (1U, 2D, 3D), (1D, 2D, 3D) define an event in which two or three lines are in the failed state. Assuming that a minimum of two lines is needed for successful system operation, this set of states also defines the system failure. The event A is, therefore, a set of system states, and the event A is said to have occurred if the system is in a state that is a member of set A.

8.3.3 Probability. A simple and useful way of looking at the probability of an occurrence of the event is by using a large number of observations.

Consider, for example, that a system is energized at time $t = 0$, and the state of the system is noted at time t. This is said to be one observation. Now, if this process is repeated N times and the system is observed in the failed state N_f times, the probability of the system being in a failed state at time t is

$$P_f(t) = N_f/N \qquad \text{(Eq 26)}$$

$$N \to \infty$$

8.3.4 Combinatorial Properties of Event Probabilities. Certain combinatorial properties of event probabilities that are useful in reliability analysis are discussed in this section.

8.3.4.1 Addition Rule of Probabilities. Two events, A_1 and A_2, are mutually exclusive if they cannot occur together. For events A_1 and A_2 that are not mutually exclusive, that is, which can happen together

$$P(A_1 \cup A_2) = P(A_1) + P(A_2) - P(A_1 \cap A_2) \qquad \text{(Eq 27)}$$

where

$P(A_1 \cup A_2)$ = Probability of A_1 or A_2, or both.
$P(A_1 \cap A_2)$ = Probability of A_1 and A_2 happening together.

When A_1 and A_2 are mutually exclusive, they cannot happen together; that is, $P(A_1 \cap A_2) = 0$, therefore Eq 27 reduces to

$$P(A_1 \cup A_2) = P(A_1) + P(A_2) \qquad \text{(Eq 28)}$$

8.3.4.2 Multiplication Rule of Probabilities. If the probability of occurrence of event A_1 is affected by the occurrence of A_2, then A_1 and A_2 are not independent. The conditional probability of event A_1, given that event A_2 has already occurred, is denoted by $P(A_1|A_2)$ and

$$P(A_1 \cap A_2) = P(A_1|A_2) P(A_2) \tag{Eq 29}$$

This formula is also used to calculate the conditional probability

$$P(A_1|A_2) = P(A_1 \cap A_2) / P(A_2) \tag{Eq 30}$$

When, however, events A_1 and A_2 are independent, that is, the occurrence of A_2 does not affect the occurrence of A_1

$$P(A_1 \cap A_2) = P(A_1) P(A_2) \tag{Eq 31}$$

8.3.4.3 Complementation. $\bar{A}_1$ is used to denote the complement of event A_1. The complement $\bar{A}_1$ is the set of states that are not members of A_1. For example, if A_1 denotes states indicating system failure, then the states not representing system failure make $\bar{A}_1$.

$$P(\bar{A}_1) = 1 - P(A_1) \tag{Eq 32}$$

8.3.5 Random Variable. A random variable can be defined as "a quantity that assumes values in accordance with probabilistic laws." A discrete random variable assumes discrete values, whereas a random variable that assumes values from a continuous interval is termed a "continuous random variable." For example, the state of a system is a discrete random variable, and the time between two successive failures is a continuous random variable.

8.3.6 Probability Distribution Function. Probability distribution function describes the variability of a random variable. For a discrete random variable X, assuming values x_i, the probability density function is defined by

$$P_X(x) = P(X = x) \tag{Eq 33}$$

The probability density function for a discrete random variable is also called the "probability mass function" and has the following properties:

(1) $P_X(x) = 0$ unless x is one of the values $x_0, x_1, x_2, \ldots$
(2) $0 \leq P_X(x_i) \leq 1$
(3) $\sum_i P_X(x_i) = 1$

Another useful function is the probability distribution function or cumulative distribution function. It is defined by

$$F_X(x) = P(X \leq x) = \sum P_X(x_i), \qquad x_i \leq x \tag{Eq 34}$$

The probability density function $f_X(x)$ [or simply $f(x)$] for a continuous random variable is defined so that

$$P(a \leq X \leq b) = \int_a^b f(y) \, dy \tag{Eq 35}$$

If, for example, X denotes the time to failure, Eq 35 gives the probability that the failure will occur in the interval (a,b). The corresponding probability distribution function for a continuous random variable is

$$F(x) = P(-\infty \leq X \leq x) = \int_{-\infty}^{x} f(y)\ \mathrm{d}y \qquad \text{(Eq 36)}$$

The function $f(x)$ has certain specific properties (see Reference [3][10]) including the following:

$$\int_{-\infty}^{\infty} f(x)\ \mathrm{d}x = 1 \qquad \text{(Eq 37)}$$

8.3.7 Expectation. The probabilistic behavior of a random variable is completely defined by the probability density function. It is often, however, desirable to have a single value characterizing the random variable. One such value is the expectation. It is defined by

$$E(X) = \sum_{i} x_i P_X(x_i) \quad \text{for a discrete random variable.}$$

$$= \int_{-\infty}^{\infty} x\, f(x)\ \mathrm{d}x \quad \text{for a continuous random variable.}$$

The expectation of X is also called the "mean value of X" and has a special relationship to the average value of X in that, if random variable X is observed many times and the arithmetic average of X is calculated, it will approach the mean value as the number of observations increases.

8.3.8 Exponential Distribution. There are several special probability distribution functions (see Reference [3]); but the one of particular interest in reliability analysis is the exponential distribution, having the probability density function of

$$f(x) = \lambda \exp(-\lambda x) \qquad \text{(Eq 38)}$$

where λ is a positive constant. The mean value of random variable X with exponential distribution is

$$d = \int_{0}^{\infty} x\lambda\ e^{-\lambda x}\ \mathrm{d}x = 1/\lambda \qquad \text{(Eq 39)}$$

Also the probability distribution is

$$F(x) = \int_{0}^{x} \lambda e^{-\lambda y}\ \mathrm{d}y = 1 - e^{-\lambda x} \qquad \text{(Eq 40)}$$

If the time between failures obeys the exponential distribution, the mean time between failures is $d = 1/\lambda$, where λ denotes the failure rate of the component. It should be noted that the failure rate for exponential distribution and only exponential distribution is constant.

---

[10]The numbers in brackets correspond to those in the references at the end of this chapter.

8.4 Reliability Measures. The term "reliability" is generally used to indicate the ability of a system to continue to perform its intended function. Several measures of reliability are described in the literature, and some of the meaningful indexes for repairable systems, especially power distribution systems, are described in this section.

(1) *Unavailability.* Unavailability is the "steady-state probability that a component or system is out of service due to failures or scheduled outages." If only the failed state is considered, this term is called "forced unavailability."

(2) *Availability.* Availability is the "steady-state probability that a component or system is in service." Numerically, availability is the complement of unavailability, that is

Availability = 1 – unavailability

(3) *Frequency of System Failure.* This index can be defined as the "mean number of system failures per unit time."

(4) *Expected Failure Duration.* This index can be defined as the "expected or long-term average duration of a single failure event."

8.5 Reliability Evaluation Methods. Numerical values for reliability measures can be obtained either by analytical methods or through digital simulation. Only the analytical techniques are discussed here (a discussion of the simulation approach can be found in Reference [3]). The three methods described in this chapter are the state-space, network reduction, and cut-set methods. The state-space method is very general but becomes cumbersome for relatively large systems. The network reduction method is applicable when the system consists of series and parallel subsystems. The cut-set method is becoming increasingly popular in the reliability analysis of transmission and distribution networks and has been primarily used in this book. The state-space and network reduction methods are discussed in this chapter for reference and for the potential benefit to the users of this book.

8.5.1 Minimal Cut-Set Method. The cut-set method can be applied to systems with simple as well as complex configurations and is a very suitable technique for the reliability analysis of power distribution systems. A cut-set is a "set of components whose failure alone will cause system failure," and a minimal cut-set has no proper subset of components whose failure alone will cause system failure. The components of a minimal cut-set are in parallel since all of them must fail in order to cause system failure and various minimal cut-sets are in series as any one minimal cut-set can cause system failure.

A simple approach for the identification of minimal cut-sets is described in Chapter 2; but more formal algorithms are also available in literature (see Reference [3]). Once the minimal cut-sets have been obtained, the reliability measures can be obtained by the application of suitable formulas (see References [1] and [2][11]). Assuming component independence and denoting the probability of failure

---

[11] IEEE publications are available from the Institute of Electrical and Electronics Engineers, IEEE Service Center, 445 Hoes Lane, Piscataway, NJ 08855-1331.

of components in cut-set C_i by $P(\bar{C}_i)$, the probability (unavailability) and the frequency of system failure for m minimal cut-sets are given by:

$$P_f = P(\bar{C}_1 \cup \bar{C}_2 \cup \bar{C}_3 \cup \cdots \cup \bar{C}_m)$$

$$= P(\bar{C}_1) + P(\bar{C}_2) + \cdots + P(\bar{C}_m) \left(\frac{m}{1}\right) \text{ terms} - [P(\bar{C}_1 \cap \bar{C}_2) + \cdots$$

$$+ [P(\bar{C}_1 \cap \bar{C}_j)] \ i \neq j \ \left(\frac{m}{2}\right) \text{ terms}$$

$$\vdots$$

$$(-1)^{m-1} P(\bar{C}_1 \cap \bar{C}_2 \cap \cdots \cap \bar{C}_m) \left(\frac{m}{m}\right) \text{ terms} \qquad \text{(Eq 41)}$$

where $\bar{C}_1 \cap \bar{C}_2$, for example, denotes the failure of components of both the minimal cut-sets 1 and 2 and, therefore, $P(\bar{C}_1 \cap \bar{C}_2)$ means the probability of failure of all the components contained in C_1 and C_2, that is

$$P(\bar{C}_1 \cap \bar{C}_2) = \Pi P_{id} \text{ and } i\epsilon(C_1 \cup C_2)$$

where

P_{id} = Probability of component i being in the failed state
  = $r_i / (d_i + r_i)$
  = $\lambda_i / (\lambda_i + \mu_i)$.
d_i = MTBF of component i.
λ_i = Failure rate of component i.
  = $1/d_i$.
r_i = MTTR of component i.
μ_i = Repair rate of component i.
  = $1/r_i$.
Π = Product.

The frequency of failure is given by

$$f_f = P(\bar{C}_1) W_1 + P(\bar{C}_2) W_2 + \cdots + P(\bar{C}_m) W_m - [P(\bar{C}_1 \cap \bar{C}_2) W_{1,2} + P(\bar{C}_1 \cap \bar{C}_3) W_{1,3}$$

$$+ \cdots + P(\bar{C}_i \cap \bar{C}_j) W_{i,j}], \quad i \neq j$$

$$\vdots$$

$$(-1)^{m-1} P(\bar{C}_1 \cap \bar{C}_2 \cap \cdots \cap \bar{C}_m) W_{1,2,\cdots, m} \qquad \text{(Eq 42)}$$

where

$$W_{i,j} = \sum_{k\epsilon C_i \cup C_j} \mu_k$$

The mean failure duration is given by

$$d_f = P_f / f_f$$

When the mean time between the failure of components is much larger than the mean time to repair (or in other words, the component availabilities approach unity), Eqs 41 and 42 can be approximated (see Reference [2]) by simpler equations

$$P_f = \sum_{i=1}^{m} P(\bar{C}_i) = \sum_{i=1}^{m} P_{cs_i} \qquad (Eq\ 43)$$

and

$$f_f = \sum_{i=1}^{m} P(\bar{C}_i)\ W_i = \sum_{i=1}^{m} f_{cs_i} \qquad (Eq\ 44)$$

where P_{cs_i} and f_{cs_i} are the probability and frequency of cut-set event i, respectively. Also

$$d_f = P_f / f_f = \sum_{i=1}^{m} P_{cs_i} / \sum_{i=1}^{m} f_{cs_i} = \sum_{i=1}^{m} f_{cs_i}\ r_{cs_i} / \sum_{i=1}^{m} f_{cs_i} \qquad (Eq\ 45)$$

where

d_f = System mean failure duration.
r_{cs_i} = Mean duration of cut-set event i.

The application of Eqs 44 and 45 to power distribution systems is discussed in Chapter 2. The components in a minimal cut-set behave like a parallel system, and f_{cs_i} (assuming n components in C_i) can be computed as follows:

$$f_{cs_i} = \prod_{j=1}^{n} P_{jd} \sum_{j=1}^{n} \mu_j \qquad (Eq\ 46)$$

and

$$r_{cs_i} = 1 / \sum_{j=1}^{n} \mu_j \qquad (Eq\ 47)$$

For example, for a cut-set having three components 1, 2, and 3:

$$f_{cs_i} = \frac{\lambda_1\ \lambda_2\ \lambda_3\ (\mu_1 + \mu_2 + \mu_3)}{(\lambda_1 + \mu_1)\ (\lambda_2 + \mu_2)\ (\lambda_3 + \mu_3)}$$

$$\simeq \lambda_1\ \lambda_2\ \lambda_3\ (r_1\ r_2 + r_2\ r_3 + r_3\ r_1),\ \text{assuming}\ \lambda_i \ll \mu_i$$

and

$$r_{cs_i} = \frac{r_1\ r_2\ r_3}{r_1\ r_2 + r_2\ r_3 + r_3\ r_1}$$

8.5.2 State-Space Method. The state-space method is a very general approach and can be used when the components are independent as well as for systems involving dependent failure and repair modes. The different steps of this approach are illustrated using a simple example of a component in series with two parallel components, as shown in Fig 19.

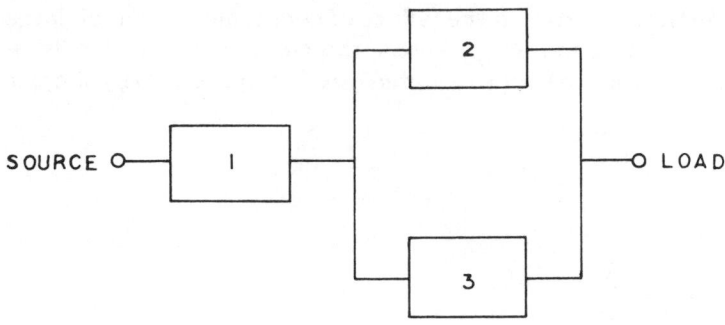

Fig 19
One Component in Series with
Two Components in Parallel

(1) *Enumerate the Possible System States.* Assuming each component can exist either in the operating state (U) or in the failed state (D) and that the components are independent, there are eight possible system states. These states are numbered 1 through 8 in Fig 20, and the description of the component states is indicated in each system state.

(2) *Determine Interstate Transition Rates.* The transition rate from s_i (that is, state i) to s_j is the mean rate of the system passing from s_i to s_j. For example, in Fig 20 the system can transit from s_1 to s_2 by the failure of component 1 and the repair of component 1, will put the system back into s_1. Therefore, the transition rate from s_1 to s_2 is λ_1, and the transition rate from s_2 to s_1 is μ_1.

(3) *Determine State Probabilities.* When the components can be assumed to be independent, state probabilities can be found by the product rule as indicated in Eq 31. When, however, statistical dependence is involved, a set of simultaneous equations needs to be solved to obtain state probabilities (see Reference [3]). Only the independent case is discussed here and for this, say the probability of being in state 2 can be determined by:

$$P_2 = P_{1d} P_{2u} P_{3u} \qquad \text{(Eq 48)}$$

where

p_{iu} = Probability of component i being in up state

$\quad = d_i/(d_i + r_i)$

$\quad = \mu_i/(\lambda_i + \mu_i).$

and

p_{id} = Probability of component i being in down state

$\quad = r_i/(d_i + r_i)$

$\quad = \lambda_i/(\lambda_i + \mu_i).$

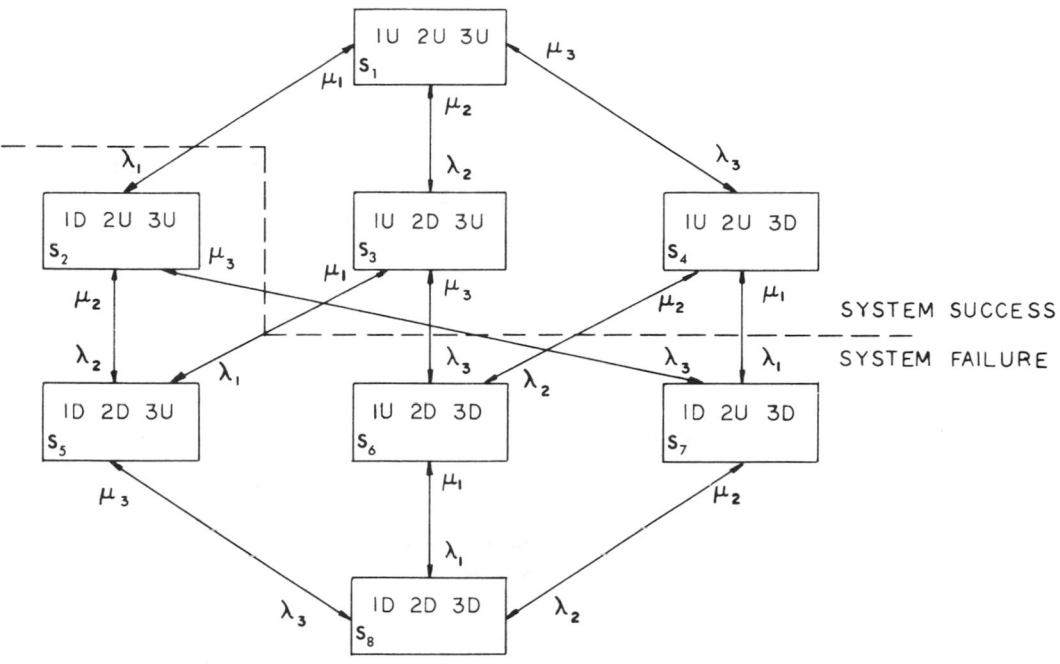

Fig 20
State Transition Diagram for the System Shown in Fig 19

(4) *Determine Reliability Measures.* The states constituting the failure, or success, or any other event of interest are identified. For the system shown in Fig 19, if the links 2 and 3 are fully redundant, system failure can occur if either component 1 fails, or components 2 and 3 fail, or if all components fail. The state space S as shown in Fig 20 is

S = {1, 2, 3, 4, 5, 6, 7, 8}

The subset A (representing failure) can be identified as:

A = {2, 5, 6, 7, 8}

and the subset representing the success states is

$S - A$ = {1, 3, 4}

Unavailability or the probability of the system being in the down state is now given by

$$P_f = \sum_{i \in A} P_i \qquad \text{(Eq 49)}$$

where $i \in A$ indicates that summation is over all states contained in subset A.

Applied to our example

$$P_f = P_2 + P_5 + P_6 + P_7 + P_8$$

where P_i can be found by the product rule as in Eq 48.

The frequency of system failure, that is, the frequency of encountering subset A, can be computed by the following relationship:

$$f_f = \sum_{i \in (S-A)} P_i \sum_{j \in A} \lambda_{ij} \qquad \text{(Eq 50)}$$

where λ_{ij} equals the transition rate from state i to state j.

Applying Eq 50 to the system in Fig 19:

$$f_f = P_1 \lambda_1 + P_3(\lambda_1 + \lambda_3) + P_4(\lambda_1 + \lambda_2)$$

The mean failure duration can be obtained from P_f and f_f using

$$d_f = P_f / f_f \qquad \text{(Eq 51)}$$

In the preceding analysis, it was assumed that the failure of a component does not alter the probability of failure of the remaining components. If, however, it is assumed that after the system failure, no further component failure will take place, the state transition diagram in Fig 20 will be modified as shown in Fig 21. Once component 1 fails or components 2 and 3 fail, no further failure is possible. The probabilities in this case cannot be calculated by simple multiplication; they can be computed by solving a set of linear equations (see Reference [3]). Once the state probabilities have been calculated, the remaining procedure is the same.

Fig 21
State Transition Diagram for the System Shown
in Fig 19 When Components Are Not Independent

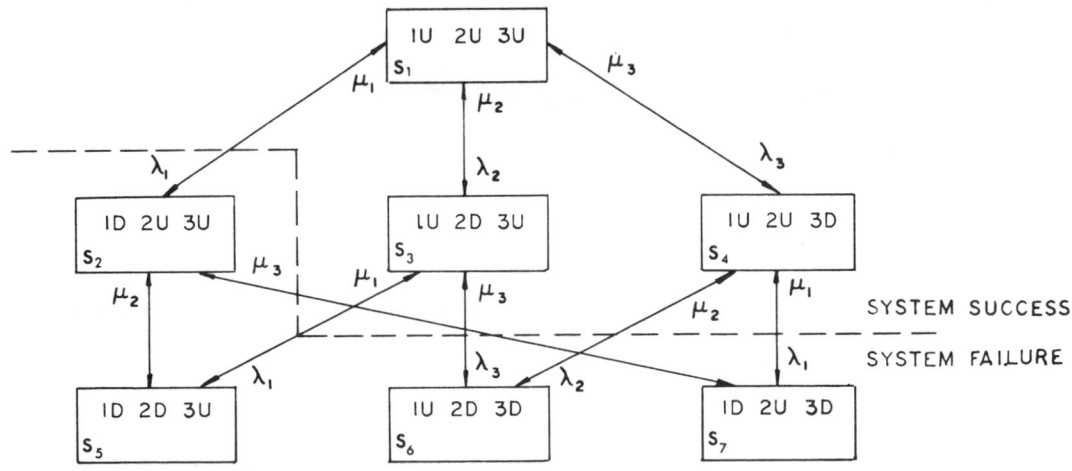

8.5.3 Network Reduction Method. The network reduction method is useful for systems consisting of series and parallel subsystems. This method consists of successively reducing the series and parallel structures by equivalent components. Knowledge of the series and parallel reduction formulas is essential for the application of this technique.

8.5.4 Series System. The components are said to be in series when the failure of any one component causes system failure. It should be noted that the components do not have to be physically in series; it is the effect of failure that is important. Two types of series systems are discussed.

8.5.4.1 Independent Components. For the series system of independent components, the failure and repair rate of the equivalent component are given by

$$\lambda_s = \sum_{i=1}^{n} \lambda_i \tag{Eq 52}$$

and

$$\mu_s = \lambda_s / \left(\prod_{i=1}^{n} (1 + \lambda_i / \mu_i) - 1 \right) \tag{Eq 53}$$

where λ_s and μ_s are the equivalent failure and repair rates of the series system and $\prod_{i=1}^{n}$ denotes the product of values 1 through n.

Assuming that λ_i is much smaller than μ_i (which, in other words, means that the MTBF is much larger than the MTTR), the quantities involving products of λ_i can be neglected. Equation 53 reduces to

$$r_s = 1/\mu_s = \sum_{i=1} r_i \lambda_i / \lambda_s \tag{Eq 54}$$

8.5.4.2 Components Involving Dependence. When it is assumed that after the system failure no more components will fail, the equivalent failure and repair parameters are

$$\lambda_s = \sum_{i=1}^{n} \lambda_i \text{ and } r_s = \sum_{i=1}^{n} r_i \lambda_i / \lambda_s \tag{Eq 55}$$

It can be seen from Eqs 54 and 55 that, for component MTBF to be much larger than MTTR, the r_s for the dependent and the independent cases should be practically equal.

8.5.5 Parallel System. Two components are considered in parallel when either can ensure system success. The equivalent failure and repair rates of a parallel system of two components are given by

$$\lambda_p = \frac{\lambda_1 \lambda_2 (r_1 + r_2)}{1 + \lambda_1 r_1 + \lambda_2 r_2} \tag{Eq 56}$$

and

$$\mu_p = \mu_1 + \mu_2 \tag{Eq 57}$$

If $\lambda_1 r_1$ and $\lambda_2 r_2$ are much smaller than 1, then Eq 56 can be written as

$$\lambda_p = \lambda_1 \lambda_2 (r_1 + r_2) \qquad \text{(Eq 58)}$$

8.6 References. The information in this chapter shall be used in conjunction with the following publications:

[1] Shooman, L. M. *Probabilistic Reliability: An Engineering Approach*, McGraw-Hill, New York, 1968.

[2] Singh, C. "On the Behavior of Failure Frequency Bounds," *IEEE Transactions on Reliability*, vol. R-26, Apr. 1977, pp. 63–66.

[3] Singh, C. and Billinton, R. *System Reliability Modelling and Evaluation*, Hutchinson Educational, London, England, 1977.

Appendixes A–M

(These appendixes are not a part of IEEE Std 493-1990, IEEE Recommended Practice for the Design of Reliable Industrial and Commercial Power Systems, and are included for information only.)

Appendix A

Report on Reliability Survey of Industrial Plants

Part 1
Reliability of Electrical Equipment

Part 2
Cost of Power Outages, Plant Restart Time, Critical Service Loss Duration Time, and Type of Loads Lost versus Time of Power Outages

Part 3
Causes and Types of Failures of Electrical Equipment, the Methods of Repair, and the Urgency of Repair

By
Reliability Subcommittee
Industrial and Commercial Power Systems Committee
IEEE Industry Applications Society

W. H. Dickinson, *Chair*

P. E. Gannon	C. R. Heising	A. D. Patton
M. D. Harris	D. W. McWilliams	W. J. Pearce
	R. W. Parisian	

Industrial and Commercial Power Systems Technical Conference
Institute of Electrical and Electronics Engineers, Inc.
Atlanta, Georgia
May 13–16, 1973

Published by
IEEE Transactions on Industry Applications
Mar./Apr. 1974

Report on Reliability Survey of Industrial Plants, Part I: Reliability of Electrical Equipment

IEEE COMMITTEE REPORT

Abstract – An IEEE sponsored survey of electrical equipment reliability in industrial plants was completed during 1972. The results are reported from this survey which included a total of 1982 equipment failures that were reported by 30 companies covering 68 plants in nine industries in the United States and Canada.

INTRODUCTION

A KNOWLEDGE of the reliability of electrical equipment is an important consideration in the design of power distribution systems for industrial plants. It is possible to make quantitative reliability comparisons between alternative designs of new systems and then use this information in cost–reliability tradeoff studies to determine which type of power distribution systems to use [1]-[10]. The cost of power outages at the various plant locations can be factored into the decision as to which type of power distribution system to use. These decisions can then be based upon total owning cost over the useful life of the equipment rather than first cost.

In 1969 a Reliability Working Group was formed under the Industrial Plants Power Systems Subcommittee, Industrial and Commercial Power Systems Committee. In 1972 the activity was changed to a Reliability Subcommittee under the same Committee. One of the major activities of the Reliability Working Group and the Reliability Subcommittee has been to conduct a survey of equipment reliability in industrial plants. This survey was conducted during the latter half of 1971 and the early part of 1972 and attempted to update a similar survey [11] which had been conducted eleven years ago. The results from the present survey contain data on failure rate and average downtime per failure for 74 equipment categories. The Reliability Subcommittee also felt that additional information was needed in the present survey beyond what was collected twelve years ago. Some of the additional information is the following:

1) cost of power outages of industrial plants;
2) plant restart time;
3) critical service loss duration time;
4) type of loads lost versus time of power outages;
5) repair or replacement time data;

Paper TOD-73-158, approved by the Industrial and Commercial Power Systems Committee of the IEEE Industry Applications Society for presentation at the 1973 Industrial and Commercial Power Systems Technical Conference, Atlanta, Ga., May 13–16. Manuscript released for publication November 5, 1973.

Members of the Reliability Subcommittee of the IEEE Industrial and Commercial Power Systems Committee are W. H. Dickinson, *Chairman*, P. E. Gannon, M. D. Harris, C. R. Heising, D. W. McWilliams, R. W. Parisian, A. D. Patton, and W. J. Pearce.

6) repair urgency information;
7) causes and types of failures;
8) maintenance data and policies.

It is not practical to publish all the results contained in the survey in a single paper. They will be presented in six separate parts. The first three parts are published at this time

Part 1: Reliability of Electrical Equipment;
Part 2: Cost of Power Outages, Plant Restart Time, Critical Service Loss Duration Time, and Type of Loads Lost Versus Time of Power Outages [11];
Part 3: Causes and Types of Failures, Methods of Repair, and Urgency of Repair [12].

A major part of the data in these three papers are presented in summary form. It is expected that the additional three papers will be presented at a later date and will contain further in-depth information where questions have been raised to point out the need for such data.

SURVEY FORM

The survey form is shown in Appendix A. Three types of cards were used for reporting the information.

Card type 1 asks for data on plant identification and other general plant information.

Card type 2 asks for data on a specific equipment class, including the total number of installed units, on their failure experience, on maintenance practices, and on estimated repair times of failed equipment.

Card type 3 asks for data on each individual failure reported on a card type 2.

It was necessary to provide definitions for "failure" and "repair time."

A *failure* is defined as any trouble with a power system component that causes any of the following to occur:

1) partial or complete plant shutdown, or below-standard plant operation;
2) unacceptable performance of user's equipment;
3) operation of the electrical protective relaying or emergency operation of the plant electrical system;
4) de-energization of any electric circuit or equipment.

A failure on a public utility supply system may cause the user to have either 1) a power interruption or loss of service, or 2) a deviation from normal voltage or frequency of sufficient magnitude or duration to disrupt plant production. A failure on an in-plant component causes a forced outage of the compo-

155

nent, and the component thereby is unable to perform its intended function until it is repaired or replaced.

Repair time of a failed component or duration of a failure is the clock hours from the time of the occurrence of the failure to the time when the component is restored to service, either by repair of the component or by substitution with a spare component. It is not the time required to restore service to a load by putting alternate circuits into operation. It includes time for diagnosing the trouble, locating the failed component, waiting for parts, repairing or replacing, testing, and restoring the component to service.

RESPONSE TO SURVEY

A total of 30 companies responded to the survey questionnaire, reporting data on 68 plants from nine industries in the United States and Canada as shown in Table I. There was a total of 1982 equipment failures reported in the survey; this included more than 620 000 unit-years of experience. Many of the plants reported data covering more than one year of experience.

Most of the data were reported to the IEEE Reliability Subcommittee during late 1971 and early 1972. Unfortunately, a downturn in the business cycle during this period of time caused many companies to reduce their work force and because of this fewer were able to participate in the survey than had been originally hoped.

SURVEY DATA PREPARATION

All of the returned survey questionnaire forms were reviewed. An attempt was made to clarify any discrepancies that were detected. Usable data were punched onto IBM cards for use in data processing.

STATISTICAL ANALYSIS OF EQUIPMENT FAILURES

Two equipment parameters are of prime importance in making system reliability studies. These parameters are 1) failure rate and 2) average outage duration or repair time. The best estimate for the failure rate of a particular type of equipment is the number of failures actually observed, divided by the total exposure time in unit-years, that is,

$$\hat{\lambda} = \frac{f}{T} \qquad (1)$$

where

$\hat{\lambda}$ best estimate of failure rate in failures per unit-year
λ true failure rate
f number of failures observed
T total exposure time in unit-years.

Statements regarding the accuracy of failure rate estimates can be made through the use of confidence limits [10], [14]–[17]. Failure rate confidence limits are upper and lower values of failure rate such that the following equations hold:

$$\Pr\left[\lambda_L \geqq \lambda\right] = \frac{1 - \gamma}{2} \qquad (2)$$

$$\Pr\left[\lambda \geqq \lambda_U\right] = \frac{1 - \gamma}{2} \qquad (3)$$

where

λ_L lower confidence limit of failure rate
λ_U upper confidence limit of failure rate
γ confidence interval (or confidence level).

A typical value often chosen for the confidence interval is 0.90. Once values for λ_L and λ_U are found, one can say that λ, whose best estimate is $\hat{\lambda}$, lies between λ_L and λ_U with 100γ percent confidence. Clearly the narrower the interval between λ_L and λ_U, the greater one's confidence that $\hat{\lambda}$ is a good estimate of λ, the true failure rate. Expressions for λ_L and λ_U are given as follows [17]:

$$\lambda_L = \frac{\chi^2(1 - \gamma)/2, 2f}{2T} \qquad (4)$$

$$\lambda_U = \frac{\chi^2(1 + \gamma)/2, 2f + 2}{2T} \qquad (5)$$

where $\chi^2 p, n$ is the p percentage point of a chi-squared distribution with n degrees of freedom. $\chi^2 p, n$ is tabled in statistical handbooks.

By substituting the value of T from (1) into (4) and (5) we get

$$\lambda_L = \frac{\chi^2(1 - \gamma)/2, 2f}{2f}(\hat{\lambda}) \qquad (6)$$

$$\lambda_U = \frac{\chi^2(1 + \gamma)/2, 2f + 2}{2f}(\hat{\lambda}). \qquad (7)$$

The deviation of the lower confidence level from $\hat{\lambda}$ in percent of $\hat{\lambda}$ is

$$\%\text{dev}_L = 100\left(1 - \frac{\lambda_L}{\hat{\lambda}}\right). \qquad (8)$$

Similarly, the deviation of the upper confidence level from $\hat{\lambda}$ in percent of $\hat{\lambda}$ is

$$\%\text{dev}_U = 100\left(\frac{\lambda_U}{\hat{\lambda}} - 1\right). \qquad (9)$$

Equations (6)–(9) were used to develop Fig. 1. These curves avoid the need of looking up $\chi^2 p, n$. Here λ_L and λ_U are plotted in terms of percent deviation from λ as a function of the observed number of failures.

The best estimate for the average outage duration or repair time for a particular type of equipment is simply the average of the observed outage durations. Confidence limit expressions for average outage durations are also available if the distributional nature of outage durations is known [17]. However, such expressions are not given here primarily because the average outage durations given in this paper are intended as a rough guide only. Equipment outage durations are believed to be more a function of the nature of a power system's operator than an inherent function of the equipment itself. Hence, average outage durations for equipment used in reliability studies should be values believed most reasonable for the particular system being studied.

The data from the survey contained information on the failure and repair characteristics of 217 categories of equipment. However, the number of observed failures for many equipment categories was too small to allow adequately accurate estimates of failure rates to be made. The Reliability Subcommittee felt that a minimum of eight to ten observed failures was required for "good" accuracy when estimating equipment failure rates (see Fig. 1). Therefore, whenever possible and reasonable from an engineering point of view, equipment categories having less than ten observed failures were combined with other categories so as to bring the number of observed failures in the combined category up to a minimum of ten. In some cases an equipment category with a large number of

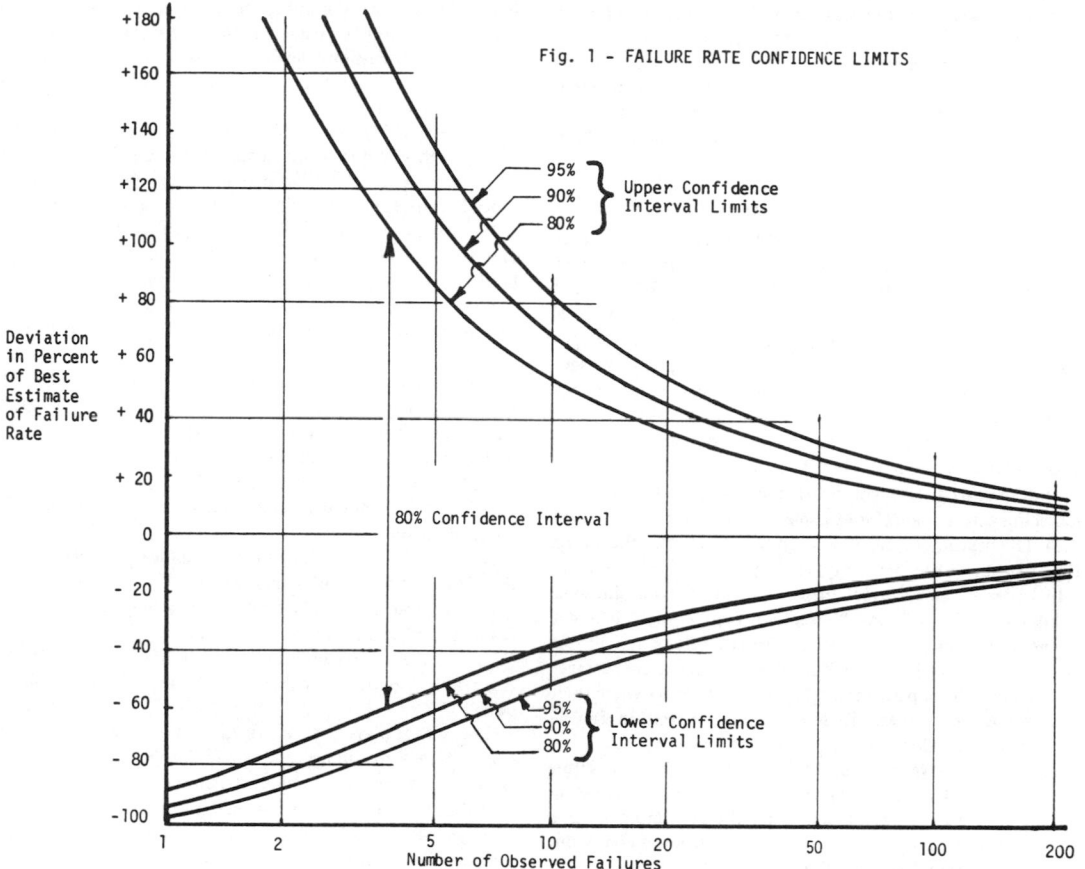

Fig. 1 - FAILURE RATE CONFIDENCE LIMITS

TABLE 1 - RESPONSE TO SURVEY QUESTIONNAIRE

Type of Industry	Number of Companies	Number of Plants
All Industry - USA & Canada.....	30*	68
Auto...........................	0	0
Cement.........................	0	0
Chemical.......................	8	21
Metal..........................	3	3
Mining.........................	0	0
Petroleum......................	5	8
Pulp and Paper.................	1	1
Rubber & Plastics..............	3	3
Textile	1	3
Other Light Manufacturing......	4	17
Other Heavy Manufacturing......	1	2
Other..........................	9	10
Foreign........................	1	1

*Some companies include more than one industry

observed failures was further subdivided. In most cases the equipment size attribute was eliminated by combining categories that were identical except for equipment size. These steps reduced the original 217 equipment categories to the 74 categories published in this paper. A total of 66 equipment categories have eight or more observed failures each; the other eight categories have between four and seven observed failures each.

Survey Results of Equipment Failures

Table 2 gives a summary of the "All Industry" equipment failure rate and equipment outage duration data for the 66 equipment categories that contain eight or more failures. The "actual hours downtime per failure" is based upon the actual outage data of the failed equipment; the "industry average" uses all equipment failures, and the "median plant average" uses all plants that reported actual outage time data on equipment failures.

The 1962 survey [11] contained equipment outage duration data on failures that have been challenged for two reasons.

1) Repairing a failed component may take much longer than replacing with a spare (for example, a large power transformer).

2) The urgency for repair is a significant factor in the outage time (low priority repairs may take days or weeks).

In order to help correct these deficiencies, two additional columns on "repair" and "replace with spare" were included in the survey and contain average estimated clock hours to fix failure during a 24-hour work day. These estimates are averaged over all the plants participating in the survey, even where there were no actual failures. These results are reported in Table 2 and are not included in the more detailed Tables 3–19.

Tables 3–19 give more detailed data on equipment failure rate and actual hours of equipment downtime per failure for 74 equipment categories; this includes the 66 equipment categories in Table 2 plus the eight equipment categories containing from four to seven failures. The additional detail includes

1) sample size in unit years;
2) number of failures;
3) number of plants reporting data;
4) additional data on actual hours of downtime per failure;
5) data for various industry groups where there were ten or more failures in that industry.

The data on average estimated clock hours to fix failure during 24-hour work day have been omitted from Tables 3–19.

The reliability data in Tables 14, 16, and 18 on cables, joints, and terminations represent a different look at the same data that are contained in Tables 13, 15, and 17. One set of tables looks at the type of insulation and the other set of tables looks at the application of the cable.

General Comments and Discussion

A survey that collects data from many plants often contains errors. Some of the errors are due to a misinterpretation of the question by the respondent, and in other cases they can be caused by omission.

Many of the respondents apparently misinterpreted the question on "number of installed units" for double- or triple-circuit electric utility power supplies. In addition, there was some confusion on the outage time after a failure of a single circuit of a double- or triple-circuit utility power supply. See the separate discussion elsewhere in this paper on these points. These are the only known major problems of misinterpretation of survey questions.

It is suspected that the failure rate estimates may be biased on the high side due to the tendency of companies to report only on equipment that has actually experienced failures. In other words, some companies may have omitted submitting unit-years of experience data on equipment that had no failures. This factor may be partially balanced out by the belief that the companies that participated in the survey may be the ones that have the best maintenance programs and keep the best records and thus may have lower failure rates than the average.

It is expected that a future paper will contain a comparison of the equipment reliability from this survey with the results from the previous survey [11] that was published in 1962. A preliminary comparison has been made and shows the following overall conclusion for 1973 versus 1962.

1) The 1973 equipment failure rates are about 0.6 times the 1962 failure rates.

2) The 1973 average downtime per failure is about 1.6 times the 1962 average downtime per failure.

3) The product of failure rate times average downtime per failure is almost the same in 1973 as 1962.

Both of these parameters are within a factor of two; and this is often the best accuracy that can be expected from reliability data.

How accurate are the failure rates shown in Tables 2–19? Fig. 1 shows the upper and lower confidence limits of the failure rate versus the number of failures observed. It can be seen that ten failures has upper and lower confidence limits of +70 percent and −46 percent for a 90 percent confidence interval. It is possible to determine the upper and lower confidence limits for the failure rate data shown in Tables 3–19.

Example of Confidence Limit Calculation

The use of Fig. 1 to determine confidence limits will be illustrated with an example. Suppose that it is desired to compute confidence limits on the failure rate of liquid-filled transformers with voltage above 15 kV in the chemical industry. The desired confidence interval is 90 percent. From Table 4, $\hat{\lambda} = 0.0119$ failures per unit-year, and the number of observed failures is 19. Entering Fig. 1 with 19 observed failures and using the 90 percent confidence interval curves yields

$$\lambda_L = \hat{\lambda} - 0.34\hat{\lambda}$$
$$= 0.0119 - 0.0041 = 0.0078 \text{ failures per unit-year}$$
$$\lambda_U = \hat{\lambda} + 0.46\hat{\lambda}$$
$$= 0.0119 + 0.0055 = 0.0174 \text{ failures per unit-year.}$$

There is a 90 percent chance that the true failure rate lies between 0.0078 and 0.0174 failures per unit-year.

TABLE 2 - SUMMARY OF "ALL INDUSTRY" EQUIPMENT FAILURE RATE AND EQUIPMENT OUTAGE DURATION DATA FOR 66 EQUIPMENT CATEGORIES CONTAINING 8 OR MORE FAILURES

Equipment	Equipment Sub Class	Failure Rate- Failures per Unit-Year	Actual Hours Downtime per Failure		Average Estimated Clock Hours to Fix Failure During 24 Hour Work Day	
			Industry Average	Median Plant Average	Repair Failed Component	Replace with Spare
Electric Utility Power Supplies..	All	0.643	1.33	1.04	–	–
=	Single Circuit	0.537	5.66	5.10	–	–
=	Double or Triple Circuit-All	0.622	0.85	1.17	–	–
=	Automatically Switched Over	0.735	0.59	0.93	–	–
=	Manual Switchover	0.458	1.87	2.00	–	–
=	Loss of All Circuits at One Time	0.119	2.00	1.58	–	–
Transformers	Liquid Filled-All	0.0041	529.	219.	378.	73.4
=	601 - 15,000 Volts - All Sizes	0.0030	174.	49.	382.	74.3
=	300-750 kVA	0.0037	61.0	10.7	49.0	3.7
=	751-2,499 kVA	0.0025	217.	64.	297.	39.7
=	2,500 kVA & up	0.0032	216.	60.0	618.	150.
=	Above 15,000 Volts	0.0130	1076.	1260.	367.	71.5
=	Dry Type; 0 - 15,000 Volts	0.0036	153.	28.	67.	39.9
=	Rectifier; Above 600 Volts	0.0298	380.	80.	300.	20.0
Circuit Breakers	Fixed Type (incl. molded case) - All	0.0052	5.8	4.0	31.7	4.5
=	0 - 600 Volts - All Sizes	0.0044	4.7	4.0	6.0	2.0
=	0 - 600 amps	0.0035	2.2	1.0	4.0	2.0
=	Above 600 amps	0.0096	9.6	8.0	8.0	2.0
=	Above 600 Volts	0.0176	10.6	3.8	44.5	12.0
=	Metalclad Drawout - All	0.0030	129.	7.6	54.2	3.9
=	0 - 600 Volts - All sizes	0.0027	147.	4.0	47.2	2.9
=	0 - 600 amps	0.0023	3.2	1.0	75.6	1.2
=	Above 600 amps	0.0030	232.	5.0	29.4	4.0
=	Above 600 Volts	0.0036	109.	168.	62.4	5.2
Motor Starters	Contact Type; 0 - 600 Volts	0.0139	65.1	24.5	8.0	4.6
=	Contact Type; 601 - 15,000 Volts	0.0153	284.	16.0	23.6	13.8

TABLE 2 (Continued)

Equipment	Equipment Sub Class	Failure Rate - Failures per Unit-Year	Actual Hours Downtime per Failure		Average Estimated Clock Hours to Fix Failure During 24 Hour Work Day	
			Industry Average	Median Plant Average	Repair Failed Component	Replace with Spare
Motors............	Induction; 0 - 600 Volts.........	0.0109	114.	18.3	50.2	13.0
"	Induction; 601 - 15,000 Volts....	0.0404	76.0	91.5	71.4	19.7
"	Synchronous; 0 - 600 Volts.......	0.0007	35.3	35.3	32.0	10.0
"	Synchronous; 601 - 15,000 Volts..	0.0318	175.	153.	146.	18.7
"	Direct Current - All.............	0.0556	37.5	16.2	69.0	5.3
Generators........	Steam Turbine Driven.............	0.032	165.	66.5	234.	201.
"	Gas Turbine driven...............	0.638	23.1	92.0	190.	400.
Disconnect Switches........	Enclosed.........................	0.0061	3.6	2.8	50.1	13.7
Switchgear Bus - Indoor & Outdoor (Unit = Number of Connected Circuit breakers or Instrument Transformer Compartments)	Insulated; 601 - 15,000 Volts....	0.00170	261.	26.8	41.0	66.0
	Bare; 0 - 600 Volts..............	0.00034	550.	24.0	41.5	24.5
	Bare; Above 600 Volts............	0.00063	17.3	13.0	20.6	7.3
Bus duct - Indoor & Outdoor...... (Unit = One Circuit Foot)	All Voltages.....................	0.000125	128.	9.5	12.9	6.0
Open Wire.............. (Unit = 1,000 Circuit Feet)...	0 - 15,000 Volts.................	0.0189	42.5	4.0	4.6	8.0
	Above 15,000 Volts)....	0.0075	17.5	12.0	8.0	-
Cable - All Types of Insulation. (Unit = 1,000 Circuit Feet)...	Above Ground & Aerial 0 - 600 Volts - All..........	0.00141	457.	10.5	20.8	39.7
"	601 - 15,000 volts - All....	0.01410	40.4	6.9	26.8	60.4
"	In Trays Above Ground.......	0.00923	8.9	8.0	49.4	119.
"	In Conduit Above Ground.....	0.04918	140.	47.5	-	19.8
"	Aerial Cable................	0.01437	31.6	5.3	10.6	28.0
"	Below Ground & Direct Burial 0 - 600 Volts...............	0.00388	15.0	24.0	-	26.8
"	601 - 15,000 Volts - All....	0.00617	95.5	35.0	20.4	26.8
"	In Duct or Conduit Below Ground....	0.00613	96.8	35.0	20.9	26.8
"	Above 15,000 Volts..........	0.00336	16.0	16.0	16.0	-

TABLE 2 (Continued)

Equipment	Equipment Sub Class	Failure Rate - Failures per Unit-Year	Actual Hours Downtime per Failure Industry Average	Actual Hours Downtime per Failure Median Plant Average	Average Estimated Clock Hours to Fix Failure During 24 Hour Work Day Repair Failed Component	Average Estimated Clock Hours to Fix Failure During 24 Hour Work Day Replace with Spare
Cable................... (Unit = 1,000 Circuit Feet)...	601 - 15,000 Volts					
"	Thermoplastic..........	0.00387	44.5	10.0	22.5	29.3
"	Thermosetting..........	0.00889	168.	26.0	27.2	55.2
"	Paper Insulated Lead Covered....	0.00912	48.9	26.8	17.3	18.3
"	Other..................	0.01832	16.1	28.5	23.2	44.8
Cable Joints -All Types of Insul.	601 - 15,000 Volts					
"	In Duct or Conduit Below Ground..	0.000864	36.1	31.2	14.7	5.5
Cable Joints...........	601 - 15,000 Volts					
"	Thermoplastic..........	0.000754	15.8	8.0	12.6	22.0
"	Paper Insulated Lead Covered.....	0.001037	31.4	28.0	30.0	-
Cable Terminations - All Types of Insulation........	Above Ground & Aerial					
"	0 - 600 Volts..........	0.000127	3.8	4.0	8.0	8.0
"	601 - 15,000 Volts - All...........	0.000879	198.	11.1	34.6	40.6
"	Aerial Cable...... in Trays Above Ground....	0.001848	48.5	11.3	15.3	18.0
"	In Duct or Conduit Below Ground	0.000333	8.0	9.0	48.8	58.3
"	601 - 15,000 Volts....	0.000303	25.0	23.4	28.8	30.0
Cable Terminations.....	601 - 15,000 Volts					
"	Thermoplastic..........	0.004192	10.6	11.5	12.0	12.0
"	Thermosetting..........	0.000307	451.	11.3	30.2	42.8
"	Paper Insulated Lead Covered...	0.000781	68.8	29.2	39.0	30.0
Miscellaneous..........	Inverters..............	1.254	107.	185.	5.0	8.0
"	Rectifiers.............	0.038	39.0	52.2	41.5	12.0

TABLE 3 - ELECTRIC UTILITY POWER SUPPLIES

Number of Plants in Sample Size	Sample Size Unit - Years	Number of Failures Reported	Industry	Equipment Sub Class	Failure Rate - Failures per Unit-Year	Actual Hours Downtime/Failure			
						Industry Average	Minimum Plant Average	Median Plant Average	Maximum Plant Average
30	314.4	202	All	All	0.643	1.33	*	1.04	24.0
7	70.8	38	"	Single Circuit	0.537	5.66	0.25	5.10	10.3
23	210.7	131	"	Double or Triple Circuit - All	0.622	0.85	*	1.17	24.0
17	140.2	103	"	Automatically Switched Over	0.735	0.59	*	0.93	6.00
6	54.6	25	"	Manual Switchover	0.458	1.87	1.82	2.00	24.0
23	210.7	25	"	Loss of All Circuits At One Time	0.119	2.00	*	1.58	6.00
7	64.8	20	Chemical	All	0.309	1.42	*	1.58	6.00
7	64.8	20	"	Double or Triple Circuit - All	0.309	1.42	*	1.58	6.00
6	60.1	20	"	Automatically Switched Over	0.333	1.42	*	1.58	6.00
3	46.5	10	Petroleum	All	0.215	6.80	0.33	4.95	9.57
2	18.5	49	Textile	All	2.649	0.28	0.014	2.17	4.33
2	18.5	49	"	Double or Triple Circuit - All	2.649	0.28	0.014	2.17	4.33
1	3.4	46	"	Automatically Switched Over	13.46	0.014	0.014	0.014	0.014
5	67.3	27	Other Light Manuf.	All	0.402	1.34	**	0.58	24.0
4	51.3	22	" "	Double or Triple Circuit - All	0.429	1.51	**	0.79	24.0
3	27.3	15	" "	Automatically Switched Over	0.549	0.51	**	0.04	1.46

* 19 cycles
** 2 seconds

TABLE 4 - TRANSFORMERS

Number of Plants in Sample Size	Sample Size Unit-Years	Number of Failures Reported	Industry		Failure Rate - Failures per Unit-Year	Actual Hours Downtime/Failure			
						Industry Average	Mini-mum Plant Average	Median Plant Average	Maxi-mum Plant Average
33	15,210	63	All	Liquid Filled - All....	0.0041	529.	2.0	219.	3744.
30	13,210	39	"	601-15,000 volts - All Sizes....	0.0030	174.	2.0	49.	840.
12	3,002	11	"	300-750 kVA....	0.0037	61.0	4.5	10.7	336.
18	6,040	15	"	751 - 2,499 kVA....	0.0025	217.	2.0	64.0	840.
11	4,036	13	"	2,500 kVA & up....	0.0032	216.	24.0	60.0	403.
12	1,848	24	"	Above 15,000 volts....	0.0130	1076.	12.8	1260.	3744.
16	4,937	18	"	Dry Type; 0-15,000 volts....	0.0036	153.	0.5	28.	720.
3	672	20	"	Rectifier, Above 600 volts....	0.0298	380.	24.0	80.	867.
14	8,598	43	Chemical	Liquid Filled - All....	0.0050	338.	8.0	168.	1800.
12	6,838	24	"	601-15,000 volts - All Sizes....	0.0035	52.3	8.0	48.5	336.
7	3,274	10	"	300-750 kVA....	0.0031	19.3	3.0	8.0	120.
9	1,601	19	"	Above 15,000 volts....	0.0119	670.	12.8	708.	3600.
2	662	16	"	Rectifier; Above 600 volts....	0.0242	425.	80.0	474.	867.
3	2,512	14	Petroleum	Liquid Filled - All....	0.0056	843.	4.5	591.	1178.
3	2,334	10	"	601-15,000 volts - All Sizes....	0.0043	244.	4.5	204.	403.

TABLE 5 – CIRCUIT BREAKERS

Number of Plants in Sample Size	Sample Size Unit-Years	Number of Failures Reported	Industry		Failure Rate - Failures per Unit-Year	Actual Hours Downtime/Failure			
						Industry Average	Minimum Plant Average	Median Plant Average	Maximum Plant Average
16	9,501	49	All	Fixed Type(includes molded case) - all	0.0052	5.8	0.5	4.0	72.0
12	8,990	40	"	0 - 600 volts - All Sizes	0.0044	4.7	0.5	4.0	11.0
9	7,643	27	"	0-600 amps	0.0035	2.2	0.5	1.0	9.0
4	1,347	13	"	Above 600 amps	0.0096	9.6	5.0	3.0	11.0
5	510	9	"	Above 600 volts	0.0176	10.6	1.5	3.8	72.0
28	40,770	124	"	Metalclad, Drawout - All	0.0030	129.	0.3	7.6	890.
18	24,490	66	"	0-600 volts - All Sizes	0.0027	147.	0.2	4.0	894.
11	11,270	26	"	0-600 amps	0.0023	3.2	0.2	1.0	4.0
13	13,220	40	"	Above 600 amps	0.0030	232.	0.2	5.0	894.
22	16,280	58	"	Above 600 volts	0.0036	109.	1.1	168.	883.
5	1,961	20	Chemical	Fixed Type(includes molded case) - All	0.0102	8.1	4.3	9.0	11.0
3	1,520	15	"	0-600 volts - All Sizes	0.0099	9.5	5.0	9.0	11.0
2	937	13	"	Above 600 amps	0.0139	9.6	5.0	8.0	11.0
7	10,850	33	"	Metalclad, Drawout - All	0.0030	83.7	5.8	97.7	576.
7	4,808	31	"	Above 600 volts	0.0064	89.3	6.3	97.7	576.
3	1,885	18	Petroleum	Fixed Type(includes molded case) - All	0.0095	5.8	1.0	4.0	72.0
2	1,817	17	"	0-600 volts - All Sizes	0.0094	1.9	1.0	2.5	4.0
2	1,817	17	"	0-600 amps	0.0094	1.9	1.0	2.5	4.0
3	10,430	28	Textile	Metalclad, Drawout - All	0.0027	289.	0.3	4.0	890.
3	9,655	25	"	0-600 volts - All Sizes	0.0026	218.	0.3	4.0	894.
2	4,943	19	"	0-600 amps	0.0038	3.8	0.3	2.2	4.0

TABLE 6 – MOTOR STARTERS

Number of Plants in Sample Size	Sample Size Unit-Years	Number of Failures Reported	Industry	Equipment Sub Class	Failure Rate – Failures per Unit-Year	Actual Hours Downtime/Failure			
						Industry Average	Minimum Plant Average	Median Plant Average	Maximum Plant Average
9	4,522	63	All	Contact Type 0-600 volts	0.0139	65.1	1.0	24.5	75.5
15	6,518	100	"	601-15,000 volts	0.0153	284.	3.0	16.0	1440.
3	854	5	"	Circuit Breaker	0.0059	2.8	2.8	2.8	2.8
7	5,340	14	Chemical	Contact Type; 601-15,000 volts	0.0026	298.	4.5	16.0	1323.
1	207	51	Metal	Contact Type; 0-600 volts	0.2470	75.5	75.5	75.5	75.5
2	626	81	Petroleum	Contact Type; 601-15,000 volts	0.1294	1440.	1440.	1440.	1440.

TABLE 7 – MOTORS

Number of Plants in Sample Size	Sample Size Unit-Years	Number of Failures Reported	Industry	Equipment Sub Class	Failure Rate – Failures per Unit-Year	Actual Hours Downtime/Failure			
						Industry Average	Minimum Plant Average	Median Plant Average	Maximum Plant Average
17	19,610	213	All	Induction 0-600 volts	0.0109	114.	0.5	18.3	312.
17	4,229	171	"	601-15,000 volts	0.0404	76.0	3.3	91.5	191.
2	13,790	10	"	Synchronous 0-600 volts	0.0007	35.3	35.3	35.3	35.3
11	4,276	136	"	601-15,000 volts	0.0318	175.	8.0	153.	360.
6	558	31	"	Direct Current	0.0556	37.5	4.0	16.2	139.
6	9,638	50	Chemical	Induction 0-600 volts	0.0052	22.5	6.	10.3	45.7
8	2,819	122	"	601-15,000 volts	0.0433	56.3	3.3	38.	191.
1	13,750	10	"	Synchronous 0-600 volts	0.0007	35.3	35.3	35.3	35.3
4	1,201	52	"	601-15,000 volts	0.0433	129.	25.8	113.	218.
3	6,467	146	Petroleum	Induction 0-600 volts	0.0226	158.	120.	139.	159.
2	1,015	34	"	601-15,000 volts	0.0335	139.	90.	119.	147.
2	2,826	78	"	Synchronous 601-15,000 volts	0.0276	207.	167.	210.	254.
3	161	12	Rubber & Plastics	Induction 601-15,000 volts	0.0748	144.	132	150.	168.
1	161	17	Textile	Direct Current	0.1056	9.4	9.4	9.4	9.4

TABLE 8 - GENERATORS

Number of Plants in Sample Size	Sample Size Unit-Years	Number of Failures Reported	Industry	Equipment Sub Class	Failure Rate - Failures per Unit-Year	Actual Hours Downtime/Failure			
						Industry Average	Minimum Plant Average	Median Plant Average	Maximum Plant Average
8	761.8	24	All............	Steam Turbine Driven......	0.032	165.	1.5	66.5	1080.
4	89.4	57	"............	Gas Turbine Driven.......	0.638	23.1	5.0	92.0	720.
4	59.4	4	"............	Driven by Motor, Diesel, or Gas Engine...........	0.067	127.	121.	133.	144.
1	5.5	54	Petroleum........	Gas Turbine Driven........	9.818	5.0	5.0	5.0	5.0

TABLE 9 - DISCONNECT SWITCHES

Number of Plants in Sample Size	Sample Size Unit-Years	Number of Failures Reported	Industry	Equipment Sub Class	Failure Rate - Failures per Unit-Year	Actual Hours Downtime/Failure			
						Industry Average	Minimum Plant Average	Median Plant Average	Maximum Plant Average
8	2,065	6	All...............	Open.............	0.0029	183.	3.0	6.0	1080.
16	15,490	94	"...............	Enclosed.........	0.0061	3.6	0.2	2.8	9.3
4	2,205	22	Chemical..........	Enclosed..........	0.0100	6.0	2.0	5.1	6.5
1	4,293	61	Metal.............	Enclosed..........	0.0142	2.8	2.8	2.8	2.8

TABLE 10 - SWITCHGEAR BUS: INDOOR & OUTDOOR
(Unit = Number of Connected Circuit Breakers or Instrument Transformer Compartments)

Number of Plants in Sample Size	Sample Size Unit-Years	Number of Failures Reported	Industry	Equipment Sub Class	Failure Rate - Failures per Unit-Year	Actual Hours Downtime/Failure			
						Industry Average	Minimum Plant Average	Median Plant Average	Maximum Plant Average
12	11,740	20	All.............	Insulated; 601-15,000 volts.....	0.00170	261.	5.0	26.8	1613.
12	32,280	11	".............	Bare 0-600 volts............	0.00034	550.	2.0	24.0	2520.
5	20,560	13	".............	Above 600 volts............	0.00063	17.3	6.9	13.0	48.
5	4,003	15	Chemical........	Insulated; 601-15,000 volts.	0.00375	340.	18.0	26.8	1613.
3	17,270	10	"............	Bare Above 600 volts...........	0.00058	19.3	6.9	42.0	48.

TABLE 11 - BUS DUCT: INDOOR & OUTDOOR
(Unit = 1 Circuit Foot)

Number of Plants in Sample Size	Sample Size Unit-Years	Number of Failures Reported	Industry	Equipment Sub Class	Failure Rate - Failures per Unit-Year	Actual Hours Downtime/Failure Industry Average	Minimum Plant Average	Median Plant Average	Maximum Plant Average
12	160,400	20	All............	All Voltages...............	0.000125	128.	0.5	9.5	2160.

TABLE 12 - OPEN WIRE
(Unit = 1,000 Circuit Feet)

Number of Plants in Sample Size	Sample Size Unit-Years	Number of Failures Reported	Industry	Equipment Sub Class	Failure Rate - Failures per Unit-Year	Actual Hours Downtime/Failure Industry Average	Minimum Plant Average	Median Plant Average	Maximum Plant Average
10	5,185	98	All............	0-15,000 volts.........	0.0189	42.5	1.0	4.0	3600.
7	1,460	11	"	Above 15,000 volts......	0.0075	17.5	0.4	12.0	48.
3	292.6	10	Chemical........	0-15,000 volts........	0.0342	606.	4.0	7.5	3600.
1	2,121	76	Petroleum.......	0-15,000 volts........	0.0358	4.1	4.1	4.1	4.1

TABLE 13 - CABLE (ALL TYPES OF INSULATION)
(Unit = 1,000 Circuit Feet)

Number of Plants in Sample Size	Sample Size Unit-Years	Number of Failures Reported	Industry	Equipment Sub Class	Failure Rate- Failures per Unit-Year	Actual Hours Downtime/Failure			
						Industry Average	Mini- mum Plant Average	Median Plant Average	Maxi- mum Plant average
			All.........	Above Ground & Aerial					
10	5,692	8	"	0-600 volts - All.........	0.00141	457.	2.0	10.5	1802.
18	5,248	74	"	601-15,000 volts - All...	0.01410	40.4	0.2	6.9	360.
7	1,517	14	"	In Trays Above Ground...	0.00923	8.9	6.0	8.0	12.7
6	183	9	"	In Conduit Above Ground...	0.04918	140.	4.0	47.5	360.
11	3,548	51	"	Aerial Cable.........	0.01437	31.6	0.2	5.3	178.
			"	Below Ground & Direct Burial					
3	2,060	8	"	0-600 volts.........	0.00388	15.0	8.0	24.0	48.0
26	19,120	118	"	601-15,000 volts - All...	0.00617	95.5	0.3	35.0	4320.
26	18,940	116	"	In Duct or Conduit Below Ground	0.00613	96.8	0.3	35.0	4320.
1	2,975	10	"	Above 15,000 volts.........	0.00336	16.0	16.0	16.0	16.0
			Chemical.........	Above Ground & Aerial					
7	1,961	44	"	601-15,000 volts - All...	0.02244	35.5	2.0	4.7	154.
3	1,137	11	"	In Trays Above Ground...	0.00968	7.8	6.0	7.0	8.0
5	737	28	"	Aerial Cable......urial	0.03800	47.1	2.0	4.7	178.
			"	Below Ground & Direct Burial					
10	11,420	70	"	601-15,000 volts - All...	0.00613	53.0	2.6	25.0	514.
10	11,420	70	"	In Duct or Conduit Below Ground	0.00613	53.0	2.6	25.0	514.
			Petroleum.........	Above Ground & Aerial					
2	2,838	15	"	601-15,000 volts - All...	0.00529	21.0	7.7	27.7	47.6
2	2,669	12	"	Aerial Cable.........	0.00450	23.1	7.7	53.8	100.
			"	Below Ground & Direct Burial					
2	981	23	"	601-15,000 volts - All...	0.02345	94.0	26.8	69.7	113.
2	981	23	"	In Duct or Conduit Below Ground	0.02345	94.0	26.8	69.7	113.
1	2,975	10	"	Above 15,000 volts.........	0.00336	16.0	16.0	16.0	16.0

TABLE 14 – CABLE (ALL APPLICATIONS)
(Unit = 1,000 Circuit Feet)

Number of Plants in Sample Size	Sample Size Unit-Years	Number of Failures Reported	Industry	Equipment Sub Class	Failure Rate- Failures per Unit-Year	Actual Hours Downtime/Failure			
						Industry Average	Mini- mum Plant Average	Median Plant Average	Maxi- mum Plant Average
			All.............	601–15,000 volts					
9	9,819	38	".............	Thermoplastic..............	0.00387	44.5	2.0	10.0	178.
15	5,960	53	".............	Thermosetting..............	0.00889	168.	0.2	26.0	4320.
10	7,126	65	".............	Paper Insulated Lead Covered..	0.00912	48.9	0.3	26.8	120.
8	1,419	26	".............	Other..............	0.01832	16.1	0.7	28.5	168.
			Chemical.............	601–15,000 volts.					
7	9,158	36	".............	Thermoplastic..............	0.00393	45.4	2.0	9.8	178.
3	2,578	26	".............	thermosetting..............	0.01009	117.	17.3	202.	387.
4	937	26	".............	Paper Insulated Lead Covered..	0.02774	10.7	2.6	25.0	120.
3	697	16	".............	Other..............	0.02297	18.3	8.0	9.0	168.
			Petroleum.............	601–15,000 volts					
2	2,520	15	".............	Thermosetting..............	0.00595	21.0	7.7	27.7	47.6
2	1,299	23	".............	Paper Insulated Lead Covered..	0.01770	94.0	26.8	69.7	113.

TABLE 15 – CABLE JOINTS (ALL TYPES OF INSULATION)

Number of Plants in Sample Size	Sample Size Unit-Years	Number of Failures Reported	Industry	Equipment Sub Class	Failure Rate- Failures per Unit-Year	Actual Hours Downtime/Failure			
						Industry Average	mini- mum Plant Average	Median Plant Average	Maxi- mum Plant Average
			All.............	601–15,000 volts					
5	7,401	6	".............	Above ground & Aerial.........	0.000811	20.3	8.0	16.5	48.0
12	40,500	35	".............	In Duct or Conduit Below Ground	0.000864	36.1	1.0	31.2	160.
			Chemical.............	601–15,000 volts					
5	24,120	21	".............	In Duct or Conduit Below Ground	0.000871	17.0	1.0	8.0	34.4

TABLE 16 - CABLE JOINTS (ALL APPLICATIONS)

Number of Plants in Sample Size	Sample Size Unit-Years	Number of Failures Reported	Industry / Equipment Sub Class	Failure Rate- Failures per Unit-Year	Actual Hours Downtime/Failure			
					Industry Average	Minimum Plant Average	Median Plant Average	Maximum Plant Average
			601-15,000 volts					
5	27,860	21	All... Thermoplastic	0.000754	15.8	3.4	8.0	36.0
4	4,857	6	" Thermosetting	0.001235	102.	14.0	60.0	160.
5	13,500	14	" Paper Insulated Lead Covered...	0.001037	31.4	1.0	28.0	75.5
			601-15,000 volts					
4	22,900	20	Chemical Thermoplastic	0.000873	14.8	3.4	8.0	34.4

TABLE 17 - CABLE TERMINATIONS (ALL TYPES OF INSULATION)

Number of Plants in Sample Size	Sample Size Unit-Years	Number of Failures Reported	Industry / Equipment Sub Class	Failure Rate- Failures per Unit-Year	Actual Hours Downtime/Failure			
					Industry Average	Minimum Plant Average	Median Plant Average	Maximum Plant Average
			Above Ground & Aerial					
4	63,120	8	All 0-600 volts - All	0.000127	3.8	0.5	4.0	5.9
13	39,840	35	" 601-15,000 volts - All	0.000879	198.	1.0	11.1	728.
4	24,010	8	" In Trays Above Ground	0.000333	8.0	7.0	9.0	11.0
3	3,920	5	" In Conduit Above Ground	0.001276	1157.	24.0	732.	1440.
7	11,910	22	" Aerial Cable	0.001848	48.5	1.0	11.3	84.4
			In Duct or Conduit Below Ground					
6	26,390	8	" 601-15,000 volts	0.000303	25.0	16.0	23.4	34.5
			Above Ground & Aerial					
7	25,790	21	chemical 601-15,000 volts - All	0.000814	284.	7.0	11.2	728.
4	1,677	9	" Aerial Cable	0.005367	14.6	9.0	13.7	24.0
			Above Ground & Aerial					
2	10,150	12	Petroleum 601-15,000 volts - All	0.001182	79.3	24.0	54.2	84.4
1	10,120	11	" Aerial cable	0.001087	84.4	84.4	84.4	84.4

TABLE 18 – CABLE TERMINATIONS (ALL APPLICATIONS)

Number of Plants in Sample Size	Sample Size Unit-Years	Number of Failures Reported	Industry	Equipment Sub Class	Failure Rate-Failures per Unit-Year	Actual Hours Downtime/Failure			
						Industry Average	Minimum Plant Average	Median Plant Average	Maximum Plant Average
				601-15,000 volts					
2	2,385	10	All...........	Thermoplastic...........	0.004192	10.6	7.0	11.5	16.0
9	42,310	13	"	Thermosetting...........	0.000307	451.	9.3	11.3	1440.
5	20,490	16	"	Paper Insulated Lead Covered.	0.000781	68.8	16.0	29.2	82.6

TABLE 19 – MISCELLANEOUS

Number of Plants in Sample Size	Sample Size Unit-Years	Number of Failures Reported	Industry	Equipment Sub Class	Failure Rate-Failures per Unit-Year	Actual Hours Downtime/Failure			
						Industry Average	Minimum Plant Average	Median Plant Average	Maximum Plant Average
5	3,164.	6	All...........	Fuses...........	0.0019	5.5	1.0	2.0	24.0
3	30,600.	6	"	Protective Relays...........	0.0002	5.0	0.5	3.8	7.2
3	11.2	14	"	Inverters...........	1.25	107.	2.1	185.	369.
3	314.	12	"	Rectifiers...........	0.0382	39.0	32.4	52.2	72.0
2	5.6	14	Chemical........	Inverters...........	2.51	107.	2.1	185.	369.
1	16.8	10	Petroleum.......	Rectifiers...........	0.5970	32.4	32.4	32.4	32.4

USER INSTRUCTIONS FOR IEEE SURVEY FORM ON
RELIABILITY OF ELECTRIC EQUIPMENT IN INDUSTRIAL PLANTS

(SPONSORED BY THE RELIABILITY WORKING GROUP,
INDUSTRIAL PLANTS POWER SYSTEMS SUBCOMMITTEE,
INDUSTRIAL AND COMMERCIAL POWER SYSTEMS COMMITTEE)

PURPOSE   This survey is intended to collect data on failures that occur in in-plant electric equipment and in public utility electric power supplies that affect operations in industrial plants. We hope that these data will determine not only accurate failure rates and repair times on major classes of equipment, but will also give an insight into the causes of these failures in such a way that remedial recommendations may be formulated to reduce failures and to improve plant performance.

MAILING INSTRUCTIONS   Mail all filled-out forms to the following address.

     IEEE-IGA Reliability Working Group
     Care of Assistant Professor A D Patton, Dept of Electrical Engineering
     Texas A&M University
     College Station, Texas 77843

DATA PROCESSING   These forms will be given a confidential company code, and will then be key punched on cards for processing by a digital computer along with data collected from others.   The computer will prepare a suitable report on failure rates, durations, and causes of failure.

ADDITIONAL INFORMATION   The reverse side of the Survey Form asks for additional information.   The following information should be filled in on the reverse side of the first page of data for each plant: company name, plant name, type and location, the name, address, and phone number of the individual submitting the data and/or the individual to whom questions about the data may be directed.

In addition, space is provided for remarks or clarifying comments on the data being reported.   These comments should be filled in on all data sheets, if needed to clarify data.

DEFINITIONS

A component is a piece of equipment, a line or circuit, or a section of a line or circuit, or a group of items which is viewed as an entity.

A system is a group of components connected or associated in a fixed configuration to perform a specified function of generating, transmitting, or distributing power.

A failure is defined as any trouble with a power system component that causes any of the following to occur.

 (1) Partial or complete plant shutdown, or below-standard plant operation
 (2) Unacceptable performance of user's equipment
 (3) Operation of the electrical protective relaying or emergency operation of the plant
   electrical system
 (4) Deenergization of any electric circuit or equipment

A failure on a public utility supply system may cause the user to have either (1) a power interruption or loss of service, or (2) a deviation from normal voltage or frequency of sufficient magnitude or duration to disrupt plant production.

A failure on an in-plant component causes a forced outage of the component, and the component thereby is unable to perform its intended function until it is repaired or replaced.

Repair time of a failed component or duration of a failure is the clock hours from the time of the occurrence of the failure to the time when the component is restored to service, either by repair of the component or by substitution with a spare component. It is not the time required to restore service to a load by putting alternate circuits into operation.

It includes time for diagnosing the trouble, locating the failed component, waiting for parts, repairing or replacing, testing, and restoring the component to service.

Revision 3-4-71

2

USER INSTRUCTIONS FOR IEEE SURVEY FORM ON
RELIABILITY OF ELECTRIC EQUIPMENT IN INDUSTRIAL PLANTS
(SPONSORED BY THE RELIABILITY WORKING GROUP,
INDUSTRIAL PLANTS POWER SYSTEMS SUBCOMMITTEE,
INDUSTRIAL AND COMMERCIAL POWER SYSTEMS COMMITTEE)

GENERAL INSTRUCTIONS

THE SURVEY FORM   The IEEE Survey Form 1`.1-70 is an input data form for a computer program. The data on these forms will be key punched onto computer cards and analyzed by the computer program.

CODED DATA   The Survey Form asks for coded and uncoded data. It is necessary to refer to the instructions in filling in either. The following shows the columns on each card type that requires filling in a code.

CARD TYPE	COLUMNS REQUIRING CODES
1	1-10, 36
2	11-18, 33-36
3	25, 29, 30-53, 57, 58

It may happen that none of the codes shown fit the particular case being reported. For such cases, the "other" code should be used, by filling a "9" or a "99" in the space provided. "Other" means not otherwise classified. If this is done, explain on reverse side of page, referring to card type and column number.

EQUIPMENT CLASS   A group of codes is used to specify an equipment class. An equipment class consists of a main code, two sub-class codes, a voltage code and a size code. These are explained in the instructions. For the example shown on the filled-out form, this code is as follows.

CLASS	CODE	DESCRIPTION
Main	20	= transformer
Sub 1	4	= power
Sub 2	34	= liquid filled
Voltage	2	= 601-15,000 volts primary
Size	3	= 300-750 kVA

The above coded equipment class covers all liquid-filled power transformers, with a primary voltage of 601-15,000 volts and rated 300-750 kVA. Any transformer in the plant that does not fit this example is a different classification and requires a different coding. Thus, a 5000 kVA power transformer, liquid filled, 13.8 kV primary voltage would be coded 20-4-34-2-5.

CARD-TYPES   The Survey Form asks for three types of information under the headings CARD-TYPE 1, CARD-TYPE 2, and CARD-TYPE 3.

In general, CARD-TYPE 1 asks for data on plant identification and other general plant information.

CARD-TYPE 2 asks for data on a specific equipment class, including the total number of installed units, on their failure experience, on maintenance practices, and on estimated repair times of failed equipment. The total installed units and their failure experience is the most essential data asked for.

CARDS-TYPE 3 asks for data on each individual failure reported on a CARD-TYPE 2.

A typical plant might have as many as, say 30 different equipment classes. These 30 equipment classes might have, for example 10 different failures. To report this information requires 30 pages of the Survey Form, one for each different equipment class. CARD-TYPE 1 is filled in completely on the first page and partly thereafter. CARD-TYPE 2 is filled in on each page. CARDS-TYPE 3 are filled in 10 times, once for each failure, if any.

CARD-TYPE 1   CARD-TYPE 1 is used to identify the reporting company and plant of that company and to give general information about that plant. The first 10 columns on this card are to be repeated by the key puncher onto CARD-TYPE 2 and CARDS-TYPE 3 for identification purposes.

Only one CARD-TYPE 1 is used by the computer program. However, we ask that on each page of the IEEE Survey Form that the first 7 columns be filled-in in case the filled-out survey forms become separated.

Fill in Items 1-8 on reverse side of first page of data for each plant.

ALL CARD TYPES   Fill in CARD-TYPE, column number, and remarks or comments on reverse side, if any, on all data cards.

3
USER INSTRUCTIONS FOR IEEE SURVEY FORM ON
RELIABILITY OF ELECTRIC EQUIPMENT IN INDUSTRIAL PLANTS
(SPONSORED BY THE RELIABILITY WORKING GROUP,
INDUSTRIAL PLANTS POWER SYSTEMS SUBCOMMITTEE,
INDUSTRIAL AND COMMERCIAL POWER SYSTEMS COMMITTEE

CARD-TYPE 2   The second or CARD-TYPE 2 is used to report on each different equipment class in the plant. A typical plant might have a one type of utility supply, and several different classes each of transformers, circuit breakers, cables, etc. These different classes are shown in Columns 11-18. These Columns 11-18 are to be repeated by the key puncher on all CARDS-TYPE 3. There will be as many CARDS-TYPE 2 as there are different equipment classes.

Each CARD-TYPE 2 is used to report (1) the total number installed of one equipment class and the total number of failures experienced (if any) of that equipment class.

In addition, each CARD-TYPE 2 is used to report on maintenance practices and estimated repair times. These are your best estimate of repair times. These estimated times will be used if actual repair times are not known, or if actual repair times are much different from the average for some special reason which is unlikely to recur. We prefer to use actual data if available.

These data are to be left blank for failures on the utility power supply, since this information is not normally available.

CARD-TYPE 3   The third or CARD-TYPE 3 is used to report on actual data for each failure reported on a corresponding CARD-TYPE 2. Thus, associated with each CARD-TYPE 2 is a set of CARDS-TYPE 3. The number of CARDS-TYPE 3 will be the same as the number of failures (column 31) reported on CARDS-TYPE 2, for example, if a CARD-TYPE 2 has a 3 in Column 31, then 3 CARDS-TYPE 3 should be filled in.

Each CARD-TYPE 3 reports specific information on one failure, such as failure duration, urgency of repair, cause of failure, loads affected by the failure, and effect of failure on plant operations.

RIGHT-ADJUSTMENT OF DATA   In filling in data, numbers should be right-adjusted, that is, they must end in the right-hand column of the assigned field. This means that if, for example, the survey form provides 3 columns to insert data but a two-digit number is to be inserted in the space available, then the number should be filled into the two right-hand columns.

SAMPLE FILLED-OUT FORM   Refer to the attached sample filled-out form. This gives an example of a report on one class of transformers with two failures.

7) DATE 3 - 4 - 71

SAMPLE

IEEE SURVEY FORM 11-1-70

PAGES 15 PAGE 4

RELIABILITY OF ELECTRIC EQUIPMENT IN INDUSTRIAL PLANTS

CARD – TYPE 1

(REFER TO SURVEY FORM INSTRUCTIONS)
(NOTE – * REFERS TO CODED DATA)

COM-PANY CODE	PLANT*					PLANT OPERATING SCHEDULE		ESTIMATED PLANT OUTAGE COST, $		PLANT MAX. DEMAND AT PLANT DESIGN CAPACITY, KW	PLANT RESTART TIME, HOURS	CRITICAL SERVICE LOSS DURATION			CARD TYPE	CARD NO.
	NO.	TYPE	LOCATION	CLIMATE	ATMOSPHERE	HR. PER DAY	DAYS PER WK.	PER FAILURE	PER HR. DOWNTIME			NO. OF UNITS	UNITS*			
1	4	6	8	9	10	11	13	15	20	25	31	33	36		79	80
G E L	1 1 0 1 5 1					8	5	4 0 0 0	2 0 0 0	5 4 0 0 0	2	1 0	4		1	1

CARD – TYPE 2

EQUIPMENT CLASS*					PERIOD COVERED BY THIS REPORT				NO. OF INSTALLED UNITS	NUMBER OF FAILURES	AVERAGE AGE*	MAIN-TENANCE		ESTIMATED CLOCK HOURS TO REPAIR A FAILURE				CARD TYPE	CARD NO.
MAIN	SUB 1	SUB 2	VOLTAGE SIZE		FROM		TO					NORMAL CYCLE, MO.	QUALITY	REPAIR FAILED COMPONENT		REPLACE WITH SPARE			
					MO.	YR.	MO.	YR.						24-HR. PER DAY	8-HR. PER DAY	24-HR. PER DAY	8-HR. PER DAY		
11	13	15	17	18	19	21	23	25	27	31	33	34	36	37	41	45	48	79	80
2 0	4 3 4 2 3				1 6 6	1 0 7 0			1 2 0	2 3	3 2			1 0 0	3 0 0	1 4	4 8	2	1

CARDS – TYPE 3

NUMBER	DATE		FOREWARNING*	DURATION		REPAIR METHOD*	REPAIR URGENCY*	MO. SINCE LAST MAINTAINED*	DAMAGED PART*	TYPE*	RESPONSI-BILITY*	INITIATING CAUSE*	CONTRIBUTING CAUSE*	CHARACTER-ISTICS*	LOADS LOST*				% PRODUCTION LOST*	PLANT OUTAGE DURATION		SERVICE RESTORED*	CARD TYPE	CARD NO.
	MO.	YR.		NO. OF UNITS	UNITS*										COMPUTER	MOTOR	LIGHTING	SOLENOID	OTHER		NO. OF UNITS	UNITS*			
19	21	23	25	26	29	30	32	34	36	38	40	42	44	46	48	49	50	51	52	53	54	57	58	79	80
1	9 6 9			6 0 2	2	1	2	9	1	1	4 9 9	1 5	1	1	1	1	2			4 4	4	3	1
2	8 7 0		1 8 0 2	1	1	3	2	1	5 9 9	1 0	6 1 1 0 9	1	2			4 4	4	3	2
3																								3	3
4																								3	4
5																								3	5
6																								3	6
7																								3	7
8																								3	8
9																								3	9
10																								3	0

WHD

5
USER INSTRUCTIONS FOR CARD-TYPE 1

(REFER TO SURVEY FORM INSTRUCTIONS)
(NOTE – * REFERS TO CODED DATA)

CARD – TYPE 1

COM-PANY CODE	PLANT*					PLANT OPERATING SCHEDULE		ESTIMATED PLANT OUTAGE COST, $		PLANT MAX. DEMAND AT PLANT DESIGN CAPACITY, KW	PLANT RESTART TIME, HOURS	CRITICAL SERVICE LOSS DURATION				CARD TYPE	CARD NO.
	NO.	TYPE	LOCATION	CLIMATE	ATMOSPHERE	HR. PER DAY	DAYS PER WK.	PER FAILURE	PER HR. DOWNTIME			NO. OF UNITS	UNITS*				
1	4	6	8	9	10	11	13	15	20	25	31	33	36			79	80
																1	1

COLUMN	NAME	CODE	DESCRIPTION
1	Company Code		Fill in on all pages a three-letter abbreviation of company name for identification of data.
4	Plant No		Fill in on all pages a sequence number starting with "1" for Plant 1, "2" for Plant 2, etc. for identification of data. A plant may consist of one or more units at the same site.
6	Plant Type		Fill in on all pages the plant type
		1	Auto Industry
		2	Cement Industry
		3	Chemical Industry
		4	Metal Industry
		5	Mining Industry
		6	Petroleum Industry
		7	Pulp and Paper Industry
		8	Rubber and Plastics Industry
		9	Textile Industry
		10	Other Light Manufacturing
		11	Other Heavy Manufacturing
		99	Other
8	Plant Location	1	USA and Canada
		2	Foreign

9 | Plant Climate (For entire plant site)

Average of daily maximums for hottest month:

	Temperature	Relative Humidity (RH) (measured at noon to 2 PM ST)
1	Hot (>90F)	High (>55 RH)
2	Hot (>90F)	Moderate (50-55 RH)
3	Hot (>90F)	Low (<50 RH)
4	Moderate (80-90F)	High (>55 RH)
5	Moderate (80-90F)	Moderate (50-55 RH)
6	MOderate (80-90F)	Low (<50 RH)
7	Low (<80F)	High (>55 RH)
8	Low (<80F)	Moderate (50-55 RH)
9	Low (<80F)	Low (<50 RH)

10	Plant Atmosphere (For entire plant site)	1	Clean to slightly polluted air
		2	With salt spray and corrosive chemicals
		3	With salt spray and dust or sand
		4	With salt spray only
		5	With corrosive chemicals and dust or sand
		6	With corrosive chemicals only
		7	With dust or sand only
		8	With conductive dust
		9	Other

	Plant Operating Schedule		
11	Hours per day		Give hours per normal working day that plant operates
13	Days per week		Give days per normal working week that plant operates
	Estimated Plant Outage Cost, Dollars		
15	Per Failure		Extra expense incurred because of a failure only (not including plant downtime), such as for damaged equipment, spoiled product, extra maintenance, or extra repair costs

EB

6
USER INSTRUCTIONS FOR CARD-TYPE 1

CARD – TYPE 1

(REFER TO SURVEY FORM INSTRUCTIONS)
(NOTE – * REFERS TO CODED DATA)

COM-PANY CODE	PLANT*					PLANT OPERATING SCHEDULE		ESTIMATED PLANT OUTAGE COST, $		PLANT MAX. DEMAND AT PLANT DESIGN CAPACITY, KW	PLANT RESTART TIME, HOURS	CRITICAL SERVICE LOSS DURATION				CARD TYPE	CARD NO.
	NO.	TYPE	LOCATION	CLIMATE	ATMOSPHERE	HR. PER DAY	DAYS PER WK.	PER FAILURE	PER HR. DOWNTIME			NO. OF UNITS	UNITS*				
1	4	6	8	9	10	11	13	15	20	25	31	33	36			79	80
																1	1

COLUMN	NAME	CODE	DESCRIPTION
20	Per hour downtime		Value of lost production in dollars per hour of plant downtime only. This is the estimated revenues (sales price) of product not made, less expenses saved in labor, material, utilities, etc. If this varies with the duration of the plant downtime, use an average value per hour.
25	Plant maximum demand at design capacity, kW		Give the maximum electric power demand when the plant is operating at its rated or design capacity in kilowatts.
31	Plant restart time, hours		Give the time required to get the plant back into operation after service is restored following a failure that has caused a complete plant shutdown, hours.
	Critical service loss duration		
33	No of units		Give the maximum time in units defined in Col 36 of loss of service to the plant which will not cause a complete plant shutdown. Any power interruption of longer duration will cause a plant shutdown. In other words, give maximum length of power failure that will not stop plant production. This time is typically in the range of cycles to minutes.
36	Units		Select code for appropriate time unit that will give accurate results.
		1	Days
		2	Hours
		3	Minutes
		4	Seconds
		5	Cycles

WHD

7
USER INSTRUCTIONS FOR CARD-TYPE 2

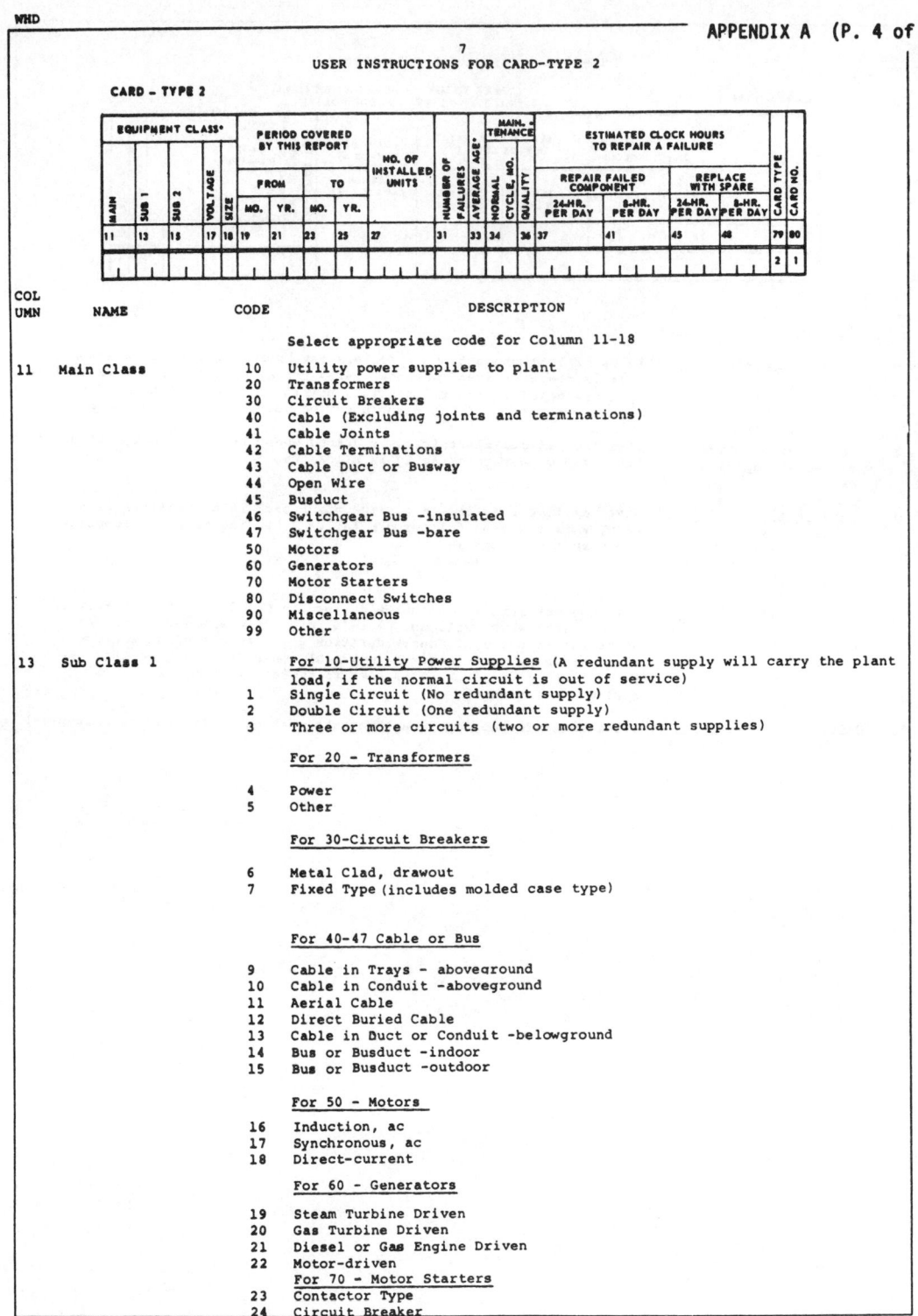

COLUMN	NAME	CODE	DESCRIPTION

Select appropriate code for Column 11-18

11   Main Class

10	Utility power supplies to plant
20	Transformers
30	Circuit Breakers
40	Cable (Excluding joints and terminations)
41	Cable Joints
42	Cable Terminations
43	Cable Duct or Busway
44	Open Wire
45	Busduct
46	Switchgear Bus -insulated
47	Switchgear Bus -bare
50	Motors
60	Generators
70	Motor Starters
80	Disconnect Switches
90	Miscellaneous
99	Other

13   Sub Class 1

For 10-Utility Power Supplies (A redundant supply will carry the plant load, if the normal circuit is out of service)

1	Single Circuit (No redundant supply)
2	Double Circuit (One redundant supply)
3	Three or more circuits (two or more redundant supplies)

For 20 - Transformers

4	Power
5	Other

For 30-Circuit Breakers

6	Metal Clad, drawout
7	Fixed Type (includes molded case type)

For 40-47 Cable or Bus

9	Cable in Trays - aboveground
10	Cable in Conduit -aboveground
11	Aerial Cable
12	Direct Buried Cable
13	Cable in Duct or Conduit -belowground
14	Bus or Busduct -indoor
15	Bus or Busduct -outdoor

For 50 - Motors

16	Induction, ac
17	Synchronous, ac
18	Direct-current

For 60 - Generators

19	Steam Turbine Driven
20	Gas Turbine Driven
21	Diesel or Gas Engine Driven
22	Motor-driven

For 70 - Motor Starters

23	Contactor Type
24	Circuit Breaker

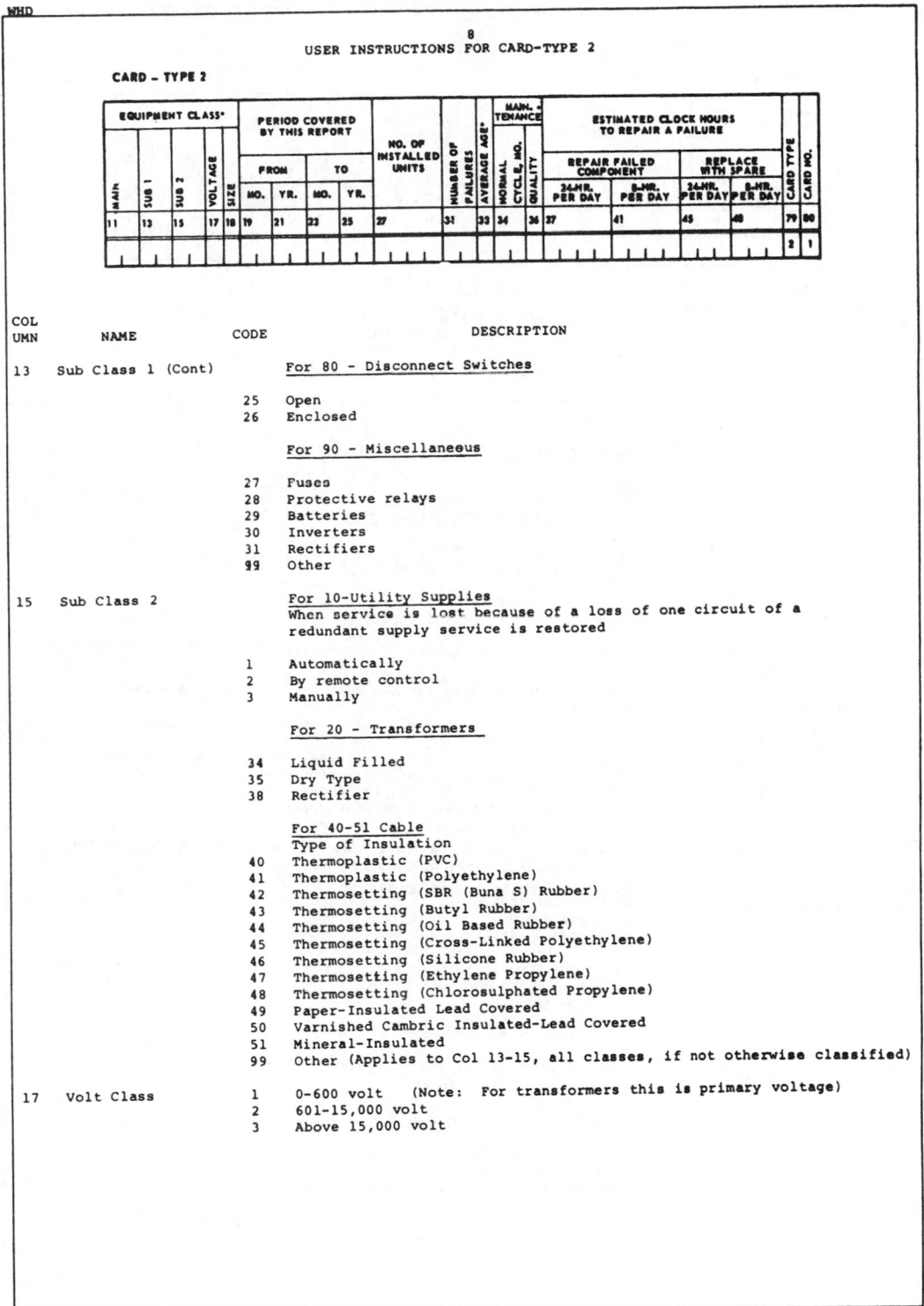

8
USER INSTRUCTIONS FOR CARD-TYPE 2

COLUMN	NAME	CODE	DESCRIPTION

13   Sub Class 1 (Cont)

For 80 - Disconnect Switches

25   Open
26   Enclosed

For 90 - Miscellaneous

27   Fuses
28   Protective relays
29   Batteries
30   Inverters
31   Rectifiers
99   Other

15   Sub Class 2

For 10-Utility Supplies
When service is lost because of a loss of one circuit of a
redundant supply service is restored

1   Automatically
2   By remote control
3   Manually

For 20 - Transformers

34   Liquid Filled
35   Dry Type
38   Rectifier

For 40-51 Cable
Type of Insulation
40   Thermoplastic (PVC)
41   Thermoplastic (Polyethylene)
42   Thermosetting (SBR (Buna S) Rubber)
43   Thermosetting (Butyl Rubber)
44   Thermosetting (Oil Based Rubber)
45   Thermosetting (Cross-Linked Polyethylene)
46   Thermosetting (Silicone Rubber)
47   Thermosetting (Ethylene Propylene)
48   Thermosetting (Chlorosulphated Propylene)
49   Paper-Insulated Lead Covered
50   Varnished Cambric Insulated-Lead Covered
51   Mineral-Insulated
99   Other (Applies to Col 13-15, all classes, if not otherwise classified)

17   Volt Class

1   0-600 volt   (Note: For transformers this is primary voltage)
2   601-15,000 volt
3   Above 15,000 volt

9
USER INSTRUCTION FOR CARD-TYPE 2

CARD – TYPE 2

EQUIPMENT CLASS*					PERIOD COVERED BY THIS REPORT				NO. OF INSTALLED UNITS	NUMBER OF FAILURES	AVERAGE AGE*	MAIN. TENANCE		ESTIMATED CLOCK HOURS TO REPAIR A FAILURE				CARD TYPE	CARD NO.
					FROM		TO					NORMAL CYCLE, MO.	QUALITY	REPAIR FAILED COMPONENT		REPLACE WITH SPARE			
MAIN	SUB 1	SUB 2	VOLTAGE	SIZE	MO.	YR.	MO.	YR.						24-HR. PER DAY	8-HR. PER DAY	24-HR. PER DAY	8-HR. PER DAY		
11	13	15	17	18	19	21	23	25	27	31	33	34	36	37	41	45	48	79	80
																		2	1

COLUMN	NAME	CODE	DESCRIPTION
18	Size Class		For Main Class 10 - Utility Supplies
			For Main Class 30 - Circuit Breakers
			For Main Class 80 - Disc Switches
			For Main Class 90 - Miscellaneous, Fuses
		1	100-600 amperes
		2	Above 600 amperes
			For Main Class 20 - Transformers
		3	300-750 kVA
		4	751-2499 kVA
		5	2500-up kVA
			For Main Class 40-45 - Cable, etc
		6	Above No 1 AWG
			For Main Class 50 - Motors
			For Main Class 70 - Motor Starters
		7	50-1500 horsepower
		8	Above 1500 horsepower
			For Main Class 60 - Generators
		9	500-up kW
	Period covered by this report		Give month and year (numerals) for period for which failure data is available
19	From: Mo		Starting Month (Try to include data from date of installation)
21	From: Yr		Starting Year
23	To: Mo		Ending Month (Try to include data to date of this report)
25	To: Yr		Ending Year
27	No of installed units		Give total number of units installed. For cable or open wire, give length of circuit or run in M ft. For cable duct or busduct, give circuit length in feet. For switchgear bus, give the number of connected circuit breakers or instrument transformer compartments. For utility power supplies, give the number of separate supplies.
31	No of Failures		Give total number of failures that occurred during period of report. If more than 10 use additional page.
			Select codes for Column 30-53
33	Average Age	1	Less than 1 year old
		2	1-10 years old
		3	More than 10 years old
	Maintenance		Give normal cycle for preventive maintenance - (even if a failure has not occurred)
34	Normal Cycle, Mo	1	Less than 12 months
		2	12-24 months
		3	More than 24 months
		4	No preventive maintenance
36	Maintenance Quality		Your estimate of quality of preventive maintenance is -
		1	Excellent (by own forces)
		2	Fair (by own forces)
		3	Poor, inadequate (by own forces)
		4	None
		5	Excellent (by contracted forces)
		6	Fair (by contracted forces)
		7	Poor inadequate (by contracted forces)

WHD

10

USER INSTRUCTIONS FOR CARD-TYPE 2

CARD – TYPE 2

EQUIPMENT CLASS*					PERIOD COVERED BY THIS REPORT				NO. OF INSTALLED UNITS	NUMBER OF FAILURES	AVERAGE AGE*	MAIN. TENANCE		ESTIMATED CLOCK HOURS TO REPAIR A FAILURE				CARD TYPE	CARD NO.
MAIN	SUB 1	SUB 2	VOLTAGE	SIZE	FROM		TO					NORMAL CYCLE, MO.	QUALITY	REPAIR FAILED COMPONENT		REPLACE WITH SPARE			
					MO.	YR.	MO.	YR.						24-HR. PER DAY	8-HR. PER DAY	24-HR. PER DAY	8-HR. PER DAY		
11	13	15	17	18	19	21	23	25	27	31	33	34	36	37	41	45	48	79	80
																		2	1

COLUMN   NAME   CODE   DESCRIPTION

Estimated clock hours   Repair time (see definitions) Fill in the clock time for diagnosing the trouble, locating the failed component, waiting for parts repairing or replacing, testing and restoring the component to service. This is your estimate of the average repair time. Please note that actual repair times are requested in CARD-TYPE 3, Col 26. Explain on reverse side how work is done if by other than own forces.

Repair failed component With repair of failed equipment

| 37 | 24-hr per day | On round-the-clock emergency basis |
| 41 | 8-hr per day | On basis of repair during normal work day |

With replacement of failed equipment with a spare by removal of failed equipment and substitution of spare equipment

Repair with spare
| 45 | 24-hr per day | On round-the-clock emergency basis |
| 48 | 8-hr per day | On basis of repair during normal work day |

181

11
USER INSTRUCTIONS FOR CARD-TYPE 3

CARDS - TYPE 3

	DATE		FOREWARNING*	DURATION		UNITS*	REPAIR METHOD*	REPAIR URGENCY*	NO. SINCE LAST MAINTAINED*	DAMAGED PART*	TYPE*	RESPONSIBILITY*	INITIATING CAUSE*	CONTRIBUTING CAUSE*	CHARACTERISTICS*	COMPUTER	MOTOR	LIGHTING	SOLENOID	OTHER	% PRODUCTION LOST*	PLANT OUTAGE DURATION		SERVICE RESTORED*	CARD TYPE	CARD NO.
NUMBER	MO.	YR.		NO. OF UNITS																		NO. OF UNITS	UNITS*			
19	21	23	25	26		29	30	32	34	36	38	40	42	44	46	48	49	50	51	52	53	54		57	58	79 80

FAILURE / LOADS LOST*

COLUMN	NAME	CODE	DESCRIPTION
19	Failure No		Fill in one card (line) for each failure. The last failure number in Col 19 should correspond with the total failures reported in Col 31 of CARD-TYPE 2. If that number was "0" then no TYPE 3 cards should be filled in.
	Failure Date		
21	Mo		Fill in month failure occured (numeral)
23	Yr		Fill in year failure occurred (numeral)
25	Failure Forewarning		For public utility power interruption only
		1	If no forewarning was given
		2	If forewarning was given For other types of failure, leave blank
	Failure Duration		Fill in duration of failure from its initiation until (1) service is restored to normal, if a power interruption, or (2) the affected component or its replacement once again becomes available to perform its intended function.
26	No of Units		Fill in the number of time units selected in Col 29.
29	Units		Select code for appropriate time unit that will give accurate results. For most cases select hours as unit.
		1	Days
		2	Hours
		3	Minutes
		4	Seconds
		5	Cycles
			Select code for Col 30-44 (Leave blank for utility failures)
30	Failure Repair Method	1	Repair of failed component in place or sent out for repair
		2	Repair by replacement of failed component with spare
32	Failure Repair Urgency	1	Requiring round-the-clock all out efforts
		2	Requiring repair work only during regular workday, perhaps with some overtime.
		3	Requiring repair work on a non-priority basis.
34	Failure, months since maintained		Failed component last had preventive maintenance -
		1	Less than 12 months ago
		2	12-24 months ago
		3	Over 24 months ago
		4	No preventive maintenance
36	Failure, Damaged Part	1	Insulation - winding
		2	Insulation - bushing
		3	Insulation - other
		4	Mechanical - bearings
		5	Mechanical - other moving parts
		6	Mechanical - other
		7	Other electrical - auxiliary device
		8	Other electrical - protective device
		9	Tap changer - no load type
		10	Tap changer - load type
		99	Other

WHD

12
USER INSTRUCTIONS FOR CARD-TYPE 3

CARDS - TYPE 3

	DATE		FOREWARNING*	DURATION		REPAIR METHOD*	REPAIR URGENCY*	MO. SINCE LAST MAINTAINED*	DAMAGED PART*	TYPE*	RESPONSI-BILITY*	INITIATING CAUSE*	CONTRIBUTING CAUSE*	CHARACTER-ISTIC*	LOADS LOST*				% PRODUCTION LOST*	PLANT OUTAGE DURATION		SERVICE RESTORED*	CARD TYPE	CARD NO.
NUMBER| MO.	YR.		NO. OF UNITS	UNITS*										COMPUTER	MOTOR	LIGHTING	SOLENOID	OTHER		NO. OF UNITS	UNITS*			
19	21	23	25	26	29	30	32	34	36	38	40	42	44	46	48	49	50	51	52	53	54	57	58	79	80
1																								3	1

COLUMN	NAME	CODE	DESCRIPTION
38	Failure Type	1	Flashover or arcing involving groun.1
		2	All other flashover or arcing
		3	Other electrical defect
		4	Mechanical defect
		99	Other

Your best estimate of suspected responsibility

40	Failure Responsibility	1	Manufacturer-defective Component
		2	Transportation to Site - defective handling
		3	Application Engineering - improper application
		4	Inadequate installation and testing prior to startup
		5	Inadequate maintenance
		6	Inadequate operating procedures
		7	Outside agency -personnel
		8	Outside agency -other
		99	Other

42	Failure Initiating Cause		Insulation breakdown caused by
		1	Transient overvoltage disturbance (lightning, switching surges, arcing ground fault in ungrounded system)
		2	Overvoltage
		3	Overheating
		4	Other insulation breakdown
		21	Mechanical breaking, cracking, loosening, abrading, or deforming of static or structural parts
		22	Mechanical burnout, friction, or seizing of moving parts
		23	Mechanically caused damage from foreign source (digging, vehicular accident, etc)
		41	Shorting by tools or metal objects
		42	Shorting by birds, snakes, rodents, etc
		51	Loss of control power
		52	Malfunction of protective relay control device, or auxiliary device
		61	Low voltage
		62	Low frequency
		99	Other

44	Failure Contributing Cause	1	Persistent overloading
		2	Above-normal temperatures
		3	Below-normal temperature
		4	Exposure to agressive chemicals or solvents
		5	Exposure to abnormal moisture or water
		6	Exposure to non-electrical fire or burning
		8	Obstruction of ventilation by foreign object or material
		9	Normal deterioration from age
		10	Severe wind, rain, snow, sleet, or other weather conditions
		11	Protective relay improperly set
		12	Loss or deficiency of lubricant
		13	Loss or deficiency of oil or cooling medium
		14	Misoperation or testing error
		15	Exposure to dust or other contaminents
		99	Other

13
USER INSTRUCTIONS FOR CARD-TYPE 3

CARDS – TYPE 3

				FAILURE											LOADS LOST*				% PRODUCTION LOST*	PLANT OUTAGE DURATION				
NUMBER	DATE		FOREWARNING*	DURATION		REPAIR METHOD*	REPAIR URGENCY*	MO. SINCE LAST MAINTAINED*	DAMAGED PART*	TYPE*	RESPONSI-BILITY*	INITIATING CAUSE*	CONTRIBUTING CAUSE*	CHARACTER-ISTICS*	COMPUTER	MOTOR	LIGHTING	SOLENOID	OTHER		NO. OF UNITS	UNITS*	SERVICE RESTORED*	CARD TYPE	CARD NO.
	MO.	YR.		NO. OF UNITS	UNITS*																				
19	21	23	25	26	29	30	32	34	36	38	40	42	44	46	48	49	50	51	52	53	54	57	58	79	80
1																								3	1

COLUMN	NAME	CODE	DESCRIPTION
46	Failure Characteristic		Utility Power Supplies (Select code)
		1	Failure of single circuit (No redundant supply)
		2	Failure of one circuit of a double-circuit redundant supply
		3	Failure of both circuits of a double-circuit redundant supply
		4	Failure of all circuits of a three or more circuit redundant supply
		5	Partial failure of a three or more circuit redundant supply
			Transformers (Select code)
		6	Automatic removal by protective equipment
		7	Partial failure reducing capacity
		8	Manual removal
			Circuit Breakers (Select code)
		9	Failed to close when it should
		10	Failed while opening
		11	Opened when it shouldn't
		12	Damaged while successfully opening
		13	Damaged while closing
		14	Failed while operating (not while opening or closing)
			General (Select code for any other class)
		15	Failed (this applies to all classes)
		16	Failed during testing or maintenance
		17	Damage discovered during testing or maintenance
		20	Partial failure
		99	Other
	Loads Lost		What loads were lost because of failure (1=yes, 0=no, 9= not known) even though power is restored promptly
48	Computer		One or more computers or solid-state control devices operated incorrectly
49	Motor		One or more motors (contactor dropout)
50	Lighting		Lighting load
51	Solenoid		One or more solenoid -operated devices dropped out, such as a solenoid-operated fuel valve
52	Other		Lost other loads, describe in remarks
53	Percent Production Lost	0	None
		1	0-30 percent
		2	Above 30 percent

WHD

14

USER INSTRUCTIONS FOR CARD-TYPE 3

CARDS - TYPE 3

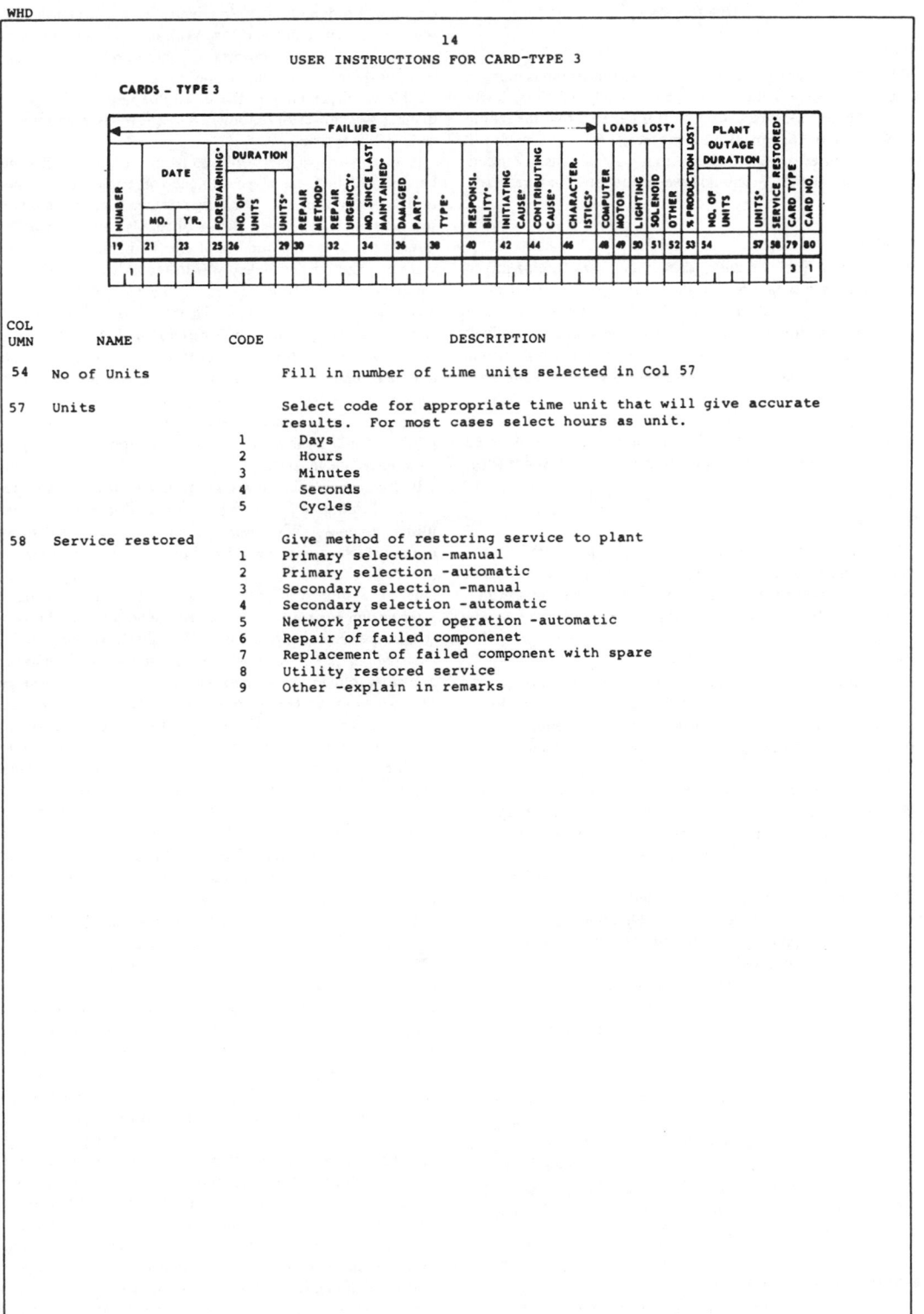

COLUMN	NAME	CODE	DESCRIPTION
54	No of Units		Fill in number of time units selected in Col 57
57	Units		Select code for appropriate time unit that will give accurate results. For most cases select hours as unit.
		1	Days
		2	Hours
		3	Minutes
		4	Seconds
		5	Cycles
58	Service restored		Give method of restoring service to plant
		1	Primary selection -manual
		2	Primary selection -automatic
		3	Secondary selection -manual
		4	Secondary selection -automatic
		5	Network protector operation -automatic
		6	Repair of failed componenet
		7	Replacement of failed component with spare
		8	Utility restored service
		9	Other -explain in remarks

185

DISCUSSION

Motors

The data in Tables 7 and 2 show that synchronous motors, 0–600 V, have a failure rate approximately 15 times lower than induction motors, 0–600 V. It is believed that the failure 0.0007 per year for synchronous motors, 0–600 V, is much too low and is in error. It is believed that synchronous and induction motors, 0–600 V, should have failure rates that are nearly the same.

Generators

The data in Tables 8 and 2 show that steam turbine driven generators have a failure rate almost 20 times lower than gas turbine driven generators. It is believed that the failure rate of 0.032 per year for steam turbine driven generators is too low; the failure rate should probably be several times higher than this value. The gas turbine data in Table 8 show that one plant in the petroleum industry had 54 failures in 5.5 unit-years; this compares with 3 failures in 83.9 unit-years for the other three plants that submitted data in the survey. It is believed that the overall failure rate of 0.638 per year for gas turbines is too high.

Open Wire

A clear definition was not given for "open wire" on the survey form (see Appendix A). It is believed that all of the respondents interpreted "open wire" to mean "bare or weatherproof conductors supported on insulators."

Cable

The data in Tables 13 and 2 show that cable above ground and aerial has a failure rate for 0–600 V that is ten times lower than 601–15 000 V. It is believed that the failure rate of 0.00141 per unit-year for 0–600 V above ground and aerial is too low.

There is a wide variation in the failure rate for cable, 601–15 000 V, based upon the application (in trays above ground, in conduit above ground, aerial cable, in duct or conduit below ground). This variation covers a range of 8 to 1. It is believed that the failure rate of 0.04918 per year is too high for cable, 601–15 000 V, in conduit above ground.

There is a wide variation in the cable failure rate shown in Table 14 (and Table 2) for the different types of insulation (601–15 000 V, all applications). These failure rates vary over a range of 5 to 1. The very low failure rate data for thermoplastic insulation and the high failure rate data for other insulation came primarily from the chemical industry.

Switchgear Bus

The failure rate in Table 10 (and Table 2) shows that insulated bus, 601–15 000 V, has a failure rate about three times higher than bare bus, above 600 V. It is believed that this is the opposite of what it should be. The data submitted by the chemical industry has caused this distortion; they had a very high failure rate for insulated bus (601–15 000 V) and a low failure rate for bare bus (above 600 V).

Electric Utility Power Supplies

The data for electric utility power supplies are shown in Tables 3 and 2. The failure rate is about the same for a single circuit and a double or triple circuit. This is evidently due to the predominance of the throwover mode of operation of multiple-circuit supplies. However, the actual downtime per failure is about three to nine times higher for a single circuit than for a double or triple circuit; the downtime depends on whether manual switchover or automatic switchover is used on a multiple-circuit system.

It appears that many respondents misinterpreted the "number of installed units" for double- or triple-circuit electric utility power supplies. What was desired was the number of separate and independent points of supply, but this was often interpreted to be the number of circuits in the utility supply system. Thus the tendency was to report two installed units for double-circuit supplies. It is believed that this error was made in almost every case. Therefore, *the Reliability Subcommittee changed the number of installed units for multiple-circuit utility supplies to 1 except in those cases where other evidence indicated the presence of more than one point of supply.* The sample size shown in Tables 3 and 2 reflects this change for double- or triple-circuit electric utility power supplies. Thus a double- or triple-circuit supply for one year is counted as one unit-year.

It also appears that a few respondents incorrectly interpreted failure duration on card type 3 for multiple-circuit electric utility supplies. What was desired was the period of time during which service was interrupted. However, in a few cases it appears that what was given was the time to repair one circuit of a multiple-circuit supply even though the supply interruption time is limited to the time required to throw over to the alternate supply circuit. The *Reliability Subcommittee changed the failure duration to the value given for plant outage duration in those cases in which such an error was believed to exist.* However, it is suspected that not all of these errors were corrected. The effect of this change was to reduce the actual hours of downtime per failure for multiple-circuit supplies. The majority of the multiple-circuit supply failures are due to loss of the normal feed, and the duration of the failure is limited to the time to switch to the alternate feed. The average outage duration in Tables 3 and 2 is shorter for automatic switching than for manual switching, as one would expect.

There were 25 recorded cases of simultaneous failure of all circuits in a double- or triple-circuit supply. This gives a failure rate of 0.119 failure per year for loss of all circuits at one time. Further details on this are given in Part 3 [13]. Thus a multiple-circuit electric utility power supply has a failure rate (loss of all circuits at one time) that is only about five times lower than the failure rate (0.537 failures per year) for a single-circuit supply and about six times lower than the all-inclusive failure rate of 0.643 failure per year. The ratio between all-inclusive failure rate and the failure rate for loss of all circuits at one time is not as large as one might suspect. Some of the reasons for this are the following.

1) Some portion of utility supply failures are due to failure of the bulk power system which feeds all the supply circuits.

2) At least some cases of loss of all circuits at one time occur when a forced outage of one circuit overlaps a scheduled or maintenance outage of the other circuit (typical utility industry data indicate that this type of overlapping outage is often more probable than overlapping forced outages).

3) The all-inclusive failure rate is, in effect, an average outage rate reflecting the performance of some throwover schemes and some normally closed breaker schemes. Thus, since throw-

over schemes are expected to have higher outage rates than normally closed breaker schemes, it follows that the computed all-inclusive outage rate is probably somewhat lower than the outage rate which would be computed for throwover schemes only. (Unfortunately we cannot compute the throwover scheme outage rate since we do not know which of the reported utility supplies are throwover schemes.)

Only point 3) reflects on the accuracy of the data; the other two points just reflect the facts of life.

A comparison of the all-inclusive failure rate (0.643 failures per year) with the failure rate for loss of all circuits at one time (0.119 failures per year) gives a rough idea of the degree of supply failure rate improvement possible by going from a throwover scheme to a scheme using normally closed circuit breakers.

References

[1] W. H. Dickinson, P. E. Gannon, C. R. Heising, A. D. Patton, and D. W. McWilliams, "Fundamentals of reliability techniques as applied to industrial power systems," in *Conf. Rec. 1971 IEEE Ind. Comm. Power Syst. Tech. Conf.* 71C18-IGA, pp. 10–31.

[2] C. R. Heising, "Reliability and availability comparison of common low-voltage industrial power distribution systems," *IEEE Trans. Ind. Gen. Appl.*, vol. IGA-6, pp. 416–424, Sept./Oct. 1970.

[3] W. H. Dickinson, "Economic evaluation of industrial power system reliability," *AIEE Trans. (Appl. Ind.)*, vol. 76, pp. 264–271, Nov. 1957.

[4] W. H. Dickinson, "Evaluation of alternative power distribution systems for refinery process units," *AIEE Trans. (Power Appl. Syst.)*, vol. 79, Apr. 1960.

[5] W. H. Dickinson, "Economic justification of petroleum industry automation and other alternatives by the revenue requirements method," *IEEE Trans. Ind. Gen. Appl.*, vol. IGA-1, pp. 39–50, Jan./Feb. 1965.

[6] D. P. Garver, F. E. Montmeat, and A. D. Patton, "Power systems reliability I—Measures of reliability and methods of calculation," *IEEE Trans. Power Appl Sys.*, vol. 83, pp. 727–737, July 1964.

[7] F. E. Montmeat, A. D. Patton, J. Zemkoski, and D. J. Cummin, "Power system reliability II—Applications and a computer program," *ibid.*, vol. PAS-84, pp. 636–643, July 1965.

[8] Z. G. Todd, "A probability method for transmission and distribution outage calculations," *IEEE Trans. Power App. Syst.*, vol. 83, pp. 695–701, July 1964.

[9] C. F. DeSieno, and L. L. Stine, "A probability method for determining the reliability of electric power systems," *IEEE Trans. Power App. Syst.*, vol. 83, pp. 174–181, Feb. 1964.

[10] "General principles for reliability analysis of nuclear power generating station protection systems," IEEE Publ. 352, ANSI N41.4, 1972.

[11] W. H. Dickinson, "Report on reliability of electric equipment in industrial plants," *AIEE Trans. (Appl. Ind.)*, vol. 81, pp. 132–151, July 1962.

[12] IEEE Committee Report, "Report on reliability survey of industrial plants, Part II: Cost of power outages, plant restart time, critical service loss duration time, and type of loads lost versus time of power outages," this issue, pp. 236–241.

[13] IEEE Committee Report, "Report on reliability survey of industrial plants; Part III: Causes and types of failures of electrical equipment, methods of repair, and urgency of repair," this issue, pp. 242–249.

[14] N. H. Roberts, *Mathematical Methods in Reliability Engineering*. New York: McGraw-Hill, 1964.

[15] I. Bazovsky, *Reliability Theory and Practice*. Englewood Cliffs, N.J.: Prentice-Hall, 1961.

[16] D. K. Lloyd and M. Lipow, *Reliability: Management, Methods, and Mathematics*. Englewood Cliffs, N.J.: Prentice-Hall, 1962.

[17] A. D. Patton, "Determination and analysis of data for reliability studies," *IEEE Trans. Power App. Syst.*, vol. PAS-87, pp. 84–100, Jan. 1968.

Report on Reliability Survey of Industrial Plants, Part II: Cost of Power Outages, Plant Restart Time, Critical Service Loss Duration Time, and Type of Loads Lost Versus Time of Power Outages

IEEE COMMITTEE REPORT

Abstract—An IEEE sponsored reliability survey of industrial plants was completed during 1972. This survey included the cost of power outages, plant restart time, critical service loss duration time, and type of loads lost versus power outage duration time. Survey results reflect data from 30 companies covering 68 plants in nine industries in the United States and Canada. This information is useful in the design of industrial power distribution systems.

INTRODUCTION

KNOWLEDGE of the cost of power outages and of plant restart time is important information for use in the design of industrial power distribution systems. In addition it is also desirable to know the critical service loss duration time and the type of loads lost versus the time of power outage.

During 1972 the Reliability Subcommittee of the IEEE Industrial and Commercial Power Systems Committee completed a reliability survey of industrial plants. This is the second part, which reports results from the survey. Included in this paper are the following results:

1) cost of power outages to industrial plants in the United States and Canada (dollars per kilowatt interrupted plus dollars per kilowatthour of undelivered energy);

2) plant restart time after a failure that has caused complete plant shutdown;

3) critical service loss duration time, that is, the maximum length of power failure that will not stop plant production;

4) type of loads lost versus the time of power outage (this

includes computer, motor, lighting, and solenoid loads, and gives plant outage duration times resulting from these failures).

Paper TOD-73-158, approved by the Industrial and Commercial Power Systems Committee of the IEEE Industry Applications Society for presentation at the 1973 Industrial and Commercial Power Systems Technical Conference, Atlanta, Ga., May 13–16. Manuscript released for publication November 5, 1973.

Members of the Reliability Subcommittee of the IEEE Industrial and Commercial Power Systems Committee are W. H. Dickinson, *Chairman*, P. E. Gannon, M. D. Harris, C. R. Heising, D. W. McWilliams, R. W. Parision, A. D. Patton, and W. J. Pearce.

SURVEY FORM

The survey form used is shown in Appendix A of Part 1 [1]. The information on the cost of power outages came from card type 1, columns 13, 20, and 25. Card type 1 also contained plant restart time (column 31) and critical service loss duration (columns 33 and 36).

The data on type of loads lost came from card type 3, columns 48, 49, 50, 51, and 52. The data on time of power outage came from columns 26 and 29 of card type 3; these data are actually the outage duration time after a failure of the electric utility power supply or a failure of electrical equipment in the power distribution system.

RESPONSE TO SURVEY

A total of 30 companies responded to the survey questionnaire reporting data on 68 plants from nine industries in the United States and Canada. Every response did not supply all the information requested on every question. Tables 22–29

188

give data on how many plants provided answers to the various questions.

The results were compiled for the United States and Canada. Data from one foreign plant are also included separately.

Cost of Power Outages

Each plant was asked to report data on the cost of power outages as follows:

1) Dollars per failure, i.e., extra expense incurred because of a failure only (not including plant downtime) such as for damaged equipment, spoiled product, extra maintenance, or extra repair costs.

2) Dollars per hour of downtime, i.e., value of lost production in dollars per hour of plant downtime only. This is the estimated revenues (sales price) of product not made, less expenses saved in labor, material, utilities, etc. If this varies with the duration of the plant downtime, an average value per hour was to be given.

3) Maximum electric power demand when the plant is operating at its rated or design capacity in kilowatts.

This made it possible to calculate an estimate of the cost of power outages in terms of the dollars per kilowatts interrupted plus the dollars per kilowatthours of undelivered energy. The average cost of power outages from the survey is given in Table 20.

Of the 41 plants that reported outage cost data in the survey, 31 had a maximum demand greater than 1000 kW and 10 had a maximum demand less than 1000 kW. Cost data for plants with maximum demands less than 1000 kW are not considered particularly reliable due to the small number of such plants represented in the data.

There is a wide spread in the cost of power outages. Consequently few plants with high outage costs can have a significant effect on the overall average cost. In such cases the median cost of power outages may be more representative than the average cost. The median cost is such that half of the plants have a cost greater than this value and half have less. Table 21 shows the median power outage costs. Additional details on the cost of power outages are given in Tables 22-27. These additional details include: 1) number of plants reporting the outage cost per failure and the outage cost per hour of downtime, 2) minimum plant cost, 3) maximum plant cost, 4) costs for various industries.

Tables 22, 24, and 26 give the cost of outage per failure per kilowatt maximum demand. Tables 23, 25, and 27 give the cost of a sustained outage per hour down per kilowatt maximum demand.

Plant Restart Time

Each plant was asked to report data on the time required to get the plant back into operation after service is restored following a failure that has caused a complete plant shutdown. A total of 43 plants reported these data. The average plant

TABLE 20 - AVERAGE COST OF POWER OUTAGES FOR INDUSTRIAL PLANTS IN THE UNITED STATES OF AMERICA AND CANADA

All Plants	$1.89 per kW + $2.68 per kWh
Plants > 1000 kW Max. Demand	$1.05 per kW + $0.94 per kWh
Plants < 1000 kW Max. Demand	$4.59 per kW + $8.11 per kWh

TABLE 21 - MEDIAN COST OF POWER OUTAGES FOR INDUSTRIAL PLANTS IN THE UNITED STATES OF AMERICA AND CANADA

All plants	$0.69 per kW + $0.83 per kWh
Plants > 1000 kW Max. Demand	$0.32 per kW + $0.36 per kWh
Plants < 1000 kW Max. Demand	$3.68 per kW + $4.42 per kWh

restart time was 17 h. The median was 4 h. Additional details are given in Table 28.

Critical Service Loss Duration Time

One of the most commonly asked questions is, What is a power failure? In particular, How long can power be lost without causing a complete plant shutdown? Each plant was asked to report data giving the maximum length of power failure that will not stop plant production. This time is typically in the range of cycles to minutes and is called "critical service loss duration time."

A total of 55 plants reported data on critical service loss duration time. The median value was 10 s, that is, half of the plants were greater than this value and half were less. Additional details are given in Table 29.

Loads Lost Versus Time of Power Outage

Each plant was asked, What loads were lost because of failure even though power was restored promptly? Five types of loads were included in the survey:

1) *computer:* one or more computers or solid-state control devices operated incorrectly;
2) *motor:* one or more motors (contactor dropout);
3) *lighting:* lighting load;
4) *solenoid:* one or more solenoid-operated devices dropped out, such as a solenoid-operated fuel valve;
5) *other:* lost other loads, to be described in remarks.

A very short outage duration time after an equipment failure (including electric utility power supply) might not result in a loss of load. Table 30 shows how short power outage duration times after an equipment failure affected the loads lost. The average plant outage duration resulting from these failures is also given in Table 30.

DISCUSSION OF RESULTS

Cost of Power Outages (Tables 20-27)

1) There is a wide spread in the cost of power outages (per kilowatt and per kilowatthour) of industrial plants. Even within a given industry, such as chemical, there is a wide spread in the cost of power outages (per kilowatt and per kilowatthour) for different plants.

2) Plants with a maximum demand of less than 1000 kW have a much higher cost of power outages (per kilowatt and per kilowatthour) than plants with a maximum demand of greater than 1000 kW. This indicates that small industrial plants have a higher cost of power outages (per kilowatt and per kilowatthour) than large industrial plants. It is suspected that this may be because the small industrial plants have more employees per kilowatt (and per kilowatthour). It is also possible that high-consumption industries tend to have a lot of electrochemical or heating processes, and these tend to have low outage costs; for example, heat not supplied now can be supplied later, providing the outage is not too long.

3) It is suggested that the "all-industry" data for the 41 and 42 plants should be compiled to show 25 percent and 75 percent in addition to the minimum median and maximum values already tabulated (Tables 22 and 23).

4) It is suggested that future surveys also include the cost of power outages (per kilowatt and per kilowatthour) of commercial buildings.

TABLE 22 - PLANT OUTAGE COST PER FAILURE PER kW OF MAXIMUM DEMAND - ALL PLANTS ($ per kW)

Industry	Number of Plants Reporting	Minimum	Median	Maximum	Average
All Industry - USA & Canada	42	.002	.69	10.00	1.89
Auto	0	-	-	-	-
Cement	0	-	-	-	-
Chemical	11	.02	.22	3.33	.75
Metal	2	.18	2.42	4.67	2.42
Mining	0	-	-	-	-
Petroleum	5	.002	.07	.31	.12
Pulp and Paper	1	.33	.33	.33	.33
Rubber and Plastics	2	.28	.50	.71	.50
Textile	2	.07	1.00	1.92	1.00
Other Light Manufacturing	6	.09	1.10	2.80	1.22
Other Heavy Manufacturing	8	1.67	3.85	10.00	5.11
Other	5	.25	.94	7.50	2.86
Foreign	1	.33	.33	.33	.33

TABLE 23 - PLANT OUTAGE COST PER HR. DOWNTIME PER kW OF MAXIMUM DEMAND - ALL PLANTS ($ per kWh)

Industry	Number of Plants Reporting	Minimum	Median	Maximum	Average
All Industry - USA & Canada	41	.0009	.83	27.00	2.68
Auto	0	-	-	-	-
Cement	0	-	-	-	-
Chemical	12	.0009	.14	2.11	.33
Metal	2	.55	.94	1.33	.94
Mining	0	-	-	-	-
Petroleum	2	.04	1.24	2.43	1.24
Pulp and Paper	1	.07	.07	.07	.07
Rubber and Plastics	3	.28	.36	1.33	.66
Textile	1	.24	.24	.24	.24
Other Light Manufacturing	6	.33	.79	2.00	.91
Other Heavy Manufacturing	8	.93	6.35	27.00	9.73
Other	6	.75	2.50	5.77	2.69
Foreign	1	.07	.07	.07	.07

TABLE 24 - PLANT OUTAGE COST PER FAILURE PER kW OF MAXIMUM DEMAND - PLANTS MORE THAN 1,000 kW MAX. DEMAND ($ per kW)

Industry	Number of Plants Reporting	Minimum	Median	Maximum	Average
All Industry - USA & Canada	32	.002	.32	7.50	1.05
Auto	0	-	-	-	-
Cement	0	-	-	-	-
Chemical	11	.02	.22	3.33	.75
Metal	1	.18	.18	.18	.18
Mining	0	-	-	-	-
Petroleum	5	.002	.07	.31	.12
Pulp and Paper	1	.33	.33	.33	.33
Rubber and Plastics	2	.28	.50	.71	.50
Textile	2	.07	1.00	1.92	1.00
Other Light Manufacturing	4	.09	1.10	2.80	1.27
Other Heavy Manufacturing	1	1.87	1.87	1.87	1.87
Other	5	.25	.94	7.50	2.86
Foreign	1	.33	.33	.33	.33

TABLE 25 - PLANT OUTAGE COST PER HR. DOWNTIME PER kW OF MAXIMUM DEMAND - PLANTS MORE THAN 1,000 kW MAX. DEMAND ($ per kWh)

Industry	Number of Plants Reporting	Minimum	Median	Maximum	Average
All Industry - USA & Canada	31	.0009	.36	5.77	.94
Auto	0	-	-	-	-
Cement	0	-	-	-	-
Chemical	12	.0009	.14	2.11	.33
Metal	1	.55	.55	.55	.55
Mining	0	-	-	-	-
Petroleum	2	.04	1.24	2.43	1.24
Pulp and Paper	1	.07	.07	.07	.07
Rubber and Plastics	3	.28	.36	1.33	.66
Textile	1	.24	.24	.24	.24
Other Light Manufacturing	4	.33	.54	1.20	.65
Other Heavy Manufacturing	1	.93	.93	.93	.93
Other	6	.75	2.50	5.77	2.69
Foreign	1	.07	.07	.07	.07

TABLE 26 – PLANT OUTAGE COST PER FAILURE PER kW OF MAXIMUM DEMAND – PLANTS LESS THAN 1,000 kW MAX. DEMAND ($ per kW)

Industry	Number of Plants Reporting	Minimum	Median	Maximum	Average
All Industry – USA & Canada	10	.50	3.68	10.00	4.59
Auto.	0	-	-	-	-
Cement	0	-	-	-	-
Chemical	0	-	-	-	-
Metal	1	4.67	4.67	4.67	4.67
Mining	0	-	-	-	-
Petroleum	0	-	-	-	-
Pulp and Paper	0	-	-	-	-
Rubber and Plastics	0	-	-	-	-
Textile	—	—	—	—	—
Other Light Manufacturing	2	.50	1.11	1.72	1.11
Other Heavy Manufacturing	7	1.67	5.00	10.00	5.57
Other	0	-	-	-	-
Foreign	0	-	-	-	-

TABLE 27 – PLANT OUTAGE COST PER HR. DOWNTIME PER kW OF MAXIMUM DEMAND – PLANTS LESS THAN 1,000 kW MAX. DEMAND ($ per kWh)

Industry	Number of Plants Reporting	Minimum	Median	Maximum	Average
All Industry – USA & Canada	10	.86	4.42	27.00	8.11
Auto.	0	-	-	-	-
Cement	0	-	-	-	-
Chemical	0	-	-	-	-
Metal	1	1.33	1.33	1.33	1.33
Mining	0	-	-	-	-
Petroleum	0	-	-	-	-
Pulp and Paper	0	-	-	-	-
Rubber and Plastics	0	-	-	-	-
Textile	—	—	—	—	—
Other Light Manufacturing	2	.86	1.43	2.00	1.43
Other Heavy Manufacturing	7	3.33	7.69	27.00	11.00
Other	0	-	-	-	-
Foreign	0	-	-	-	-

TABLE 28 - PLANT RESTART TIME (After Service is Restored Following a Failure that has Caused Complete Plant Shutdown)

Industry	Number of Plants Reporting	Average (Hours)	Median (Hours)
All Industry - USA & Canada..	43	17.4	4.0
Auto.	0	-	-
Cement.	0	-	-
Chemical.	19	20.7	20
Metal.	1	4	4
Mining.	0	-	-
Petroleum.	3	37.3	24
Pulp and Paper.	1	10	10
Rubber & Plastics.	3	2.33	2
Textile.	3	58.3	72
Other Light Manufacturing.	7	2.14	2
Other Heavy Manufacturing.	1	2	2
Other.	5	2.6	2
Foreign.	1	48	48

TABLE 29 - CRITICAL SERVICE LOSS DURATION (Maximum Length of Power Failure that Will Not Stop Plant Production)

Industry	Number of Plants Reporting	Average	Median
All Industry - USA & Canada....	55	12.6 min.	10.0 sec.
Auto.	0	-	-
Cement.	0	-	-
Chemical.	20	4.56 min.	1.25 sec.
Metal.	2	15.0 min.	15.0 min.
Mining.	0	-	-
Petroleum.	1	1.0 sec.	1.0 sec.
Pulp and Paper.	1	10.0 cycles	10.0 cycles
Rubber & Plastics.	3	30.0 sec.	20.0 sec.
Textile.	3	3.34 min.	30.0 cycles
Other Light Manufacturing.	7	10.3 min.	10.0 sec.
Other Heavy Manufacturing.	10	47 min.	45 min.
Other.	8	1.9 min.	20.0 cycles
Foreign.	1	15.0 cycles	15.0 cycles

TABLE 30 - LOADS LOST VERSUS TIME OF POWER OUTAGE
(Tabulation of the Percentage of Equipment Failures
for Which the Designated Load was Lost and Average
Plant Outage Duration Resulting from these Failures)

Type of Load	For Equipment Failures 1 Cycle or less in Duration			For Equipment Failures Between 1 and 10 Cycles in Duration			For Equipment Failures 10 Cycles or More in Duration		
	Yes	No	Not Known	Yes	No	Not Known	Yes	No	Not Known
Computer	0%	0%	0%	4%	96%	0%	9%	91%	0%
Motor	0%	0%	0%	33%	67%	0%	67%	33%	0%
Lighting	0%	0%	0%	22%	78%	0%	38%	61%	2%
Solenoid	0%	0%	0%	22%	74%	4%	25%	66%	9%
Other	0%	0%	0%	7%	15%	78%	25%	62%	13%
Average Plant Outage Duration	0.0 Hours			1.39 Hours			22.6 Hours		

Only non-zero data was used in computing the average plant outage duration

5) Additional information on the cost of power outages in Sweden, Norway, and the United States is contained in [2].

Plant Restart Time (Table 28)

The textile, petroleum, and chemical industries have a much longer plant restart time than the other industries included in the survey.

Critical Service Loss Duration (Table 29)

1) There is a wide spread in critical service loss duration time for the 55 plants in the survey.

2) It is suggested that the data from the 55 plants should be compiled to show several percentiles (10, 25, 75, and 90 percent) in addition to the median value already tabulated.

Loads Lost Versus Time of Power Outage (Table 30)

1) An outage between 1 to 10 cycles resulted in 33 percent of the plants losing motor loads and 22 percent losing a solenoid and only 4 percent losing a computer load. An outage greater than 10 cycles resulted in 67 percent of the plants losing motor loads and 25 percent losing a solenoid and only 9 percent losing a computer load; many plants must not have

had computer loads to give such a low value. In fact, many plants must not have had motor loads or solenoid loads either. The important parameter to look at is the change in these percentages from 0 to the maximum value as the length of power outage time is increased.

2) It is suggested that loss of load data be compiled for the following additional categories of outage duration time:

a) 10 to 15 cycles,
b) 15+ to 30 cycles,
c) 0.5 + to 2.0 s,
d) 2.0+ to 4.0 s,
e) greater than 4.0 s.

The average plant outage duration should also be determined for these categories.

REFERENCES

[1] IEEE Committee Report, "Report on reliability survey of industrial plants; Part I: Reliability of electrical equipment," this issue, pp. 213–235.
[2] R. B. Shipley, A. D. Patton, and J. S. Denison "Power reliability cost vs worth," *IEEE Trans. Power App. Syst.*, vol. PAS-91, pp. 2204–2212, Sept./Oct. 1972.

Report on Reliability Survey of Industrial Plants, Part III: Causes and Types of Failures of Electrical Equipment, the Methods of Repair, and the Urgency of Repair

IEEE COMMITTEE REPORT

Abstract—An IEEE sponsored reliability survey of industrial plants was completed during 1972. This included the causes and types of failures of electrical equipment, the methods of repair, and the urgency of repair. The results are reported from the survey of 30 companies covering 68 plants in nine industries in the United States and Canada. This information is useful in the design of industrial power distribution systems.

INTRODUCTION

A KNOWLEDGE of the causes and types of failures of electrical equipment is useful in the design of industrial power distribution systems. In addition it is also useful to know the failure repair method, whether or not the repair was urgent, and how long it had been since the previous maintenance had been performed. During 1972 the Reliability Subcommittee of the IEEE Industrial and Commercial Power Systems Committee completed a reliability survey of industrial plants. This is the third paper reporting results from the survey. Included in this paper are the results for 14 main classes of electrical equipment on

1) failure repair method;
2) failure repair urgency;
3) failure, months since maintained;
4) failure, damaged part;
5) failure type;
6) suspected failure responsibility;
7) failure initiating cause;
8) failure contributing cause;
9) failure characteristic.

The failure repair method includes either the repair of the failed component or the replacement of the failed component with a spare. This can have a significant effect on the average downtime per failure, and thus is an important factor in reliability and availability calculations.

Paper TOD-73-158, approved by the Industrial and Commercial Power Systems Committee of the IEEE Industry Applications Society for presentation at the 1973 Industrial and Commercial Power Systems Technical Conference, Atlanta, Ga., May 13–16. Manuscript released for publication November 5, 1973.

Members of the Reliability Subcommittee of the IEEE Industrial and Commercial Power Systems Committee are W. H. Dickinson, *Chairman*, P. E. Gannon, M. D. Harris, C. R. Heising, D. W. McWilliams, R. W. Parisian, A. D. Patton, and W. J. Pearce.

The failure repair urgency also has a significant effect on the average downtime per failure and thus is an important factor in reliability and availability calculations.

A preventive maintenance program can have an effect on the failure rate of electrical equipment. Thus a knowledge of whether or not maintenance has been performed recently prior to the failure is a significant factor in helping to determine whether or not the maintenance program is adequate.

The damaged part from a failure is of interest. In addition, a knowledge is also desirable of the type of failure, initiating cause, contributing cause, and suspected responsibility. This information is useful for correcting deficiencies in electrical equipment and electrical systems.

The failure characteristic can be defined as the effect that the failure has on the electrical system. Thus this information is very important.

SURVEY FORM

The survey form used is shown in Appendix A of Part 1 [1]. All of the information reported on in this paper came from card type 3, columns 30–46. The definitions of *failure* and *repair time* are given in Part 1 [1].

RESPONSE TO SURVEY

A total of 30 companies responded to the survey questionnaire, reporting data on 68 plants from nine industries in the United States and Canada. Every failure report on card type 3 did not have filled in all the information called for in columns 30–46. Tables 31 and 32 give the data for each main equipment class on how many failures had the information called for in columns 30–46. Each main equipment class contains 18 or more failures; this is believed to be an adequate statistical sample size.

STATISTICAL ANALYSIS

The results were compiled for 14 main equipment classes. The number of failures were tabulated for each category of each column (30–46, card type 3). This was then divided by the total failures in each column so as to give the percentage for each category for each column (for each main equipment class).

SURVEY RESULTS

The results are tabulated for the 14 main equipment classes in Tables 33–41. Each table represents one column (of 30–46, card type 3).

SUMMARY OF CONCLUSIONS

Transformers

In the cases reported, there were approximately an equal number of incidences of repairing the failed transformer and replacing it with a spare. The repair urgency slightly favored a round-the-clock repair over the regular work-day schedule. Inadequate preventive maintenance did not seem to have much influence on the reported failures since no preventive maintenance was reported on only 5 percent of the failures; 11 percent of the failures were blamed on inadequate maintenance. Damaged insulation both in the windings and bushings accounted for the majority of the transformer damage, with the majority of failures being flashovers involving ground. 24 percent of the reported cases considered normal deterioration from age as the contributing cause of the failure, yet 39 percent reported that they felt the manufacturer was primarily responsible. Transient overvoltages, from lightning or switching surges, and other insulation breakdown account for 41 percent of the reported failures. In 90 percent of the reported cases the transformers were removed from the system by automatic protective devices; only 7 percent had manual removal.

Circuit Breakers

About the same number of circuit breakers were repaired in place as were replaced by spares. The relative importance of circuit breakers was indicated by 73 percent of the survey respondents making repairs on a round-the-clock basis. The bulk of the reported failures involved flashovers to ground with damage primarily to the protective device components and the device insulation. Transient overvoltages, insulation breakdowns, and protective device malfunctions were considered a major initiating cause with normal deterioration from age and misoperation or testing errors considered as contributing causes. However, 33 percent of the respondents could not classify the initiating cause into any of the survey classes, and 55 percent could not classify the contributing cause into any of the survey classes. In addition, 36 percent of the suspected causes of failure were blamed on "other." 42 percent of the reported failures involved circuit breakers opening when they should not; it is possible that several of these failures were external to the circuit breaker and of unknown cause and were blamed on the circuit breaker. 32 percent of the reported failures involved circuit breakers that failed during a load-carrying condition.

23 percent of the failures were blamed on the manufacturer and another 23 percent on inadequate maintenance, but 36 percent were blamed on "other." Inadequate preventive maintenance (PM) could be a factor of some significance since no PM was reported on 16 percent of the failures.

Motor Starters

Of the reported motor starter failures, about two thirds were repaired by replacing the starter with a spare and two thirds were repaired on a round-the-clock basis. About half of the cases reported indicate that the damage was other than the classes listed in the survey, primarily resulting from flashovers or electrical defects. 64 percent felt that a malfunction of a protective relay control device initiated the failure with 40 percent of the respondents reporting that normal deterioration from age was a contributing cause. Over half of the respondents felt that improper application was primarily responsible for the failure. In the cases reported 36 percent had been discovered during testing or maintenance, and 20 percent were only partial failures. Lack of preventive maintenance was not a big problem. Those starters that had been maintained less than 12 months prior to the failure accounted for 67 percent of the cases reported.

Motors

Of the reported motor failures, about three quarters were repaired versus about one fourth being replaced by a spare. About three quarters were repaired on a regular work-day basis. The types of failures varied from flashovers to electrical defects, to mechanical defects, with winding insulation and bearings sustaining the majority of the damage. Insulation breakdown, overheating, and mechanical seizing were blamed as the primary initiating causes with normal deterioration from age, loss or deficiency of lubricant, exposure to abnormal moisture, and exposure to aggressive chemicals ranking high on the list of contributing causes. 30 percent of the failures were discovered during testing or maintenance, which probably resulted in less actual damage in those cases. Inadequate maintenance, improper application, and defective equipment were listed as having primary responsibility. However, over half of the respondents could not assign responsibility into one of the survey classes. The motors that had been maintained between 12 and 24 months prior to the failure accounted for 57 percent of the reported cases with less than 12 months and more than 24 months accounting for 22 percent and 19 percent, respectively. No preventive maintenance accounted for only 2 percent, yet this does not correlate well with inadequate maintenance being listed as having primary responsibility in 17 percent of the reported cases.

Generators

Of the reported generator failures 84 percent were repaired in place. About the same number were repaired on a round-the-clock basis as were repaired on a regular work-day basis. 69 percent of the respondents reported damage other than the survey classes with electrical auxiliaries, winding insulation, and moving parts sustaining some damage. Mechanical breaking, transient overvoltages, and about half unclassified items were considered the primary initiating causes with normal deterioration from age and persistent overloading considered contributing causes. Responsibility was spread between inadequate maintenance and defective components with about half of the respondents unable to place primary responsibility into any of the survey classes. Infrequent or no preventive maintenance were not involved in any of the reported cases, a point that does not correlate with the fact that some of the respondents felt inadequate maintenance was the primary responsibility.

Disconnect Switches

Of the reported disconnect switch failures, 70 percent were repaired by replacement with a spare, with work in 80 percent of the cases being performed on a regular work-day schedule. Electrical defects, mechanical defects, and flashovers to ground resulted in damage to mechanical components and insulation. Some form of mechanical breaking or contact from foreign

TABLE 31 - NUMBER OF FAILURES FOR ELECTRIC UTILITY
POWER SUPPLIES THAT CONTAINED THE
INFORMATION CALLED FOR IN COLUMNS 30-46,
CARD - TYPE 3

Card Type 3 Column	Title	Number of Failures
30	Failure Repair Method.............	28
32	Failure Repair Urgency............	35
34	Failure, Months Since Maintained..	25
36	Failure, Damaged Part.............	39
38	Failure Type.....................	49
40	Suspected Failure Responsibility..	43
42	Failure Initiating Cause..........	53
44	Failure Contributing Cause........	53
46	Failure Characteristic............	145

TABLE 32 - NUMBER OF FAILURES FOR EACH MAIN EQUIPMENT
CLASS THAT CONTAINED THE INFORMATION CALLED
FOR IN COLUMNS 30-46, CARD-TYPE 3

Main Equipment Class	Maximum	Minimum	Avg.
Transformers	101	97	100
Circuit Breakers	176	161	171
Motor Starters	88	88	88
Motors	561(col.36)	493(col.40)	517
Generators	83(col.36)	31(all other)	37
Disconnect Switches	101	100	101
Swgr. Bus-Insulated	20	20	20
Swgr. Bus-Bare	24	20	23
Bus Duct	20	18	20
Open Wire	109	104	108
Cable	223	211	218
Cable Joints	45	44	45
Cable Terminations	51	47	50

sources accounted for about half of the initiating causes, with exposure to dust and contaminants and a large number of unclassified items considered contributing causes. Inadequate operating procedures, inadequate maintenance, and defective components were considered primarily responsible, which seems to correlate with over 66 percent of the reported cases not having any preventive maintenance and 21 percent not having any preventive maintenance 24 months prior to the failure.

Switchgear Bus, Bare

Of the reported uninsulated switchgear bus failures, about two thirds were repaired in place, with a little more than half of them being repaired on a round-the-clock basis. 79 percent of the respondents report some form of insulation damage all resulting from flashovers either to ground (79 percent) or between phases (21 percent). Mechanical failure, shorting by metal objects, and insulation breakdown were the predominant initiating causes with exposure to abnormal moisture, exposure to dust, exposure to aggressive chemicals, and normal deterioration due to age listed as contributing causes. Interestingly, 15 percent of the respondents listed misoperation or testing errors as a contributing cause. 39 percent felt that an outside agency was responsible for the failure, while 22 percent blamed inadequate maintenance.

Switchgear Bus, Insulated

Of the reported insulated switchgear bus failures, essentially all were repaired in place with over two thirds of the repairs being completed on a round-the-clock basis. 90 percent of the respondents reported insulation damage resulting primarily from flashovers to ground and between phases. Insulation breakdown was considered to have initiated the failure in about half of the cases, with exposure to contaminants, moisture, severe weather, and normal deterioration from age being considered as contributing factors. Improper application (45 percent) and inadequate maintenance (35 percent) were held responsible for the failures.

Bus Duct

Of the reported bus duct failures, 65 percent were repaired in place with the majority of them being repaired on a round-the-clock basis. 90 percent of the respondents reported some form of damaged insulation resulting from a flashover to ground. Mechanical failure, insulation breakdown, and overheating were blamed as initiating factors, with normal deterioration due to age being listed as a contributing factor in half of the cases. Responsibility for the reported failures varied from defective components (26 percent), improper application (16 percent), to inadequate maintenance (16 percent).

Open Wire

Of the reported open-wire failures, 70 percent were repaired in place with a little over half involving a round the clock effort. About half of the failures involved flashovers either to ground or between phases and about 25 percent involved other electrical defects. In the reported failures, transient overvoltages, overheating, or shorting by metal objects were considered the most significant initiating causes, with severe weather and exposure to aggressive chemicals being the predominant contributing causes. 81 percent of the respondents indicated that no preventive maintenance had been performed in over two years, which supports the fact that over a third of them blamed inadequate maintenance as being responsible.

Cables

The relative importance of primary cable was again indicated by about two thirds of the reported cases making repairs on a round-the-clock basis. There were a few more reported cases where repairs to cables were made by complete replacement rather than by in-place repairs. About three quarters of the failures involved flashovers to ground, resulting in insulation damage.

TABLE 33 — FAILURE REPAIR METHOD
TABLE 34 — FAILURE REPAIR URGENCY

Table, Title, Category	CABLE TERMINATIONS	CABLE JOINTS	CABLE	OPEN WIRE	BUS DUCT	SWITCHGEAR BUS-BARE	SWITCHGEAR BUS-INSULATED	DISCONNECT SWITCHES	GENERATORS	MOTORS	MOTOR STARTERS	CIRCUIT BREAKERS	TRANSFORMERS	ELECTRIC UTILITY POWER SUPPLIES
	%	%	%	%	%	%	%	%	%	%	%	%	%	%
TABLE 33 — FAILURE REPAIR METHOD (Col. 30)														
1. Repair of failed component in place or sent out for repair	60	87	47	70	65	71	95	30	84	78	33	51	47	50
2. Repair by replacement of failed component with spare	34	13	53	9	35	29	5	70	16	22	67	49	53	46
99. Other	6	0	0	21	0	0	0	0	0	0	0	0	0	4
TABLE 34 — FAILURE REPAIR URGENCY (Col. 32)														
1. Requiring round-the-clock all out efforts	53	56	66	55	80	58	70	20	48	23	66	73	51	91
2. Requiring repair work only during regular workday, perhaps with some overtime	31	22	28	26	15	33	25	80	52	74	34	22	45	9
3. Requiring repair work on a non-priority basis	16	22	6	0	5	8	5	0	0	2	0	5	4	0
99. Other	0	0	0	19	0	0	0	0	0	0	0	0	0	0

TABLE 35 – FAILURE, MONTHS SINCE MAINTAINED
TABLE 36 – FAILURE, DAMAGED PART

TABLE 35 – FAILURE, MONTHS SINCE MAINTAINED (Col. 34)

1. Less than 12 months ago
2. 12-24 months ago
3. Over 24 months ago
4. No preventive maintenance
99. Other

Equipment	1	2	3	4	99
CABLE TERMINATIONS	12	12	36	40	0
CABLE JOINTS	18	20	2	60	0
CABLE	11	13	10	66	0
OPEN WIRE	1	8	81	9	0
BUS DUCT	25	45	10	20	0
SWITCHGEAR BUS – BARE	35	30	13	22	
SWITCHGEAR BUS – INSULATED	10	35	55	0	0
DISCONNECT SWITCHES	8	5	21	66	0
GENERATORS	58	42	0	0	0
MOTORS	22	57	19	2	0
MOTOR STARTERS	67	17	16	0	0
CIRCUIT BREAKERS	18	60	5	16	0
TRANSFORMERS	34	38	22	5	0
ELECTRIC UTILITY POWER SUPPLIES	56	40	4	0	0

TABLE 36 – FAILURE, DAMAGED PART (Col. 36)

1. Insulation – winding
2. Insulation – bushing
3. Insulation – other
4. Mechanical – bearings
5. Mechanical – other moving parts
6. Mechanical – other
7. Other electrical – auxiliary device
8. Other electrical – protective device
9. Tap changer – no load type
10. Tap changer – load type
99. Other

Equipment	1	2	3	4	5	6	7	8	9	10	99
CABLE TERMINATIONS	0	12	75	0	0	4	0	0	0	0	10
CABLE JOINTS	0	0	91	0	0	0	0	0	0	0	9
CABLE	5	0	84	3	0	1	1	0	0	0	6
OPEN WIRE	0	1	6	0	4	3	3	0	0	0	84
BUS DUCT	15	10	65	0	0	0	0	0	0	0	10
SWITCHGEAR BUS – BARE	0	8	71	0	0	0	0	0	0	0	21
SWITCHGEAR BUS – INSULATED	0	5	90	0	0	5	0	0	0	0	0
DISCONNECT SWITCHES	0	1	14	9	30	1	0	0	0	0	38
GENERATORS	7	0	0	2	7	4	10	1	0	0	69
MOTORS	50	0	3	29	3	1	3	0	0	0	11
MOTOR STARTERS	5	0	0	0	16	2	13	2	0	0	52
CIRCUIT BREAKERS	0	2	19	1	11	6	6	28	1	0	26
TRANSFORMERS	68	13	3	0	0	3	1	7	1	7	3
ELECTRIC UTILITY POWER SUPPLIES	0	8	10	0	3	15	10	0	0	0	44

TABLE 37 – FAILURE TYPE

Table, Title, Category TABLE 37 – FAILURE TYPE (col. 38)	ELECTRIC UTILITY POWER SUPPLIES	TRANSFORMERS	CIRCUIT BREAKERS	MOTOR STARTERS	MOTORS	GENERATORS	DISCONNECT SWITCHES	SWITCHGEAR BUS-INSULATED	SWITCHGEAR BUS-BARE	BUS DUCT	OPEN WIRE	CABLE	CABLE JOINTS	CABLE TERMINATIONS
1. Flashover or arcing involving ground	43	58	33	14	28	19	15	65	79	70	34	73	70	55
2. All other flashover or arcing	4	13	10	20	4	3	4	35	21	30	23	1	9	4
3. Other electrical defect	14	12	19	55	32	29	47	0	0	0	25	7	20	37
4. Mechanical defect	8	10	11	11	31	32	14	0	0	0	6	5	—	4
99. Other	31	7	27	0	6	16	21	0	0	0	12	14	0	0

TABLE 38 – SUSPECTED FAILURE RESPONSIBILITY

Category	CABLE TERMINATIONS %	CABLE JOINTS %	CABLE %	OPEN WIRE %	BUS DUCT %	SWITCHGEAR BUS-BARE %	SWITCHGEAR BUS-INSULATED %	DISCONNECT SWITCHES %	GENERATORS %	MOTORS %	MOTOR STARTERS %	CIRCUIT BREAKERS %	TRANSFORMERS %	ELECTRIC UTILITY POWER SUPPLIES %
1. Manufacturer-defective Component	0	0	16	0	26	9	5	29	19	15	18	23	39	8
2. Transportation to Site – defective handling	0	0	0	0	0	0	0	0	0	0	0	0	0	0
3. Application Engineering – improper application	18	0	8	2	16	4	45	6	0	9	51	4	2	0
4. Inadequate installation and testing prior to start-up	39	50	14	9	5	17	10	4	3	1	0	3	3	0
5. Inadequate maintenance	22	18	10	30	16	22	35	13	19	17	8	23	11	0
6. Inadequate operating procedures	0	0	3	2	0	0	0	40	3	4	3	6	9	6
7. Outside agency – personnel	0	5	4	5	5	22	0	1	0	0	0	5	2	17
8. Outside agency – other	8	2	6	21	0	17	0	0	6	0	0	1	4	32
99. Other	14	25	38	31	32	9	5	8	48	53	19	36	30	38

Table, Title, Category

TABLE 39 – FAILURE INITIATING CAUSE

Table, Title, Category

TABLE 39 – FAILURE INITIATING CAUSE (Col. 42)

1. Transient overvoltage disturbance (lightning, switching surges, arcing ground fault in ungrounded system)
2. Overvoltage
3. Overheating
4. Other insulation breakdown
21. Mechanical breaking, cracking, loosening, abrading or deforming of static or structural parts
22. Mechanical burnout, friction, or seizing of moving parts.
23. Mechanically caused damage from foreign source (digging, vehicular, accident, etc.)
41. Shorting by tools or metal objects
42. Shorting by birds, snakes, rodents, etc.
51. Loss of control power
52. Malfunction of protective relay control device, or auxiliary device.
61. Low voltage
62. Low frequency
99. Other

Equipment	No.	1	2	3	4	21	22	23	41	42	51	52	61	62	99
CABLE TERMINATIONS	12	0	2	51	24	0		4	2	2	0	0	0	0	4
CABLE JOINTS	11	0	0	40	31	0			0	0	0	0	0	0	18
CABLE	26	0	1	29	24	0		7	2	0	0	0	0	0	10
OPEN WIRE	26	0	21	8	7	0		10	14	3	0	0	0	0	11
BUS DUCT	0	0	30	20	45	0			5	0	0	0	0	0	0
SWITCHGEAR BUS - BARE	5	0	5	18	23	0			23	9	0	0	0	0	18
SWITCHGEAR BUS - INSULATED	5	0	0	50	10	0			0	0	0	0	0	0	35
DISCONNECT SWITCHES	4	0	4	5	17	2		20	0	0	0	0	3	0	45
GENERATORS	10	0	3	3	29	3		3	0	0	0	3	0	0	45
MOTORS	6	0	26	30	4	20	3		0	0	0	5	0	0	5
MOTOR STARTERS	1	0	1	8	8	6			5	1	0	64	0	0	7
CIRCUIT BREAKERS	13	0	3	18	13	5		1	2	1	1	11	0	0	33
TRANSFORMERS	23	0	11	18	17	0		1	1	2	0	1	0	0	25
ELECTRIC UTILITY POWER SUPPLIES	33	0	0	5	7	2		14	12	2	0	2	0	2	21

TABLE 40 – FAILURE CONTRIBUTING CAUSE

TABLE 40 – FAILURE CONTRIBUTING CAUSE (Col. 44)

1. Persistent overloading
2. Above-normal temperatures
3. Below-normal temperature
4. Exposure to aggressive chemicals or solvents
5. Exposure to abnormal moisture or water
6. Exposure to non-electrical fire or burning
8. Obstruction of ventilation by foreign objects or material
9. Normal deterioration from age
10. Severe wind, rain, snow, sleet, or other weather conditions
11. Protective relay improperly set
12. Loss or deficiency of lubricant
13. Loss or deficiency of oil or cooling medium
14. Misoperation or testing error
15. Exposure to dust or other contaminants
99. Other

Equipment	1	2	3	4	5	6	8	9	10	11	12	13	14	15	99
CABLE TERMINATIONS	0	0	0	10	12	0	0	24	16	0	0	0	8	0	29
CABLE JOINTS	0	2	0	13	22	0	0	29	2	0	0	0	0	0	31
CABLE	2	0	0	14	8	2	1	30	15	0	0	0	3	1	24
OPEN WIRE	0	0	0	28	1	3	0	3	30	1	0	0	2	2	31
BUS DUCT	6	0	0	0	17	0	0	50	11	0	0	0	6	0	11
SWITCHGEAR BUS – BARE	0	0	0	10	20	5	0	10	5	0	0	0	15	20	15
SWITCHGEAR BUS – INSULATED	0	5	0	0	15	0	0	20	20	0	0	0	0	40	0
DISCONNECT SWITCHES	8	3	1	0	4	0	0	5	0	0	0	0	0	26	53
GENERATORS	10	6	0	0	6	3	0	32	3	6	0	0	0	0	32
MOTORS	5	1	0	7	10	0	2	34	2	0	15	1	0	5	18
MOTOR STARTERS	0	0	0	0	0	0	0	40	0	0	2	0	3	1	53
CIRCUIT BREAKERS	4	1	0	2	3	0	0	17	0	2	1	0	3	3	55
TRANSFORMERS	13	0	0	0	6	0	0	24	6	0	0	0	3	3	44
ELECTRIC UTILITY POWER SUPPLIES	2	4	0	0	2	0	0	4	38	2	0	0	0	4	45

TABLE 41 – FAILURE CHARACTERISTIC

Table, Title, Category	%	Utility Power Supplies (Select code)					Transformers (Select Code)		
		1. Failure of single circuit (no redundant supply)	2. Failure of one circuit of a double-circuit redundant supply	3. Failure of both circuits of a double-circuit redundant supply	4. Failure of all circuits of a three or more circuit redundant supply	5. Partial failure of a three or more circuit redundant supply	6. Automatic removal by protective equipment	7. Partial failure reducing capacity	8. Manual removal
CABLE TERMINATIONS	%	12	2	0	0	0	2	0	0
CABLE JOINTS	%	0	0	0	0	0	0	0	0
CABLE	%	17	7	0	0	0	4	0	0
OPEN WIRE	%	0	0	0	0	0	0	0	0
BUS DUCT	%	10	0	0	0	0	0	0	0
SWITCHGEAR BUS - BARE	%	8	0	0	8	0	0	0	0
SWITCHGEAR BUS - INSULATED	%	30	5	0	0	0	0	0	0
DISCONNECT SWITCHES	%	0	0	0	0	0	4	0	0
GENERATORS	%	0	0	0	0	3	0	0	0
MOTORS	%	0	0	0	0	0	0	0	0
MOTOR STARTERS	%	0	0	0	0	0	0	0	0
CIRCUIT BREAKERS	%	0	1	0	1	0	0	0	0
TRANSFORMERS	%	1	0	0	0	0	90	1	7
ELECTRIC UTILITY POWER SUPPLIES	%	10	71	15	2	0	0	0	0

TABLE 41 – FAILURE CHARACTERISTIC (Col. 46)

TABLE 41 – FAILURE CHARACTERISTIC

Table, Title, Category	CABLE TERMINATIONS	CABLE JOINTS	CABLE	OPEN WIRE	BUS DUCT	SWITCHGEAR BUS – BARE	SWITCHGEAR BUS – INSULATED	DISCONNECT SWITCHES	GENERATORS	MOTORS	MOTOR STARTERS	CIRCUIT BREAKERS	TRANSFORMERS	ELECTRIC UTILITY POWER SUPPLIES
Circuit Breakers (Select Code)														
9. Failed to close when it should	0	0	0	0	0	0	0	0	0	0	1	5	0	0
10. Failed while opening	0	0	0	0	15	0	0	0	0	0	1	9	0	1
11. Opened when it shouldn't	0	0	0	0	0	0	0	0	6	0	1	42	0	0
12. Damaged while successfully opening	0	0	0	0	0	0	0	0	0	0	0	7	0	0
13. Damaged while closing	0	0	0	0	0	0	0	0	0	0	0	2	0	0
14. Failed while operating (not while opening or closing)	0	0	0	0	0	0	0	0	0	0	0	32	0	0
General (Select Code for any other class)														
15. Failed (this applies to all classes)	65	96	65	69	65	71	65	68	65	68	34	0	1	0
16. Failed during testing or maintenance	2	4	2	2	5	13	0	3	0	1	5	1	0	0
17. Damage discovered during testing or maintenance	12	0	2	1	0	0	0	18	0	30	36	1	0	0
20. Partial failure	6	0	3	6	5	0	0	6	16	0	20	0	0	0
99. Other	0	0	1	23	0	0	0	1	10	0	1	0	0	0

207

An interesting point is that in over two thirds of the failures there had been no preventive maintenance, yet inadequate maintenance was only listed in 10 percent of the cases as being responsible for the failure. 16 percent placed the responsibility with the manufacturer, 14 percent with inadequate installation and testing prior to start-up, with 38 percent of the cases reporting reasons for the failure in classes other than those listed in the survey.

The initiating causes varied from transient overvoltage disturbances to insulation breakdown, to mechanical failures, with 30 percent reporting normal deterioration from age as a contributing cause.

Cable Joints

Of the failures reported, 87 percent were repaired in place, with just over half being repaired on a round-the-clock basis. Almost all of the failures resulted in damaged insulation, primarily from flashovers to ground, which were initiated by insulation breakdowns, transient overvoltages, or mechanical failure.

29 percent of the respondents felt that normal deterioration from old age contributed to the failure, while 35 percent blamed abnormal moisture or exposure to aggressive chemicals. Inadequate installation and testing were considered responsible for 50 percent of the failures. 60 percent of the respondents reported that no preventive maintenance had been performed, but only 18 percent blamed the failure on inadequate maintenance.

Cable Terminations

Of the reported cable termination failures, 60 percent were repaired in place with just over half of the repairs being made on a round-the-clock basis. The primary damage was insulation involving either a flashover to ground or other electrical defect. About half of the respondents felt that the failure was initiated by an insulation breakdown, with normal deterioration due to age, severe weather, and exposure to abnormal moisture or aggressive chemicals contributing significantly to the problem. 39 percent felt that inadequate installation and testing prior to start-up was primarily responsible, while 22 percent felt that inadequate maintenance should be blamed. This also seems to correspond to the reporting that in 40 percent of the cases no preventive maintenance had been performed in over two years.

GENERAL CONCLUSIONS

Electrical Equipment

The general picture from Tables 38 and 35 spotlights inadequate maintenance as a significant factor in the suspected responsibility for failures. Yet the owner appears willing to work round the clock to fix failures after they have occurred. Lack of cleaning and lubrication is apparent on disconnect switches, buses, open wire, cable, cable joints, cable terminations, and motors.

Electric Utility Power Supplies

Many of the results shown in Tables 33–38 are not really applicable for electric utility power supplies because the questions asked are not well suited. The importance of the utility supply was indicated by 91 percent of respondents making repairs on a round-the-clock basis. The failures were predominantly flashovers involving ground, caused by lightning during severe weather or by dig-ins or vehicular accident. Outside agencies, probably the local utility, were predominantly responsible for the failure with preventive maintenance having no apparent effect on the cases reported.

The data reported under "failure characteristic" in Table 41 are of special significance in the case of double- or triple-circuit electric utility power supplies. In particular, the failure rate can

TABLE 42 - SIMULTANEOUS FAILURE OF ALL CIRCUITS IN ELECTRIC UTILITY POWER SUPPLIES

% of 145 Failures from Table 41	Number of Failures	Utility Power Supplies - Failure Characteristic from Table 41
15%	22	3. Failure of both circuits of a double-circuit redundant supply
2%	3	4. Failure of all circuits of a three or more circuit redundant supply
17%	25	Total number of simultaneous failures of all circuits in a double or more circuit redundant supply

be calculated for the simultaneous failure of all circuits in a double- or triple-circuit electric utility power supply.

From Table 3 of Part 1 [1] the sample size is 210.7 unit-years for a double- or triple-circuit electric utility power supply. A double- or triple-circuit supply operating for one year is counted as one unit-year. It is possible to calculate a failure rate from these data as follows:

$$\frac{25}{210.7} = 0.119 \text{ failures per year for simultaneous failure of all circuits in a double- or triple-circuit electric utility power supply.}$$

Some discrepancies were found in the data on the number of installed units for double- and triple-circuit electric utility power supplies. See the discussion in Part 1 [1] on this point.

Discrepancies

A survey such as this one often obtains some data that appear to contain errors. Sometimes the results look ridiculous. However, some of the ridiculous looking results may actually be correct. Some of the errors are believed due to a misinterpretation of the question by the respondent.

The data in Tables 31–41 have been published without attempting to correct discrepancies or errors. A brief list of some possible discrepancies is given.

Table 36: The damaged part of one percent of failed circuit breakers is a tap changer. The damaged part of three percent of failed cables is a bearing. Winding insulation is shown as the damaged part in failures of cables, bus ducts, and motor starters.

Table 39: Three percent of the failures in disconnect switches were initiated by low voltage.

REFERENCES

[1] IEEE Committee Report, "Report on reliability survey of industrial plants, Part I: Reliability of electrical equipment," this issue, pp. 213–235.

Discussion

J. Krasnodebski, N. M. Thompson, D. H. Cooke, A. W. W. Cameron, S. Basu, and T. J. Ravishanker (Ontario Hydro, Toronto, Ont., Canada):

1) *Quality of Input Data:* The confidence level of data in a survey of this kind cannot be assessed by mathematics only. One key problem is the adequacy of records and completeness of data. Some of the apparent discrepancies noted in the paper seem to indicate quite substantial omissions in records. Unless the industries involved keep much better failure records than we have done to date, this is not surprising. The first requirement of a useful reliability program is an adequately complete and accurate system for recording failures and consequences (in outage terms).

TABLE A
GENERATORS

Forced Outages

EEI Report

Sample Size (unit-years)	Number of Occurrences per Unit-Year	Outage Hours per Occurrence
204	0.142	91.8
404	0.839	126.5
705	0.521	54.4
483	0.393	125.6

IEEE Reliability Survey

Type of Drive	Sample Size (unit-years)	Number of Occurrences per Unit-Year	Outage Hours per Occurrence
Steam turbines* Jet engines	761.8	0.032	165.0
Gas turbines	89.4	0.638	23.1
Diesel engines	59.4	0.067	127.0

*EEI results are for generators 60–89 MW.

The requirements for better records, along with the detail involved in the report forms, indicate that acquiring useful data of this kind is time consuming.

It is suggested that, if a choice is necessary, it might be preferable to have a limited (but statistically adequate) number of plants establish a reliably complete recording and reporting system rather than increase the size of the sample under current record systems.

2) *Survey Results on Equipment Failures:* The failure rate is given in failures per unit-year. Is year in this context a calendar year or 8760 hours of plant or equipment operating time? If the failure rate is given per calendar year, were adjustments made for plants operating for 40 hours per week against those operating for up to 168 hours per week?

3) *Discussion of Equipment:*

Motors: It is suspected that the discrepancy in failure rates results from the different application of the two types of motors. Synchronous motors are usually applied only in engineered situations and are carefully designed for the application. Large synchronous motors are usually slow speed. Induction motors are mass produced, purchased off the shelf at the lowest cost, and usually operated to take advantage of any service factor. The survey figures are probably correct but cannot be used for comparison of reliability, leading to a conclusion that synchronous motors are more reliable. It is a comparison of apples and oranges.

Switchgear Bus: The paper states that the reported data are the opposite to what they should be. The reported figures may be correct. Manufacturers regularly reduce the spacing between buses and the spaces between phases and ground when they use insulated bus. As the conductor insulation is usually also reduced by design and occasionally by inferior material standards compared to that on insulated cables, and workmanship is frequently less than perfect, failures on this type of gear are probably at least as common as those on air-insulated equipment.

Circuit Breakers: The failure rate for circuit breakers appears much too low. It must of course be a function of the frequency of operation as well as lapsed time. We did not find a definition of circuit breaker failure, which we believe should differ from cable, transformer, or other static device failures. Circuit breaker failures should be based on failure to operate satisfactorily either to remain closed or to open or to close when called upon. It should be clear whether these figures include failures caused by auxiliaries such as instrument transformers, relays, and control switches. Since any calculation of the reliability of a power system would be made unreasonably complex by attempts to treat all these devices individually, a figure for circuit breaker failures which includes them is usually required by the designer.

Generators: For the generators in the electrical power industry a good source of data exists in the EEI "Report on Equipment Availability for Twelve-Year Period 1960–1971." The comparison between the failure rates and average repair time contained in that report and the survey discussed are shown in Table 43. EEI data quoted for steam turbine driven generators are for the size class 60–89 MW, which is probably larger than the average size of a corresponding generator included in the industrial survey.

It can be seen that the EEI failure rate for steam turbine driven generators based on forced outages is higher by a factor of 5 than in the industrial survey. For gas turbines, failure rates contained in both reports are of the same order, while the outage duration quoted in the EEI report is higher. 54 failures in 5.5 unit years in the petroleum industry can probably be explained by the start-up troubles.

In summary, experience in the utility industry seems to explain results obtained in the industrial survey to a large degree.

4) *Causes of Failure:*

a) How important is the age of equipment? It is mentioned only as a "contributing cause," second in frequency only to "other." Are there economic replacement times, or does obsolescence usually come first?

b) Should the inference be drawn that reliability of industrial equipment, which is reasonably well suited to its job, depends mainly on 1) stringent acceptance testing, especially overvoltage testing, 2) adequate cleaning, and 3) proper lubrication of bearings?

5) *Additional Suggestions for Analysis:* Consideration should be given to add the manufacturer of the main class of equipment to provide information on reliability of different manufacturers.

Carl Becker (Cleveland Electric Illuminating Company, Cleveland, Ohio 44101): The Reliability Subcommittee did an outstanding job in as-

sembling and correlating the mountainous volume of data in a simple, easy to understand tabulation. I would like to add some discussion that I feel would help the value of these tables and add to the accuracy of future studies. My two main points are 1) the downtime per failure on a single-circuit utility supply is extremely high (possibly by a factor of five), and 2) the equation for the dollars lost per interruption may be improved by using other than the kilowatt demand and kilowatt-hour usage as bases.

My company gathers, codes, and analyzes by computer all interruptions to our three quarter million customers. The average downtime per customer on our distribution system (which is a single-circuit radial supply) has been between 51 and 61 min for five of the past six years. Our service area experienced a catastrophic storm during 1969 which caused the average downtime per customer to jump to 124 min. In addition, my company is of the opinion that no plant should be down for more than 4 h (barring major catastrophies). A report is therefore written for each interruption exceeding 4 h in duration, and these reports are extremely few in number. Furthermore, 13 utilities have polled their reliability statistics for customers fed from the distribution system and found the average downtime per interruption for 1971 to be approximately 1½ h long. The average downtimes ranged from 0.75 to 3.2 h.

This information shows that the downtime per failure for industrial plants is probably outside the predicted tolerance on the IEEE data. This variance may be due to either a major long disturbance affecting a majority of those industrial plants participating or to misinterpretation of the information required.

For over five years I have worked with our customers in regard to reliability problems. My experience has shown that the plant investment, labor cost, and value of product is a better gauge of the cost per minute down than would be either maximum kilowatthour demand or usage. For example, I worked with a manufacturer of magnesium parts for military aircraft (I will call this plant *A*) and another manufacturer of parts for conveyor systems (plant *B*). The dollar loss for *A* per minute down was 100 times greater than that for *B*. However, plant *B*'s demand is 2500 kW and *A*'s demand is 500 kW, which is an indication that the kilowatthour consumptions in these particular cases are not related at all to the economical loss due to a power interruption. In general I find that the cost of downtime is tied heavily to one of the following: 1) the number of employees, 2) the cost of the product in production (piecework), or 3) the dollar output per hour (high production). A combination of these three items would indicate that loss is tied to the dollars out of the plant per unit of time. Therefore I feel that future studies should relate downtime to dollars per minute of plant production, gross plant, etc.

J. W. Beard (Union Carbide Corporation, South Charleston, W. Va. 25303): The report format and the manner in which the information is presented is generally quite adequate. Appendix A (Part I) is somewhat difficult to read because of the reduced print, but I am not suggesting it be upgraded for this report. Because of the many and various pieces of data used for the report, it is understandable that the reader must spend a great deal of time in studying and analyzing the information in order to properly apply it. The "readily" understandable factor should perhaps be given more consideration in defining the criteria for future surveys.

It is my opinion that the most useful types of information presented are:

1) failure rate and failure rate confidence limits;
2) failure, damaged part;
3) failure type;
4) failure initiating cause;
5) failure contributing cause;
6) failure characteristics.

I believe it is a good assumption that the raw data submitted for many of the other types of information represented were of much lesser accuracy than for these. For example, most plants reporting data for information types such as plant outage cost, critical service loss duration, and loads lost versus time of power outage probably had to draw on someone's memory of each failure and then apply the "best estimate" principle. This factor alone raises the question as to whether these types of information can ever be constructed to have useful meaning. Except for near catastrophic failures, which result in heavy financial losses, it is doubtful that most plants will spend the money to document this type of data. Furthermore, in a practical sense, when configuring systems and applying electrical equipment, the reliability requirement must be carefully considered for each producing unit served inasmuch as there are many variables that enter into the calculation of downtime losses.

The following suggestions are offered for consideration in any future surveys.

1) Basically concentrate on failure rates and failure causes.

2) Simplify and reduce scope of the survey questionnaire forms (present forms tend to scare users from contributing).

3) Omit asking for types of information such as cost of outage, repair time, plant start-up time, etc.

4) Instruct users *not to* report failures of equipment where reasonable preventative maintenance is not performed.

5) Instruct users *not to* report failures of misapplied equipment.

6) Instruct users *not to* include equipment installed prior to January 1, 1968.

7) Instruct users to give "in-service" date (energized) of all equipment units, not just on the reported failures.

8) Define "failure" as "damage to equipment sufficiently severe to force an outage by either manual or automatic removal of voltage." (Keep in mind that failures caused by the conditions in 4) and 5) are not to be reported.)

Part I: There seemed to be a great deal of confusion by the respondents on the information desired for electric power supplies. Thus the published failure rates may be questionable. It is my opinion that the questionnaire form for this was too nondescript. Perhaps one way to clearly describe the power supplies on which information is desired would be to include on the form simple single-line diagrams of the more common types of utility services.

It is my opinion that the lack of response by many companies was due primarily to poor and/or nonexistent records. A major contributing cause may have been the massive amount of information asked for.

The Reliability Subcommittee's judgement that a minimum of 8 to 10 observed failures was required for "good" accuracy when estimating equipment failure rates seems reasonable.

The value chosen for the confidence interval (0.90) was a good choice. The inclusion of confidence limits curves (Fig. 1) adds measurably to the report.

I generally concur with the Subcommittee's discussion comments. Their discussion of some of the results presented in the tables reinforces my feeling that the survey was too broad in scope, and the information submitted by the plants too ambiguous for meaningful interpretation.

While the sample sizes would be made smaller, as a general rule I feel that equipment should be grouped by voltage class. For example, in Table 2 one grouping of cable terminations is for 601–15 000 V. In this instance it would be especially helpful to know the failure rate on 15-kV cable terminations alone.

Part II: As stated in my general comments, I feel that it is not practical to generate reasonably accurate information of these types.

The bases for the units used in cost calculations, dollars per kilowatt plus dollars per killowatthour, are somewhat confusing. Clarification of this would be helpful.

In the Subcommittee's discussion of the cost of power outages, item 2), I must disagree with their thought that electrochemical or heating processes tend to have low outage costs because heat not supplied now can be supplied later.

In the discussion of loads lost versus time of power outage the "time" factor is questionable. Most plants are not equipped to measure short-duration power outages (cycles or even seconds).

Part III: Many of the information types in this part are very important. Some, I feel, are not. I suggest that the questions on failure repair method; failure repair urgency; failure, months since maintenance; and suspected failure responsibility be omitted from future surveys. The remaining types of information may be refined using knowledge gained from this survey.

In the Subcommittee's Summary of Conclusions they report that transient overvoltages were a major cause of failure in equipment such as, for example, transformers and circuit breakers; but I got the impression that much of this was speculation on the part of those responding. The possibility of transient overvoltage should be considered

in the investigation of most equipment failures, and IEEE could perform an important service to industry by developing a socalled "evaluation of possibility of transient overvoltage contribution to equipment failures" guide.

Stanley Wells (Union Carbide Corporation, Port Lavaca, Tex. 77979): The Reliability Subcommittee should be congratulated for performing such a comprehensive reliability survey of industrial plants and for providing a very thorough report.

I would like to limit my discussion to Part 3 and, in particular, the preventive maintenance effect on the failure rate. A preventive maintenance program can very definitely have a direct effect on the failure rate of electrical equipment. In the modern automated plant of today, production demands and losses associated with downtime influence maintenance schedules. Equipment is often allowed to remain in operation for periods that exceed desired preventive maintenance time schedules. It is interesting to note that the survey indicates that preventive maintenance can be performed, yet equipment failures occur within a time period which is less than 12 months since preventive maintenance was performed. Our first attempt at a preventive maintenance program met with the same results. The program was reviewed in depth and it was found that it was inadequate and that the preventive maintenance procedures and time schedules should be reviewed and correlated with our failure experience. As experience was gained, the equipment preventive maintenance program developed into a very useful tool to practically eliminate electrical equipment failure. We soon recognized that where preventive maintenance periods were over 24 months or where no preventive maintenance at all was performed, chances of failure were extremely high. This fact is born out in the results of this survey. Table 35, "Failure–Months Since Maintained," has been rearranged to show that a large reduction in failures may be possible if preventive maintenance periods are on a 12- to 18-month basis (Table B).

Let's define preventive maintenance. Preventive maintenance is a system of routine inspections designed to minimize or forestall future equipment operating problems or failures, and which may, depending upon equipment type, require equipment exercising or proof testing. From this definition, the four following items listed under Table 38, "Suspected Failure Responsibility," can be considered a definite part of a maintenance program:

1) manufacture, defective components (locate by inspection or test);

2) application engineering, improper application;

3) inadequate installation and testing prior to start-up (proof test);

4) inadequate maintenance.

It is interesting to note that the survey indicates that these four items are responsible for a very large percentage of failures. The total for each category is listed below.

	Percent
Transformers	55
Circuit breakers	53
Motor starters	77
Motors	42
Generators	41
Disconnect switches	52
Switchgear bus insulated	95
Switchgear bus uninsulated	52
Bus duct	63
Open wire	41
Cable	48
Cable joints	68
Cable terminations	79

To increase the electrical system reliability, each failure should be very carefully analyzed to determine the failure cause, and corrective action to prevent additional failures should be applied to all applicable equipment.

TABLE B
FAILURES

	Less than 12 Months Ago Preventive Maintenance	12 Months or More or No Preventive Maintenance
Transformers	34	65
Circuit breakers	18	81
Motor starters	67	33
Motors	22	78
Generators	58	42
Disconnect switch	8	92
Switchgear bus insulated	10	90
Switchgear bus uninsulated	35	65
Bus duct	25	75
Open wire	1	98
Cable	11	89
Cable joint	18	82
Cable terminations	12	88

R. E. Kuehn (IEEE Reliability Group): The reliability, maintainability, and downtime logistics in the power area is very important and should lend itself to cost analysis, which is the ultimate judge of the value of reliability and maintainability programs. A great deal of data have been analyzed with all the obvious advantages and disadvantages that are entailed in such a data base. Parts 1 and 2 present me with a severe problem as a reliability professional and manager. In both papers a large effort was spent indicating that the survey results do not agree with what the engineering judgment says the results should be; for example, the discussion of Part 1 on motors, generators, cable, and switchgear bus. My quandary is that if I accept your judgment in all logic, I must question the validity of all the data collected, not just for motors, generators, ~able, and switchgear bus. A possible procedure would have been to test the hypothesis that a part of the data was significantly different enough from the total grouped data to justify its rejection as part of the group data.

I would like to recommend analysis of variance or multiple regression in analyzing the data. It would appear that a number of possible variables exist and their effects are suitable for quantization. These procedures are covered in [1]–[4].

REFERENCES

[1] R. G. Stokes and F. N. Sehle, "Some life-cycle cost estimates for electronic equipments," in *Proc. 1968 Annu. Symp. on Reliability*, pp. 169–183.
[2] B. L. Retterer, "State of art assessment of reliability and maintainability as applied to ship systems," in *Proc. 1969 Annu. Symp. on Reliability*, pp. 133–145.
[3] H. Dagen, "Multiple regression," in *Proc. 1972 Annu. Symp. on Reliability*," pp. 51–58.
[4] "Cost effectiveness evaluation procedures for shipboard electronic equipment," ARINC Research Publ. 509-01-2-564 and 541-01-1-766.

Tai C. Wong (American Electric Power Service Corporation, New York, N.Y. 10004): The members of the Reliability Subcommittee are to be commended for conducting and analyzing the results of a survey that covers so many elements in industrial power systems.

Perhaps the authors want to clarify why the chi-squared distribution was used in fitting the data and what kind of statistical testing technique was employed to ensure the adequacy of the distribution chosen. The authors did compare the results of the recent survey against those obtained in 1962. The readers should be warned that this is only an observation based on empirical data and that any inference of a trend in the equipment reliability may not be valid. The paper indicates that many of the reported data cover more than one year of operating experience. Because the first survey was conducted twelve years ago, it is felt that the number of years that the different equipments were in service should be published (or the data collected during the next survey if they are not yet available) so that the reader can have a better understanding of the data background when he has to draw further conclusions, beyond the tables presented.

The authors indicated that the purpose of this survey is to make possible the quantitative reliability comparisons between alternative designs of new systems and then use this information in cost–reliability tradeoff studies to determine which type of power distribution system to use. It appears that the authors focus on making the economic tradeoff comparisons based on the available system components at a given time. However, the authors pointed out that the product of failure rate times the average downtime per failure is almost the same in 1973 as in 1962. Perhaps the equipment manufacturers and the industries can establish more dialogues, leading to an answer to the following two questions.

1) Should the equipments have a lower failure rate, but when failing, take longer to repair? or

2) Should the equipments have a higher failure rate, but when failing, need shorter repair time?

In a few instances during the survey, the respondents misinterpreted either the question(s) and/or the definition of the terms, thus leading to unreliable or biased results. This is especially true in the area of preventive maintenance. I might suggest that during the next survey 1) the definition of all terms that are likely to cause confusion in the questionnaire be included, 2) a pilot survey be instituted and any necessary modifications be made to the questionnaire before a full-scale survey is launched, or 3) the survey form be sent out without requesting data, but instead requesting the respondent's interpretations of the questions and the terms used. Then the survey form may be redesigned and data requested.

I. O. Sunderman (Lincoln Electric System, Lincoln, Nebr.): The authors have presented an interesting cross section of costs involved with industrial electric equipment downtime as accumulated by the computer. The data are to be utilized by interested parties in the choice of a reliability design for industrial power distribution systems. The wide range of costs as split into the two parts over 1000 kW and under 1000 kW suggests consideration of other kW brackets at 500, 2500, 5000, 7500, 10 000 kW, etc. The sufficiency of data will dictate breaking points, as the author already questions the cost data below 1000 kW.

In Part 3 the authors have reviewed and presented in excellent tables the results of electric equipment outage reports and repair. It must have been disturbing to note the numerous "other than categories classified." Perhaps further reporting on the "other" category comments, if available, would bring additional results to light.

IEEE Reliability Subcommittee: The authors wish to thank those who presented discussions on these three papers. Some of the suggestions given can be considered for incorporation into future surveys and they can also be used in the analysis of the results.

Several discussers have raised the question about the effect of "in service date" or age on the reliability of electrical equipment. Population data were collected on the average age of equipment in service; these will be published in Part 4. However, the Reliability Subcommittee did not request these data in the survey questionnaire on equipment failures. This subject was considered by the Subcommittee when making up the questionnaire; it was not included because this would have added additional complications to a questionnaire that was already considered too long. This meant that the assumption was made that the failure rate was constant with age. Thus a chi-squared distribution is appropriate for use in calculating the confidence limits of

the failure rate. The assumption of a constant failure rate with age can be justified for most electrical equipment based upon reliability surveys made by others.

Mr. Becker and Mr. Beard have raised questions about the accuracy of the cost of power outage data and the attempt to relate it to kilowatts and kilowatthours. Information was collected but not published on the estimated plant outage costs 1) per failure and 2) per hour of downtime. The authors consider that the cost of power outages is an important factor that should be considered in the design of power distribution systems for industrial plants. Since power distribution systems are designed on the basis of kilowatt capacity and kilowatthour of delivered energy, it was felt that it is necessary to attempt to relate the cost of power outages to these two parameters. The approach used by the Reliability Subcommittee is the same as that which has been used by electric power companies in several European countries. The survey result of the median cost of 83¢ per kilowatthour of undelivered energy is in the same range as values obtained from surveys that have been made in Sweden, Norway, France, Italy, and West Germany. The authors agree that the published data of the cost of power outages are more meaningful if related to specific types of plants.

The authors acknowledge Mr. Beard's suggestion that a one-line diagram should be used in the survey of the electric utility supply. A new survey of the electric utility supply is being started, and Mr. Beard's suggestion will be included. This new survey should clear up the problem of the questionable accuracy mentioned by Mr. Beard. The authors acknowledge Mr. Beard's comment questioning the accuracy of the "time" factor in loads lost versus time of power outage in Table 30.

In answer to several questions raised by Mr. Krasnodebski, the authors make the following comments.

1) The failure rates are based upon a calendar year of 8760 h, not upon an operating time, which could be less and would thus result in a higher failure rate than reported in the survey.

2) The failures of circuit breakers are meant to include the auxiliaries.

3) The failure modes of circuit breakers are included in Table 41; this includes "fail to close," "fail to open," etc. However, data were not collected on the number of circuit breaker operations.

4) The Reliability Subcommittee does not consider that it would be appropriate for a technical society such as IEEE to collect and publish reliability data by name of manufacturer.

5) The authors agree that better record keeping of failures would improve survey results. It is expected that future surveys will cover only a few categories of electrical equipment that are considered trouble areas.

6) The authors acknowledge the logic in the very interesting comments made on synchronous motors and switchgear bus and generators.

7) The steam turbine generators in industrial plants probably have constant operation and thus could be expected to have a much lower failure rate than 60–89 MW units in utility applications where the operation was cyclical.

The authors wish to thank Mr. Kuehn for his suggestions in analyzing the data. These suggestions included 1) test hypothesis that part of data can be rejected, and 2) analysis of variance or multiple regression. Mr. Becker has raised a point where this approach for analyzing the data could possibly be tried. Mr. Becker feels that the survey results are too high on the downtime per failure of a single-circuit electric utility supply. This may be true for his system, but perhaps other utilities are not as good as his company's system.

Mr. Wong has raised a warning about drawing the conclusion that equipment reliability has improved since the previous survey conducted 11 to 12 years earlier. A separate paper has been prepared on this subject and will be published in the near future. This paper contains the conclusion that the failure rate of electrical equipment has shown a definite trend of improvement during the 12-year interval.

The authors wish to thank Mr. Wells for his discussion on preventive maintenance. A lot more data on preventive maintenance are being processed and will be included in Part 4. Mr. Wells' Table B shows more failures in the "12 months or more" category than for the "less than 12 months ago" category. The authors would like to point out that the electrical equipment has more unit-years of exposure in the "12 months or more" category and thus could be expected to have more failures. Thus it is not possible to conclude that more frequent preventive maintenance will reduce the failure rate. The Reliability Subcommittee is investigating this subject in further detail and will publish the results in Part 4.

Appendix B
Report on Reliability Survey of Industrial Plants

Part 4
Additional Detailed Tabulation of Some Data
Previously Reported in the First Three Parts

Part 5
Plant Climate, Atmosphere, and Operating Schedule,
the Average Age of Electrical Equipment,
Percent Production Lost, and the Method of
Restoring Electrical Service After a Failure

Part 6
Maintenance Quality of Electrical Equipment

By
Reliability Subcommittee
Industrial and Commercial Power Systems Committee
IEEE Industry Applications Society

A. D. Patton, *Chair*

C. E. Becker
W. H. Dickinson

P. E. Gannon
C. R. Heising
D. W. McWilliams

R. W. Parisian
S. J. Wells

Industrial and Commercial Power Systems Technical Conference
Institute of Electrical and Electronics Engineers, Inc.
Denver, Colorado
Jun. 3-6, 1974

Also Published
IEEE Transactions on Industry Applications
Jul./Aug. and Sep./Oct. 1974
Part 4, pp. 456-462
Part 5, pp. 463-466
Part 6, pp. 467-468, 681, 469-476

Reprinted from pp. 113-136
in 74CHO855-71A, 1974 I&CPS Technical Conference

Report on Reliability Survey of Industrial Plants, Part IV: Additional Detailed Tabulation of Some Data Previously Reported in the First Three Parts

IEEE COMMITTEE REPORT

Abstract—An IEEE sponsored reliability survey of industrial plants was completed during 1972. This survey included 30 companies covering a total of 68 industrial plants in the United States and Canada. Additional detailed results are reported on some data that were previously reported in the first three parts. This includes failure modes of circuit breakers, cost of power outages, critical service loss duration time, loss of motor load versus time of power outage, and the effect of failure repair method and repair urgency on the average downtime per failure of electrical equipment. This information is useful in the design of industrial power distribution systems.

INTRODUCTION AND RESULTS

DURING 1972 the Reliability Subcommittee of the Industrial and Commercial Power Systems Committee completed a reliability survey of industrial plants. This paper presents Part IV of the results from the survey. The first three parts [1]–[3] were published previously. Some of the data in the first three parts caused questions to be raised about the possibility of obtaining additional details. These additional details are being reported in this paper and include the following results.

Table 43 gives failure modes of circuit breakers, including

1) metalclad drawout
   a) 0–600 V
   b) 601–15 000 V
   c) all voltages
2) fixed type (includes molded case)
   a) 0–600 V
   b) all voltages.

Tables 44, 45 give cost of power outages, adding 25 and 75 percentile data to what was previously published.

Table 46 gives critical service loss duration time (maximum length of power failure that will not stop plant production), adding 10, 25, 75, and 90 percentile data to what was previously published.

Paper TOD-74-33, approved by the Industrial and Commercial Power Systems Committee of the IEEE Industry Applications Society for presentation at the 1974 Industrial and Commercial Power Systems Technical Conference, Denver, Colo., June 2–6. Manuscript released for publication April 15, 1974.

Members of the Reliability Subcommittee of the IEEE Industrial and Commercial Power Systems Committee are A. D. Patton, *Chairman*, C. E. Becker, W. H. Dickinson, P. E. Gannon, C. R. Heising, D. W. McWilliams, R. W. Parisian, and S. Wells.

Table 47 lists loss of motor load versus time of power outage, adding the following length of power outage categories:

1) 10 to 15 cycles
2) 15+ to 30 cycles
3) 0.5+ to 2.0 s
4) 2+ to 4.0 s
5) >4.0 s.

Tables 48 through 56 report the effect of failure repair method and failure repair urgency on the average downtime per failure for the following equipment categories:

1) transformers—liquid filled
   a) 601–15 000 V
   b) above 15 000 V
2) circuit breakers—metalclad drawout
   a) 0–600 V
   b) above 600 V
3) motors
   a) induction, 0–600 V
   b) induction, 601–15 000 V
   c) synchronous, 601–15 000 V
4) cable
   a) above ground and aerial, 601–15 000 V
   b) below ground and direct burial, 601–15 000 V

In each of the Tables 43 through 56 reference is made to the tables in Parts I, II, and III where previous results had been reported.

DISCUSSION—FAILURE MODES OF CIRCUIT BREAKERS

The data on failure modes of circuit breakers given in Table 43 show some very interesting results.

Circuit Breakers, 0–600 V

71 percent of the failures of metalclad drawout circuit breakers were "opened when it shouldn't" versus 5 percent of the failures for fixed-type circuit breakers (includes molded case). 77 percent of the failures of fixed-type circuit breakers (includes molded case) were "failed while operating (not while opening or closing)," and only 10 percent of the metalclad drawout failures included this failure mode.

None of the failures reported for either type of circuit breaker were "failed while opening." Only 9 percent and

TABLE 43 - FAILURE MODES OF CIRCUIT BREAKERS - Percent of Total Failures in Each Failure Mode
(Data Previously Reported in Tables 5 and 41)

All Circuit Breakers	Metalclad Drawout-All	Metalclad Drawout-601-15,000 Volts	Metalclad Drawout-0-600 Volts All Sizes	*Fixed Type 0-600 Volts All Sizes	*Fixed Type-All	Card-Type 3, Col. 46 FAILURE CHARACTERISTIC
%	%	%	%	%	%	
5	5	2	7	8	6	Failed to close when it should
9	12	21	0	0	2	Failed while opening
42	58	49	71	5	4	Opened when it shouldn't
7	6	4	9	5	4	Damaged while successfully opening
2	1	0	0	0	4	Damaged while closing
32	16	24	10	77	73	Failed while operating (not while opening or closing)
1	0	0	0	0	2	Failed during testing or maintenance
1	2	0	3	0	0	Damage discovered during testing or maintenance
1	0	0	0	5	5	Other
100%	100%	100%	100%	100%	100%	Total Percent
-	117	53	59	39	48	Number of Failures in Total Percent
-	7	0	7	1	1	Number Not Reported in Col. 46, Card-Type 3
-	124	-	66	40	49	Total Failures in Table 5

*Includes molded case

5 percent, respectively, of the failures were "damaged while successfully opening." Only 7 to 8 percent of the failures were "failed to close when it should."

It appears that the dominate failure mode for metalclad drawout circuit breakers, 0–600 V, is "opened when it shouldn't." It is possible that some of these failures were external to the breaker and of unknown cause and were blamed on the breaker. Some of these may have been due to improper-setting of the trip current.

The dominate failure mode for fixed-type circuit breakers (includes molded case), 0–600 V, is "failed while operating (not while opening or closing)."

Metalclad Drawout Circuit Breakers, 601–15 000 V

Metal drawout circuit breakers, 601–15 000 V, had 21 percent of the failures classified as "failed while opening" and 4 percent classified as "damaged while successfully opening." Another 24 percent of the failures were classified as "failed while operating (not while opening or closing)." 49 percent of the failures were classified as "opened when it shouldn't;" it is suspected that some of these may have been due to improper setting of the trip current.

It appears that metalclad drawout circuit breakers, 601–15 000 V, have about half of their failures as "opened when it shouldn't" and the other half as "failed while operating or while opening."

DISCUSSION—LOSS OF MOTOR LOAD VERSUS TIME OF POWER OUTAGE

The data on loss of motor load shown in Table 47 indicate that for power outages greater than 10 cycles duration most of the plants lose the motor load. However, for power outages between 1 to 10 cycles duration, only about half as many lose the motor load. Thus, power outages of less than 10 cycles duration may often not result in losing the motor load.

There were many power outages of more than 4.0 s duration, and 35 percent did not lose motor load. It is suspected that many of these did not have a motor load. Some may have had a duplicate feed and thus did not lose the motor load.

DISCUSSION—EFFECT OF FAILURE REPAIR METHOD AND FAILURE REPAIR URGENCY ON AVERAGE HOURS DOWNTIME PER FAILURE

Data were given in Part I on the average hours downtime per failure for 74 categories of electrical equipment. It is known that the downtime after a failure can be affected to a large extent by the failure repair method and the failure repair urgency. The failure repair method includes either repair of the failed component or else replacement with a spare. Some data were given in Tables 33 and 34 of Part III on the failure repair method and the failure repair urgency for whole classes of electrical equipment.

A more detailed study is reported in Tables 48–56 of this paper on the effect of the failure repair method and the failure repair urgency on the average hours downtime per failure. This is only reported for 9 electrical equipment categories, rather than the 74 categories given in Part I. These 9 electrical equipment categories were selected because an adequate sample size existed of the number of failures and because the average downtime per failure was effected significantly by the failure repair method and/or the failure repair urgency.

TABLE 44 – PLANT OUTAGE COST PER FAILURE PER kW OF MAXIMUM DEMAND ($ per kW)
All Industry – USA & Canada
(Data Previously Reported in Tables 22, 24 and 26)

Plant Size	Number of Plants Reporting	Minimum	25% Percentile	Median	75% Percentile	Maximum	Average
All Plants	42	.002	.17	.69	2.55	10.00	1.89
Plants > 1000 kW Max. Demand	32	.002	.09	.32	1.31	7.50	1.05
Plants < 1000 kW Max. Demand	10	.50	1.71	3.68	8.27	10.00	4.59

TABLE 45 – PLANT OUTAGE COST PER HR. DOWNTIME PER kW OF MAXIMUM DEMAND ($ per kWh)
All Industry – USA & Canada
(Data Previously Reported in Tables 23, 25 and 27)

Plant Size	Number of Plants Reporting	Minimum	25% Percentile	Median	75% Percentile	Maximum	Average
All Plants	41	.0009	.18	.83	2.71	27.00	2.68
Plants > 1000 kW Max. Demand	31	.0009	.12	.36	1.20	5.77	.94
Plants < 1000 kW Max. Demand	10	.86	1.83	4.42	12.50	27.00	8.11

TABLE 46 – CRITICAL SERVICE LOSS DURATION (Maximum Length of Power Failure
that Will Not Stop Plant Production)
(Data Previously Reported in Table 29)

Industry	Number of Plants Reporting	Average	10% Percentile	25% Percentile	Median	75% Percentile	90% Percentile
All Industry - USA & Canada	55	12.6 min.	5.0 cycles	10.0 cycles	10.0 sec.	15.0 min.	60.0 min.
Chemical	20	4.56 min.	3.2 cycles	8.5 cycles	1.25 sec.	5.0 min.	28.5 min.

TABLE 47 – LOSS OF MOTOR LOAD VERSUS TIME OF POWER OUTAGE
Tabulation of the Percentage of Equipment Failures
for Which the Motor Load was Lost
(Data Previously Reported in Table 30)

Length of Equipment Failure	Number of Failures Reported	TYPE OF LOAD LOST Motor		
		Yes	No	Not Known
1 cycle or less	0	0%	0%	0%
1+ to 10- cycles	–	33%	67%	0%
10 to 15 cycles	7	86%	14%	0%
15+ to 30 cycles	28	96%	4%	0%
0.5+ to 2.0 sec.	30	77%	13%	10%
2.0+ to 4.0 sec.	10	100%	0%	0%
>4.0 second	998	64%	35%	0%

TABLE 48   TRANSFORMERS–LIQUID FILLED, 601-15,000 VOLTS
EFFECT OF FAILURE REPAIR METHOD AND FAILURE REPAIR URGENCY
ON THE AVERAGE HOURS DOWNTIME PER FAILURE
(Previous Data Given in Tables 4, 33 and 34)

FAILURE REPAIR METHOD			FAILURE REPAIR METHOD		FAILURE REPAIR URGENCY
Repair	Replace with Spare	Total	Repair	Replace with Spare	
Number of Failures			Average Hours Downtime per Failure		
4	22	26	*	130	1. Requiring round-the-clock all out efforts
10	3	13	342	*	2. Requiring repair work only during regular workday, perhaps with some overtime
0	0	0	-	-	3. Requiring repair work on a non-priority basis
14	25	39	Average 174. Hours		Total

*Small Sample Size

TABLE 49 – TRANSFORMERS–LIQUID FILLED, ABOVE 15,000 VOLTS
EFFECT OF FAILURE REPAIR METHOD AND FAILURE REPAIR URGENCY
ON THE AVERAGE HOURS DOWNTIME PER FAILURE
(Previous Data Given in Tables 4, 33 and 34)

FAILURE REPAIR METHOD			FAILURE REPAIR METHOD		FAILURE REPAIR URGENCY
Repair	Replace with Spare	Total	Repair	Replace with Spare	
Number of Failures			Average Hours Downtime per Failure		
2	5	7	*	*	1. Requiring round-the-clock all out efforts
12	4	16	1842	*	2. Requiring repair work only during regular workday, perhaps with some overtime
0	1	1	-	*	3. Requiring repair work on a non-priority basis
14	10	24	Average 1076. Hours		Total

*Small Sample Size

CIRCUIT BREAKERS – METALCLAD DRAWOUT, 0-600 VOLTS
EFFECT OF FAILURE REPAIR METHOD AND FAILURE REPAIR URGENCY
ON THE AVERAGE HOURS DOWNTIME PER FAILURE
(Previous Data Given in Tables 5, 33 and 34)

FAILURE REPAIR URGENCY	Number of Failures			Average Hours Downtime per Failure	
	Repair	Replace with Spare	Total	Repair	Replace with Spare
1. Requiring round-the-clock all out efforts	31	19	50	3.3	3.8
2. Requiring repair work only during regular workday, perhaps with some overtime	6	1	7	*	*
3. Requiring repair work on a non-priority basis	8	1	9	*	*
Total	45	21	66	Average 147. Hours	

*Small Sample Size

TABLE 51 – CIRCUIT BREAKERS – METALCLAD DRAWOUT, ABOVE 600 VOLTS
EFFECT OF FAILURE REPAIR METHOD AND FAILURE REPAIR URGENCY
ON THE AVERAGE HOURS DOWNTIME PER FAILURE
(Previous Data Given in Tables 5, 33 and 34)

FAILURE REPAIR URGENCY	Number of Failures			Average Hours Downtime per Failure	
	Repair	Replace with Spare	Total	Repair	Replace with Spare
1. Requiring round-the-clock all out efforts	34	12	46	83.1	2.1
2. Requiring repair work only during regular workday, perhaps with some overtime	3	9	12	*	*
3. Requiring repair work on a non-priority basis	0	0	0	-	-
Total	37	21	58	Average 109. Hours	

*Small Sample Size

TABLE 52 - MOTORS - INDUCTION, 0-600 VOLTS
EFFECT OF FAILURE REPAIR METHOD AND FAILURE REPAIR URGENCY
ON THE AVERAGE HOURS DOWNTIME PER FAILURE
(Previous Data Given in Tables 7, 33 and 34)

FAILURE REPAIR URGENCY	Number of Failures			Average Hours Downtime per Failure	
	Repair	Replace with Spare	Total	Repair	Replace with Spare
1. Requiring round-the-clock all out efforts	12	19	31	44.7	6.6
2. Requiring repair work only during regular workday, perhaps with some overtime	175	2	177	123	*
3. Requiring repair work on a non-priority basis	0	5	5	-	*
Total	187	26	213	Average 114. Hours	

*Small Sample Size

TABLE 53 - MOTORS - INDUCTION, 601-15,000 VOLTS
EFFECT OF FAILURE REPAIR METHOD AND FAILURE REPAIR URGENCY
ON THE AVERAGE HOURS DOWNTIME PER FAILURE
(Previous Data Given in Tables 7, 33 and 34)

FAILURE REPAIR URGENCY	Number of Failures			Average Hours Downtime per Failure	
	Repair	Replace with Spare	Total	Repair	Replace with Spare
1. Requiring round-the-clock all out efforts	14	10	24	88.1	*
2. Requiring repair work only during regular workday, perhaps with some overtime	93	48	141	83.6	34.7
3. Requiring repair work on a non-priority basis	6	0	6	*	-
Total	113	58	171	Average 76. Hours	

*Small Sample Size

TABLE 54 – MOTORS – SYNCHRONOUS, 601 – 15,000 VOLTS
EFFECT OF FAILURE REPAIR METHOD AND FAILURE REPAIR URGENCY
ON THE AVERAGE HOURS DOWNTIME PER FAILURE
(Previous Data Given in Tables 7, 33 and 34)

FAILURE REPAIR URGENCY	FAILURE REPAIR METHOD Number of Failures			FAILURE REPAIR METHOD Average Hours Downtime per Failure	
	Repair	Replace with Spare	Total	Repair	Replace with Spare
1. Requiring round-the-clock all out efforts	28	2	30	198	*
2. Requiring repair work only during regular workday, perhaps with some overtime	55	8	63	201	*
3. Requiring repair work on a non-priority basis	1	0	1	*	–
Total	84	10	94	Average 175. Hours	

*Small Sample Size

TABLE 55 – CABLE – ABOVE GROUND & AERIAL, 601-15,000 VOLTS
EFFECT OF FAILURE REPAIR METHOD AND FAILURE REPAIR URGENCY
ON THE AVERAGE HOURS DOWNTIME PER FAILURE
(Previous Data Given in Tables 13, 33 and 34)

FAILURE REPAIR URGENCY	FAILURE REPAIR METHOD Number of Failures			FAILURE REPAIR METHOD Average Hours Downtime per Failure	
	Repair	Replace with Spare	Total	Repair	Replace with Spare
1. Requiring round-the-clock all out efforts	46	4	50	9.0	*
2. Requiring repair work only during regular workday, perhaps with some overtime	11	8	19	*	*
3. Requiring repair work on a non-priority basis	2	2	4	*	*
Total	59	14	73	Average 40.4 Hours	

*Small Sample Size

TABLE 56 - CABLE - BELOW GROUND & DIRECT BURIAL, 601-15,000 VOLTS
EFFECT OF FAILURE REPAIR METHOD AND FAILURE REPAIR URGENCY
ON THE AVERAGE HOURS DOWNTIME PER FAILURE
(Previous Data Given in Tables 13, 33 and 34)

FAILURE REPAIR METHOD			FAILURE REPAIR METHOD		FAILURE REPAIR URGENCY
Repair	Replace with Spare	Total	Repair	Replace with Spare	
Number of Failures			Average Hours Downtime per Failure		
17	57	74	26.5	19.0	1. Requiring round-the-clock all out efforts
2	33	35	*	77.8	2. Requiring repair work only during regular workday, perhaps with some overtime
3	3	6	*	*	3. Requiring repair work on a non-priority basis
22	93	115	Average 95.5 Hours		Total

*Small Sample Size

In several cases there is a disparity in the downtime between the "average" and the cases where work is done "round the clock." When making availability calculations, this should be considered when deciding what value to use for the downtime after a failure.

Transformers, Liquid Filled

Transformers, above 15 000 V, had an average downtime per failure of 1842 h when sent out for repair without round-the-clock urgency. This compares with an overall average of 1076 h for all outage times, which included several cases of replacement with a spare. Thus it can be concluded that repair gives a much longer outage time than replacement with a spare for transformers, above 15 000 V.

Transformers, 601-15 000 V, had an average downtime per failure of 342 h when sent out for repair without round-the-clock urgency. This compares with 130 h for replacement with a spare while working round the clock. Thus it can be concluded that repair gives a much longer outage time for transformers, 601-15 000 V, than replacement with a spare while working round the clock.

Circuit Breakers, Metalclad Drawout

Metalclad drawout circuit breakers, 0-600 V, had an average downtime per failure of 3.3 h to 3.8 h when fixing the failure with round-the-clock efforts. This compares with an overall average of 147 h for all outage times. Thus it can be concluded that 24 percent of the outages of metalclad drawout circuit breakers, 0-600 V, had low urgency for fixing the failure, and that these 24 percent of the failures resulted in increasing the average downtime per failure from 3.8 h to 147 h.

Metalclad drawout circuit breakers above 600 V, had an average downtime per failure of 109 h for all outages. However, when round-the-clock effort was applied it only took 83 h for repair and only took 2.1 h for replacement with a spare. This shows that it is possible to reduce the downtime by having a spare and working round the clock when fixing metalclad drawout circuit breakers, above 600 V.

Motors

Most users of synchronous motors, 601-15 000 V, did not have a spare. Thus the average downtime per failure was 175 h for all failures.

Induction motors, 601-15 000 V, had an average downtime per failure of 35 h for replacement with a spare, compared to 84 to 88 h for repair. Induction motors, 0-600 V, had an average downtime per failure of 6.6 h for replacement with a spare while working round the clock. This compares with 123 h for repair and not working round the clock.

Cables

Cables, above ground and aerial, 601-15 000 V, had an average downtime per failure of 9 h for repair when working round the clock. This compares with 40 h for all failures. This shows that it is possible to reduce the downtime by working round the clock when fixing cables, above ground and aerial, 601-15 000 V.

Cables, below ground and direct burial, 601-15 000 V, had an average downtime per failure of 96 h for all failures. However, this was only 19 to 27 h when working round the clock. This shows that it is possible to reduce the downtime by working round the clock when fixing cables, below ground and direct burial, 601-15 000 V.

DISCUSSION—COST OF POWER OUTAGES

Data are given in Tables 44 and 45 on the cost of power outages to industrial plants. This has added 25th and 75th percentile data to what had previously been reported in Part II. These were added because of the wide spread in the cost of power outages to industrial plants.

REFERENCES

[1] W. H. Dickinson et al., "Report on reliability survey of industrial plants, part I: Reliability of electrical equipment," IEEE Trans. Ind. Appl., vol. IA-10, pp. 213-235, Mar./Apr. 1974.
[2] W. H. Dickinson et al., "Report on reliability survey of industrial plants, part II: Cost of power outages, plant restart time, critical service loss duration time, and type of loads lost versus time of power outages," IEEE Trans. Ind. Appl., vol. IA-10, pp. 236-241, Mar./Apr. 1974.
[3] W. H. Dickinson et al., "Report on reliability survey of industrial plants, part III: Causes and types of failures of electrical equipment, the methods of repair, and the urgency of repair," IEEE Trans. Ind. Appl., vol. 14-10, pp. 242-252, Mar./Apr. 1974.

Report on Reliability Survey of Industrial Plants, Part V: Plant Climate, Atmosphere, and Operating Schedule, the Average Age of Electrical Equipment, Percent Production Lost, and the Method of Restoring Electrical Service after a Failure

IEEE COMMITTEE REPORT

Abstract—An IEEE sponsored reliability survey of industrial plants was completed during 1972. This survey included the plant climate, atmosphere, and operating schedule, the average age of electrical equipment, percent production lost, and the method of restoring electrical service after a failure. The results are reported from the survey of 30 companies covering 68 plants in nine industries in the United States and Canada. This information is useful in the design of industrial power distribution systems.

INTRODUCTION AND RESULTS

DURING 1972 the Reliability Subcommittee of the Industrial and Commercial Power Systems Committee completed a reliability survey of industrial plants. This paper presents Part V of the results from the survey. The first three parts [1]–[3] were published previously; some of the data of lesser importance were not published at that time but are presented in this paper. Included in Part V are

Table 57—Failure Forewarning for Public Utility Power Interruption Only,
Table 58—Percent Production Lost,
Table 59—Method of Service Restoration,
Table 60—Average Age of Electrical Equipment,
Table 61—Plant Climate,
Table 62—Plant Atmosphere,
Table 63—Plant Operating Schedule.

These data are useful when using the results published in Parts I, II, III, IV [4], and VI [5]. This information is also useful in the design of industrial power distribution systems. The data on average age of electrical equipment and plant operating schedule provide answers to some points raised in the written discussion to Part I.

Paper TOD-74-33, approved by the Industrial and Commercial Power Systems Committee of the IEEE Industry Applications Society for presentation at the 1974 Industrial and Commercial Power Systems Technical Conference, Denver, Colo., June 2–6. Manuscript released for publication April 15, 1974.
Members of the Reliability Subcommittee of the IEEE Industrial and Commercial Power Systems Committee are A. D. Patton, *Chairman*, C. E. Becker, W. H. Dickinson, P. E. Gannon, C. R. Heising, D. W. McWilliams, R. W. Parisian, and S. Wells.

TABLE 57 – FAILURE FOREWARNING for PUBLIC UTILITY POWER INTERRUPTION ONLY

Percent	Col. 25 Card-Type 3
97% 3% —	1. If no forewarning was given 2. If forewarning was given For other types of failure, leave blank
100%	Total Percent
172	Total Interruptions Reported

SURVEY FORM

The survey form is shown in Appendix A of Part I [1]. The information reported in this paper came from 1) card type 3, columns 25, 53, and 58; 2) card type 2, column 33; and 3) card type 1, columns 9–11 and 13. The definition of *failure* is given in Part I.

RESPONSE TO SURVEY

A total of 30 companies responded to the survey questionnaire, reporting data covering 68 plants in nine industries in the United States and Canada. For the purpose of reporting results in this paper, Part V, the number of industries were reduced from nine down to five plus an "all other" category. The five industries selected were the ones for which equipment failure rate data were reported in Tables 3 through 19, Part I. All of the remaining industries were combined into an "all other" category in Tables 61–63 on plant climate, plant atmosphere, and plant operating schedule.

DISCUSSION—FOREWARNING FOR PUBLIC UTILITY POWER INTERRUPTION

Only 3 percent of the time was a failure forewarning given for a public utility power interruption to the industrial plant. Data from Table 3, Part I, and Table 57, Part V, indicate that a large percentage of these interruptions were on double- or triple-circuit supplies. Forewarning can be important to plants with a single circuit. It can also be important to plants containing a double circuit with manual switchover.

TABLE 58 - PERCENT PRODUCTION LOST

Col. 53 Card Type 3 / Percent Production Lost	CABLE TERMINATIONS %	CABLE JOINTS %	CABLE %	OPEN WIRE %	BUS DUCT %	SWITCHGEAR BUS-BARE %	SWITCHGEAR BUS-INSULATED %	DISCONNECT SWITCHES %	GENERATORS %	MOTORS %	MOTOR STARTERS %	CIRCUIT BREAKERS %	TRANSFORMERS %	ELECTRIC UTILITY POWER SUPPLIES %
0 None	47	33	28	62	30	17	20	20	80	24	85	19	22	41
1 0-30 Percent	35	58	60	25	55	33	60	75	5	73	13	73	63	32
2 Above 30 Percent	18	9	13	13	15	50	20	5	15	3	2	8	15	27
Total Percent	100	100	101	100	100	100	100	100	100	100	100	100	100	100
Total Failures Reported	51	45	223	108	20	24	20	101	85	561	168	177	101	202

TABLE 59 – METHOD OF SERVICE RESTORATION

Col. 58 Card Type 3

Give method of restoring service to plant:
1 Primary selection – manual
2 Primary selection – automatic
3 Secondary selection – manual
4 Secondary selection – automatic
5 Network protector operation – automatic
6 Repair of failed component
7 Replacement of failed component with spare
8 Utility restored service
9 Other – explain in remarks

Method	TOTAL %	ELECTRIC UTILITY POWER SUPPLIES %	TRANSFORMERS %	CIRCUIT BREAKERS %	MOTOR STARTERS %	MOTORS %	GENERATORS %	DISCONNECT SWITCHES %	SWITCHGEAR BUS – INSULATED %	SWITCHGEAR BUS – BARE %	BUS DUCT %	OPEN WIRE %	CABLE %	CABLE JOINTS %	CABLE TERMINATIONS %
1 Primary selection – manual	7	8	3	6	0	5	20	0	58	25	20	13	14	28	19
2 Primary selection – automatic	2	1	0	1	0	0	0	0	0	5	0	4	5	8	0
3 Secondary selection – manual	11	1	25	6	0	14	33	0	17	10	10	2	20	32	23
4 Secondary selection – automatic	2	0	3	8	0	0	0	0	0	0	0	1	0	8	4
5 Network protector operation – automatic	0+	0	0	0	0	0	0	0	0	5	0	0	0	0	0
6 Repair of failed component	22	5	25	11	12	30	20	3	17	20	35	31	42	24	27
7 Replacement of failed component with spare	22	2	39	38	10	29	14	77	0	10	35	6	2	0	12
8 Utility restored service	12	81	0	1	0	0	13	0	0	0	0	1	1	0	0
9 Other – explain in remarks	22	1	5	29	78	22	0	20	8	25	0	42	16	0	15
Total Percent	100	100	100	100	100	100	100	100	100	100	100	100	100	100	100
TOTAL NUMBER REPORTED	1204	171	75	160	68	318	15	69	12	20	20	103	122	25	26

TABLE 60 – AVERAGE AGE OF ELECTRICAL EQUIPMENT

NUMBER OF INSTALLED UNITS

Age:
1 Less than 1 year old
2 1-10 years od
3 More than 10 years old

Age	TRANSFORMERS	CIRCUIT BREAKERS	MOTOR STARTERS	MOTORS	GENERATORS	DISCONNECT SWITCHES	SWITCHGEAR BUS – INSULATED	SWITCHGEAR BUS – BARE	BUS DUCT	OPEN WIRE	CABLE	CABLE JOINTS	CABLE TERMINATIONS
1 Less than 1 year old	9	989	101	104	0	0	0	0	0	30	15	0	12
2 1-10 years od	694	3691	3162	1884	9	909	646	1998	1206	12	1019	1385	3314
3 More than 10 years old	835	1944	608	3643	77	552	691	555	13640	472	1831	2338	5712

TABLE 61 – PLANT CLIMATE (for entire plant site)
TABLE 62 – PLANT ATMOSPHERE (for entire plant site)

Table, Title, Card-Type 1 Column No.

TABLE 61 – PLANT CLIMATE (Col. 9)

Average of Daily Maximums for Hottest Month

Temperature	Relative Humidity (RH) (measured at noon to 2 PM ST)	ALL INDUSTRY	CHEMICAL	METAL	PETROLEUM	RUBBER AND PLASTICS	TEXTILE	ALL OTHER
1 Hot (>90F)	High (>55 RH)	14	8	1	3	0	1	1
2 Hot (>90F)	Moderate (50-55 RH)	3	3	0	0	0	0	0
3 Hot (>90F)	Low (<50 RH)	12	0	0	0	0	0	12
4 Moderate (80-90F)	High (>55 RH)	14	4	1	2	0	0	7
5 Moderate (80-90F)	Moderate (50-55RH)	16	5	1	0	1	0	8
6 Moderate (80-90F)	Low (<50 RH)	6	1	0	1	2	1	1
7 Low (<80F)	High (>55 RH)	1	0	0	0	0	1	1
8 Low (<80F)	Moderate (50-55 RH)	2	0	0	2	0	0	0
9 Low (<80F)	Low (<50 RH)	0	0	0	0	0	0	0

(NUMBER OF PLANTS)

TABLE 62 – PLANT ATMOSPHERE (Col. 10)

Card-Type 1 Column No.	ALL INDUSTRY	CHEMICAL	METAL	PETROLEUM	RUBBER AND PLASTICS	TEXTILE	ALL OTHER
1 Clean to slightly polluted air	34	2	1	7	0	2	22
2 With salt spray and corrosive chemicals	5	4	0	1	0	0	0
3 With salt spray and dust or sand	0	0	0	0	0	0	0
4 With salt spray only	0	0	0	0	0	0	0
5 With corrosive chemicals and dust or sand	13	8	0	0	1	1	3
6 With corrosive chemicals only	4	4	0	0	0	0	0
7 With dust or sand only	2	0	0	0	0	0	2
8 With conductive dust	5	0	2	0	2	0	1
9 Other	1	0	0	0	0	0	1

(NUMBER OF PLANTS)

TABLE 63 - PLANT OPERATING SCHEDULE

ALL INDUSTRY	CHEMICAL	METAL	PETROLEUM	RUBBER AND PLASTICS	TEXTILE	ALL OTHER	Title, Card-Type 1 Column No.
NUMBER OF PLANTS							HOURS PER DAY (Col. 11)
0	0	0	0	0	0	0	Less than 8
9	2	0	1	0	0	6	8
0	0	0	0	0	0	0	9 to 15
19	0	2	0	0	0	17	16
0	0	0	0	0	0	0	17 to 23
40	19	1	7	3	3	7	24
							DAYS PER WEEK (Col. 13)
0	0	0	0	0	0	0	Less than 5
30	1	2	1	2	0	24	5
3	1	0	0	0	0	2	6
35	19	1	7	1	3	4	7

DISCUSSION—PERCENT PRODUCTION LOST

The most severe category of failure in an industrial plant is where above 30 percent of the production is lost. Data from Table 58 show that the following percent of equipment class failures resulted in losing above 30 percent of the production.

Switchgear bus—bare	50 percent
Electric utility power supplies	27 percent
Switchgear bus—insulated	20 percent
Cable terminations	18 percent
Bus duct	15 percent
Transformers	15 percent
Generators	15 percent
Open wire	13 percent
Cable	13 percent
Cable joints	9 percent
Circuit breakers	8 percent
Motors	3 percent
Motor starters	2 percent

It can be seen that failures of switchgear bus and electric utility power supplies often result in losing above 30 percent of the production.

DISCUSSION—METHOD OF SERVICE RESTORATION

The data on method of electrical service restoration to plant is shown in Table 59. A percentage breakdown of the total shows the following results.

Replacement of failed component with spare	22 percent
Repair of failed components	22 percent
Other	22 percent
Utility service restored	12 percent
Secondary selection—manual	11 percent
Primary selection—manual	7 percent
Primary selection—automatic	2 percent
Secondary selection—automatic	2 percent
Network protector operation—automatic	0+ percent

The most common methods of service restoration are replacement of failed component with a spare or repair of failed component. Only 22 percent of the time is primary selection or secondary selection used; this would indicate that most power distribution systems are radial.

DISCUSSION—AVERAGE AGE OF ELECTRICAL EQUIPMENT

Many respondents to the reliability survey of industrial plants submitted data covering a ten-year period. Thus it is not surprising to see that Table 60 shows a large population that is more than ten years old. The following percent of installed units are classified as more than ten years old.

Bus duct	92 percent
Open wire	92 percent
Generators	90 percent
Motors	65 percent
Cable	64 percent
Cable joints	63 percent
Cable terminations	63 percent
Transformers	54 percent
Switchgear bus—insulated	52 percent

Motor starters, disconnect switches, switchgear bus—bare, and circuit breakers had over 50 percent of the installed units one to ten years old.

15 percent of the circuit breakers were less than one year old. All other equipment classes had less than 6 percent of the installed units less than a year old.

DISCUSSION—PLANT CLIMATE AND ATMOSPHERE

Data on plant climate and plant atmosphere are given in Tables 61 and 62. 43 percent of the plants were in a hot climate, 53 percent in a moderate climate, and only 4 percent in a low climate (cold climate). 43 percent of the plants had high relative humidity, 31 percent had moderate relative humidity, and 26 percent had low rela-

tive humidity. 53 percent of the plants had a plant atmosphere classified as "clean to slightly polluted air." The other 47 percent had an atmosphere with some contamination.

DISCUSSION—PLANT OPERATING SCHEDULE

The data on plant operating schedule are given in Table 63. 52 percent of the plants operated 7 days per week, 4 percent for 6 days, and 44 percent for 5 days. 59 percent of the plants operated 24 h per day, 28 percent for 16 h, and 13 percent for 8 h.

REFERENCES

[1] W. H. Dickinson et al., "Report on reliability survey of industrial plants, part I: Reliability of electrical equipment," IEEE Trans. Ind. Appl., vol. IA-10, pp. 213–235, Mar./Apr. 1974.
[2] W. H. Dickinson et al., "Report on reliability survey of industrial plants, part II: Cost of power outages, plant restart time, critical service loss duration time, and type of loads lost versus time of power outages," IEEE Trans. Ind. Appl., vol. IA-10, pp. 236–241, Mar./Apr. 1974.
[3] W. H. Dickinson et al., "Report on reliability survey of industrial plants, part III: Causes and types of failures of electrical equipment, the methods of repair, and the urgency of repair," IEEE Trans. Ind. Appl., vol. IA-10, pp. 242–249, Mar./Apr. 1974.
[4] A. D. Patton et al., "Report on reliability survey of industrial plants, part IV: Additional detailed tabulation of some data previously reported in the first three parts," this issue, pp. 456–462.
[5] A. D. Patton et al., "Report on reliability survey of industrial plants, part VI: Maintenance quality of electrical equipment," this issue, pp. 467–476.

Report on Reliability Survey of Industrial Plants, Part VI: Maintenance Quality of Electrical Equipment

IEEE COMMITTEE REPORT

Abstract—An IEEE sponsored reliability survey of industrial plants was completed during 1972. This included maintenance quality, the frequency of schedule maintenance, and the failures caused by inadequate maintenance. The results are reported from the survey of 30 companies covering 68 plants in nine industries in the United States and Canada. This information is useful in the design of industrial power distribution systems.

INTRODUCTION

A KNOWLEDGE of maintenance quality of electrical equipment in industrial plants is useful information when planning the maintenance program of industrial power distribution systems. In addition it is useful to know how this correlates with the normal maintenance cycle and the failures blamed on inadequate maintenance. During 1972 the Reliability Subcommittee of the Industrial and Commercial Power Systems Committee completed a reliability survey of industrial plants. This paper presents Part VI of the results from the survey. The first three parts [1]–[3] were published previously. Table 38 from Part III reported that inadequate maintenance was blamed for between 8 to 30 percent of the failures of electrical equipment. This information has caused the Reliability Subcommittee to make a further study of the failure data; the results from this study are being reported in this paper. Included in Part VI are the results for 12 main classes of electrical equipment on

1) equipment population versus a) maintenance quality and b) normal maintenance cycle;
2) failures due to all causes versus a) failure, months since maintained, and b) maintenance quality;
3) failures due to inadequate maintenance versus a) failure, months since maintained, and b) maintenance quality.

The "maintenance quality" is an opinion that was reported by each participant in the survey. The four classifications used were "excellent," "fair," "poor," and "none." The "normal maintenance" cycle is the frequency of performing preventive maintenance.

Paper TOD-74-33, approved by the Industrial and Commercial Power Systems Committee of the IEEE Industry Applications Society for presentation at the 1974 Industrial and Commercial Power Systems Technical Conference, Denver, Colo., June 2–6. Manuscript released for publication April 15, 1974.
Members of the Reliability Subcommittee of the IEEE Industrial and Commercial Power Systems Committee are A. D. Patton, *Chairman*, C. E. Becker, W. H. Dickinson, P. E. Gannon, C. R. Heising, D. W. McWilliams, R. W. Parisian, and S. Wells.

SURVEY FORM

The survey form is shown in Appendix A of Part I [1]. The information reported in this paper came from 1) card type 2, col. 34 (maintenance, normal cycle); 2) card type 2, col. 36 (maintenance quality); 3) card type 3, col. 34 (failure, months since maintained); 4) card type 3, col. 40 (suspected failure responsibility). The definition of *failure* is given in Part I.

RESPONSE TO SURVEY

A total of 30 companies responded to the survey questionnaire, reporting data from nine industries in the United States and Canada. Every plant did not report all the information called for in card type 2, columns 34 and 36. Every failure report did not have filled out all of the information called for in card type 3, columns 34 and 40; a total of 1469 failures had this information filled in and are reported here in Part VI, and 240 of these failures were blamed on inadequate maintenance. Differences in the number of failures and unit-years reported here in Part VI and those previously reported in Part I and Part III can be explained from the preceding.

STATISTICAL ANALYSIS

The subject of statistical analysis of equipment failures is discussed in Part I [1]. Confidence limits for the failure rate are shown in Fig. 1 of Part I. The Reliability Subcommittee concluded that eight failures is an adequate sample size for reporting failure rates in the summary in Table 2, Part I. In a few cases, failure rate data were reported in Tables 3 through 19, Part I, where there were less than eight failures.

In this paper several cases are reported in Tables 67 through 78, where the failure rate contains less than eight failures; these cases have been marked "small sample size."

SURVEY RESULTS

Results are tabulated for 12 main equipment classes in Table 64 where the equipment population is given versus 1) maintenance quality and 2) normal maintenance cycle.

Table 65 summarizes the percent of each electrical equipment class population versus the maintenance quality. Table 66 summarizes the percent of each electrical equipment class population versus the normal maintenance cycle.

Results are tabulated for each of the 12 main equipment classes in Tables 67 through 78, where the number of failures is given for 1) failures due to all causes and 2)

Correction to "Report on Reliability Survey of Industrial Plants, Part VI: Maintenance Quality of Electrical Equipment"

IEEE COMMITTEE REPORT

TABLE 64 - POPULATION OF ELECTRICAL EQUIPMENT
VERSUS MAINTENANCE QUALITY & NORMAL MAINTENANCE CYCLE

MAINTENANCE QUALITY Card-Type 2 Col. 36	Less Than 12 Months	12-24 Months	More Than 24 Months	No Preventive Maintenance	Total
			MAINTENANCE, NORMAL CYCLE Card-Type 2 Col. 34 — Population: Unit-Years		
SWITCHGEAR BUS - INSULATED**					
Excellent	0	364	12,160	0	12,524
Fair	0	1,706	0	0	1,706
Poor	0	0	0	0	0
None	0	0	0	1,541	1,541
Total	0	2,070	12,160	1,541	15,771
SWITCHGEAR BUS - BARE**					
Excellent	0	1,854	27,580	0	29,434
Fair	0	19,440	2,826	0	22,266
Poor	0	769	0	0	769
None	0	0	0	369	369
Total	0	22,063	30,406	369	52,838
OPEN WIRE (Unit = 1,000 Circuit Feet)					
Excellent	0	2,217	1,014	0	3,231
Fair	0	103	2,630	0	2,733
Poor	0	0	0	0	0
None	0	0	0	680	680
Total	0	2,320	3,644	680	6,644
CABLE (Unit = 1000 Circuit Feet)					
Excellent	600	329	400	0	1,329
Fair	7	7,900	8,519	135	16,561
Poor	0	23	563	35	621
None	0	0	203	9,920	10,123
Total	607	8,252	9,685	10,090	28,634
CABLE JOINTS					
Excellent	0	9,374	311	0	9,685
Fair	12	2,800	23,530	0	26,342
Poor	0	0	1,483	0	1,483
None	0	0	0	12,110	12,110
Total	12	12,174	25,324	12,110	49,620
CABLE TERMINATIONS					
Excellent	2,500	14,290	15,650	0	32,440
Fair	0	1,452	35,200	1,170	37,822
Poor	0	0	0	845	845
None	0	0	0	54,280	54,280
Total	2,500	15,742	51,695	55,450	125,387

**Unit - Number of Connected Circuit Breakers or Instrument Transformer Compartments

TABLE 64 - POPULATION OF ELECTRICAL EQUIPMENT
VERSUS MAINTENANCE QUALITY & NORMAL MAINTENANCE CYCLE

MAINTENANCE QUALITY Card-Type 2 Col. 36	Less Than 12 Months	12-24 Months	More Than 24 Months	No Preventive Maintenance	Total
			MAINTENANCE, NORMAL CYCLE Card-Type 2 Col. 34 — Population: Unit-Years		
TRANSFORMERS					
Excellent	19	8,904	2,314	0	11,237
Fair	292	3,081	5,961	0	9,334
Poor	0	130	210	0	340
None	0	0	0	39	39
Total	311	12,115	8,485	39	20,950
CIRCUIT BREAKERS					
Excellent	297	11,640	5,014	0	16,951
Fair	1	12,620	11,860	0	24,481
Poor	0	0	1,810	0	1,810
None	0	0	0	7,608	7,608
Total	298	24,260	18,684	7,608	50,850
MOTOR STARTERS					
Excellent	126	2,724	0	0	2,850
Fair	68	4,348	3,435	0	7,851
Poor	0	680	427	0	1,107
None	0	0	0	70	70
Total	194	7,752	3,862	70	11,878
MOTORS					
Excellent	14,650	1,372	1,259	17	17,298
Fair	121	21,930	2,958	0	25,009
Poor	0	0	74	70	144
None	0	0	0	13	13
Total	14,771	23,302	4,291	100	42,464
GENERATORS					
Excellent	104.4	380.7	0	0	485.1
Fair	74.4	279.8	354.2	0	354.2
Poor	0	0	0	0	0
None	0	0	0	0	0
Total	178.8	660.5	0	0	839.3
DISCONNECT SWITCHES					
Excellent	0	6,287	1,435	0	7,722
Fair	58	426	2,642	0	3,126
Poor	0	402	0	0	402
None	0	0	0	7,365	7,365
Total	58	7,115	4,077	7,365	18,615

(see pp. 681 for the second part of Table 64)

failures due to inadequate maintenance, versus 1) failure, months since maintained, and 2) maintenance quality. Failure rate calculations are also given in Tables 67 through 78; these calculations used the population data from Table 64.

Table 79 summarizes the number of failures for all equipment classes combined versus the maintenance quality. Table 80 summarizes the number of failures for all equipment classes combined versus the months since maintained.

GENERAL CONCLUSIONS—MAINTENANCE QUALITY

The maintenance quality is an opinion that was reported by each participant in the survey. The major portion of the electrical equipment population in the survey had a maintenance quality that was classified as excellent or fair. Less than 5 percent of the population in each equipment class (except for motor starters) were classified as poor. Four equipment categories had between 24 percent to 43 percent of the population classified as "none" under maintenance quality; this included cable termination (43 percent), disconnect switches (40 percent), cable (35 percent), and cable joints (24 percent).

Maintenance quality had a significant effect on the percent of all failures that were blamed on inadequate maintenance. In the "poor" category 33 percent of all failures were blamed on inadequate maintenance. This compares with 18 percent for fair maintenance and 12 percent for excellent maintenance. The "none" category for maintenance quality also had 12 percent of all failures blamed on inadequate maintenance; but 82 percent of these failures were for equipment classes that do not require much maintenance (cable, cable terminations, cable joints,

TABLE 65 - PERCENT OF ELECTRICAL EQUIPMENT
POPULATION VERSUS MAINTENANCE QUALITY
(All Data Taken from Table 64)

MAINTENANCE QUALITY Card-Type 2 Col. 36	TRANSFORMERS	CIRCUIT BREAKERS	MOTOR STARTERS	MOTORS	GENERATORS	DISCONNECT SWITCHES	SWITCHGEAR BUS-INSULATED	SWITCHGEAR BUS-BARE	OPEN WIRE	CABLE	CABLE JOINTS	CABLE TERMINATIONS
	%	%	%	%	%	%	%	%	%	%	%	%
Excellent	54	33	24	41	58	41	79	56	49	5	20	26
Fair	44	48	66	59	42	17	11	42	41	58	53	30
Poor	2	4	10	0+	0	2	0	1	0	2	3	1
None	0+	15	0	0+	0	40	10	1	10	35	24	43
Total	100	100	100	100	100	100	100	100	100	100	100	100

TABLE 66 - PERCENT OF ELECTRICAL EQUIPMENT
POPULATION VERSUS NORMAL MAINTENANCE CYCLE
(All Data Taken from Table 64)

MAINTENANCE, NORMAL CYCLE Card-Type 2 Col. 34	TRANSFORMERS	CIRCUIT BREAKERS	MOTOR STARTERS	MOTORS	GENERATORS	DISCONNECT SWITCHES	SWITCHGEAR BUS-INSULATED	SWITCHGEAR BUS-BARE	OPEN WIRE	CABLE	CABLE JOINTS	CABLE TERMINATIONS
	%	%	%	%	%	%	%	%	%	%	%	%
Less than 12 Months	1	1	2	35	21	0+	0	0	0	2	0+	2
12-24 Months	58	47	65	55	79	38	13	42	35	29	25	13
More than 24 Months	41	37	32	10	0	22	77	57	55	34	51	41
No Preventive Maintenance	0+	15	1	0+	0	40	10	1	10	35	24	44
Total	100	100	100	100	100	100	100	100	100	100	100	100

TABLE 67 - NUMBER OF TRANSFORMER
FAILURES VERSUS MONTHS SINCE MAINTAINED AND MAINTENANCE QUALITY

MAINTENANCE QUALITY Card-Type 2 Col. 36	FAILURE, MONTHS SINCE MAINTAINED Card-Type 3, Col. 34					Failures per Unit-Year ALL CAUSES
	Less Than 12 Months Ago	12 - 24 Months Ago	More Than 24 Months Ago	No Preventive Maintenance	Total	
	Number of Failures Due to ALL CAUSES					
Excellent	22	11	5	0	38	
Fair	10	26	16	1	53	
Poor	2	1	1	1	5	
None	0	0	0	3	3	
Total	34	38	22	5	99	.00473
	Number of Failures Due to INADEQUATE MAINTENANCE (Card-Type 3 Col. 40)					INADEQUATE MAINTENANCE
Excellent	0	1	2	0	3	.00027*
Fair	1	0	6	0	7	.00075*
Poor	0	0	0	1	1	.00294*
None	0	0	0	0	0	.00000*
Total	1	1	8	1	11	.00053

* Small Sample Size

TABLE 68 - NUMBER OF CIRCUIT BREAKER
FAILURES VERSUS MONTHS SINCE MAINTAINED AND MAINTENANCE QUALITY

MAINTENANCE QUALITY Card-Type 2 Col. 36	FAILURE, MONTHS SINCE MAINTAINED Card-Type 3, Col. 34					Failures per Unit-Year ALL CAUSES
	Less Than 12 Months Ago	12 -24 Months Ago	More Than 24 Months Ago	No Preventive Maintenance	Total	
	Number of Failures Due to ALL CAUSES					
Excellent	13	60	3	1	77	
Fair	18	42	4	1	65	
Poor	0	2	2	0	4	
None	1	0	0	26	27	
Total	32	104	9	28	173	.00340
	Number of Failures Due to INADEQUATE MAINTENANCE (Card-Type 3 Col. 40)					INADEQUATE MAINTENANCE
Excellent	2	1	3	1	7	.00041*
Fair	2	18	2	0	22	.00090
Poor	0	1	2	0	3	.00166*
None	0	0	0	4	4	.00053*
Total	4	20	7	5	36	.00071

* Small Sample Size

TABLE 69 - NUMBER OF MOTOR STARTER
FAILURES VERSUS MONTHS SINCE MAINTAINED AND MAINTENANCE QUALITY

MAINTENANCE QUALITY Card-Type 2 Col. 36	FAILURE, MONTHS SINCE MAINTAINED Card-Type 3 Col. 34					Failures per Unit-Year ALL CAUSES
	Less Than 12 Months Ago	12 - 24 Months Ago	More Than 24 Months Ago	No Preventive Maintenance	Total	
	Number of Failures Due to ALL CAUSES					
Excellent	13	1	4	0	18	
Fair	45	13	8	0	66	
Poor	1	1	2	0	4	
None	0	0	0	0	0	
Total	59	15	14	0	88	.00741
	Number of Failures Due to INADEQUATE MAINTENANCE (Card-Type 3 Col. 40)					INADEQUATE MAINTENANCE
Excellent	1	0	0	0	1	.00035*
Fair	0	1	3	0	4	.00051*
Poor	1	0	1	0	2	.00170*
None	0	0	0	0	0	---
Total	2	1	4	0	7	.00059*

* Small Sample Size

TABLE 70 - NUMBER OF MOTOR
FAILURES VERSUS MONTHS SINCE MAINTAINED AND MAINTENANCE QUALITY

MAINTENANCE QUALITY Card-Type 2 Col. 36	FAILURE, MONTHS SINCE MAINTAINED Card-Type 3 Col. 34					Failures per Unit-Year ALL CAUSES
	Less Than 12 Months Ago	12 - 24 Months Ago	More Than 24 Months Ago	No Preventive Maintenance	Total	
	Number of Failures Due to ALL CAUSES					
Excellent	56	14	7	0	77	
Fair	58	280	90	11	439	
Poor	0	0	2	0	2	
None	0	0	0	0	0	
Total	114	294	99	11	518	.01221
	Number of Failures Due to INADEQUATE MAINTENANCE (Card-Type 3 Col. 40)					INADEQUATE MAINTENANCE
Excellent	8	1	1	0	10	.00058
Fair	2	25	41	2	70	.00280
Poor	0	0	2	0	2	.01390*
None	0	0	0	0	0	.00000*
Total	10	26	44	2	82	.00194

* Small Sample Size

TABLE 71 - NUMBER OF GENERATOR
FAILURES VERSUS MONTHS SINCE MAINTAINED AND MAINTENANCE QUALITY

MAINTENANCE QUALITY Card-Type 2 Col. 36	FAILURE, MONTHS SINCE MAINTAINED Card-Type 3 Col. 34					Failures per Unit-Year ALL CAUSES
	Less Than 12 Months Ago	12 - 24 Months Ago	More Than 24 Months Ago	No Preventive Maintenance	Total	
	Number of Failures Due to ALL CAUSES					
Excellent	14	9	0	0	23	
Fair	1	4	0	0	5	
Poor	0	0	0	0	0	
None	0	0	0	0	0	
Total	15	13	0	0	28	.03360
	Number of Failures Due to INADEQUATE MAINTENANCE (Card-Type 3 Col. 40)					INADEQUATE MAINTENANCE
Excellent	3	0	0	0	3	.00618*
Fair	0	2	0	0	2	.00565*
Poor	0	0	0	0	0	
None	0	0	0	0	0	
Total	3	2	0	0	5	.00596*

* Small Sample Size

TABLE 72 - NUMBER OF DISCONNECT SWITCH
FAILURES .VERSUS MONTHS SINCE MAINTAINED AND MAINTENANCE QUALITY

MAINTENANCE QUALITY Card-Type 2 Col. 36	FAILURE, MONTHS SINCE MAINTAINED Card-Type 3 Col. 34					Failures per Unit-Year ALL CAUSES
	Less Than 12 Months Ago	12 - 24 Months Ago	More Than 24 Months Ago	No Preventive Maintenance	Total	
	Number of Failures Due to ALL CAUSES					
Excellent	4	0	1	0	5	
Fair	4	5	4	0	13	
Poor	0	0	16	·0	16	
None	0	0	0	67	67	
Total	8	5	21	67	101	.00542
	Number of Failures Due to INADEQUATE MAINTENANCE (Card-Type 3 Col. 40)					INADEQUATE MAINTENANCE
Excellent	0	0	1	0	1	.00013*
Fair	0	4	1	0	5	.00160*
Poor	0	0	0	0	0	.00000*
None	0	0	0	7	7	.00095*
Total	0	4	2	7	13	.00070

* Small Sample Size

TABLE 73 - NUMBER OF SWITCHGEAR BUS-INSULATED
FAILURES VERSUS MONTHS SINCE MAINTAINED AND MAINTENANCE QUALITY

MAINTENANCE QUALITY Card-Type 2 Col. 36	FAILURE, MONTHS SINCE MAINTAINED Card-Type 3 Col. 34					Failures per **Unit-Year ALL CAUSES
	Less Than 12 Months Ago	12 - 24 Months Ago	More Than 24 Months Ago	No Preventive Maintenance	Total	
	Number of Failures Due to ALL CAUSES					
Excellent	2	3	10	0	15	
Fair	0	4	1	0	5	
Poor	0	0	0	0	0	
None	0	0	0	0	0	
Total	2	7	11	0	20	.00127
	Number of Failures Due to INADEQUATE MAINTENANCE (Card-Type 3 Col. 40)					INADEQUATE MAINTENANCE
Excellent	0	0	6	0	6	.00048*
Fair	0	0	1	0	1	.00059*
Poor	0	0	0	0	0	
None	0	0	0	0	0	.00000*
Total	0	0	7	0	7	.00044*

* Small Sample Size
**Unit = Number of Connected Circuit Breakers or Instrument Transformer Compartments

TABLE 74 - NUMBER OF SWITCHGEAR BUS-BARE
FAILURES VERSUS MONTHS SINCE MAINTAINED AND MAINTENANCE QUALITY

MAINTENANCE QUALITY Card-Type 2 Col. 36	FAILURE, MONTHS SINCE MAINTAINED Card-Type 3 Col. 34					Failures per **Unit-Year ALL CAUSES
	Less Than 12 Months Ago	12 - 24 Months Ago	More Than 24 Months Ago	No Preventive Maintenance	Total	
	Number of Failures Due to ALL CAUSES					
Excellent	2	1	1	0	4	
Fair	4	6	2	2	14	
Poor	2	0	0	0	2	
None	0	0	0	3	3	
Total	8	7	3	5	23	.00044
	Number of Failures Due to INADEQUATE MAINTENANCE (Card-Type 3 Col. 40)					INADEQUATE MAINTENANCE
Excellent	0	0	0	0	0	.00000*
Fair	1	1	2	0	4	.00018*
Poor	0	0	0	0	0	.00000*
None	0	0	0	1	1	.00271*
Total	1	1	2	1	5	.00009*

* Small Sample Size
**Unit = Number of Connected Circuit Breakers or Instrument Transformer Compartments

TABLE 75 - NUMBER OF OPEN WIRE
FAILURES VERSUS MONTHS SINCE MAINTAINED AND MAINTENANCE QUALITY

MAINTENANCE QUALITY Card-Type 2 Col. 36	FAILURE, MONTHS SINCE MAINTAINED Card-Type 3 Col. 34					Failures per **Unit-Year ALL CAUSES
	Less Than 12 Months Ago	12 - 24 Months Ago	More Than 24 Months Ago	No Preventive Maintenance	Total	
	Number of Failures Due to ALL CAUSES					
Excellent	0	1	3	0	4	
Fair	1	8	85	0	94	
Poor	0	0	0	0	0	
None	0	0	0	10	10	
Total	1	9	88	10	108	.01628
	Number of Failures Due to INADEQUATE MAINTENANCE (Card-Type 3 Col. 40)					INADEQUATE MAINTENANCE
Excellent	0	1	1	0	2	.00062*
Fair	0	1	30	0	31	.01132
Poor	0	0	0	0	0	*
None	0	0	0	0	0	.00000*
Total	0	2	31	0	33	.00497

* Small Sample Size
** Unit = 1,000 Circuit Feet

TABLE 76 - NUMBER OF CABLE
FAILURES VERSUS MONTHS SINCE MAINTAINED AND MAINTENANCE QUALITY

MAINTENANCE QUALITY Card-Type 2 Col. 36	FAILURE, MONTHS SINCE MAINTAINED Card-Type 3 Col. 34					Failures per **Unit-Year ALL CAUSES
	Less Than 12 Months Ago	12 - 24 Months Ago	More Than 24 Months Ago	No Preventive Maintenance	Total	
	Number of Failures Due to ALL CAUSES					
Excellent	5	6	2	21	34	
Fair	18	19	16	6	59	
Poor	0	3	2	21	26	
None	0	0	2	95	97	
Total	23	28	22	143	216	.00755
	Number of Failures Due to INADEQUATE MAINTENANCE (Card-Type 3 Col. 40)					INADEQUATE MAINTENANCE
Excellent	0	0	0	0	0	.00000*
Fair	0	2	0	0	2	.00012*
Poor	0	0	2	6	8	.01290
None	0	0	0	12	12	.00119
Total	0	2	2	18	22	.00077

* Small Sample Size
** Unit = 1,000 Circuit Feet

TABLE 77 - NUMBER OF CABLE JOINT
FAILURES VERSUS MONTHS SINCE MAINTAINED AND MAINTENANCE QUALITY

MAINTENANCE QUALITY Card-Type 2 Col. 36	FAILURE, MONTHS SINCE MAINTAINED Card-Type 3 Col. 34					Failures per Unit-Year ALL CAUSES
	Less Than 12 Months Ago	12 - 24 Months Ago	More Than 24 Months Ago	No Preventive Maintenance	Total	
	Number of Failures Due to ALL CAUSES					
Excellent	2	4	0	0	6	
Fair	6	5	1	5	17	
Poor	0	0	0	7	7	
None	0	0	0	15	15	
Total	8	9	1	27	45	.00091
	Number of Failures Due to INADEQUATE MAINTENANCE (Card-Type 3 Col. 40)					INADEQUATE MAINTENANCE
Excellent	0	0	0	0	0	.00000*
Fair	1	0	0	0	1	.00004*
Poor	0	0	0	6	6	.00405*
None	0	0	0	1	1	.00008*
Total	1	0	0	7	8	.00016

* Small Sample Size

TABLE 78 - NUMBER OF CABLE TERMINATION
FAILURES VERSUS MONTHS SINCE MAINTAINED AND MAINTENANCE QUALITY

MAINTENANCE QUALITY Card-Type 2 Col. 36	FAILURE, MONTHS SINCE MAINTAINED Card-Type 3 Col. 34					Failures per Unit-Year ALL CAUSES
	Less Than 12 Months Ago	12 - 24 Months Ago	More Than 24 Months Ago	No Preventive Maintenance	Total	
	Number of Failures Due to ALL CAUSES					
Excellent	3	3	4	0	10	
Fair	3	3	14	3	23	
Poor	0	0	0	1	1	
None	0	0	0	16	16	
Total	6	6	18	20	50	.00040
	Number of Failures Due to INADEQUATE MAINTENANCE (Card-Type 3 Col. 40)					INADEQUATE MAINTENANCE
Excellent	1	1	1	0	3	.00009*
Fair	0	0	5	0	5	.00013*
Poor	0	0	0	0	0	.00000*
None	0	0	0	3	3	.00006*
Total	1	1	6	3	11	.00008

* Small Sample Size

TABLE 79 - NUMBER OF FAILURES VERSUS
MAINTENANCE QUALITY FOR ALL EQUIPMENT
CLASSES COMBINED

MAINTENANCE QUALITY Card-Type 2 Col. 36	Number of Failures in Tables 67 thru 78		PERCENT of Failures Due to Inadequate Maintenance
	ALL CAUSES	INADEQUATE MAINTENANCE	
Excellent	311	36	11.6%
Fair	853	154	18.1%
Poor	67	22	32.8%
None	238	28	11.8%
Total	1,469	240	16.4%

TABLE 80 - NUMBER OF FAILURES VERSUS
MONTHS SINCE MAINTAINED FOR ALL
EQUIPMENT CLASSES COMBINED

FAILURE, MONTHS SINCE MAINTAINED Card-Type 3, Col. 34	Number of Failures in Tables 67 thru 78		PERCENT of Failures Due to Inadequate Maintenance
	ALL CAUSES	INADEQUATE MAINTENANCE	
Less than 12 Months Ago	310	23	7.4%
12-24 Months Ago	535	60	11.2%
More Than 24 Months Ago	308	113	36.7%
No Preventive Maintenance	316	44	13.9%
Total	1,469	240	16.4%

and disconnect switches). Thus this 12 percent for "none" is explainable and is not inconsistent with what could be expected.

As maintenance quality decreases from "excellent" to "fair" to "poor," the following equipment classes showed an increasing failure rate from failures blamed on inadequate maintenance: transformers, circuit breakers, motor starters, motors, disconnect switches, switchgear bus—bare, open wire, cable, and cable joints. In some of these cases the sample size is smaller than desirable (less than eight failures) in order to conclusively prove this general statement.

OTHER CONCLUSIONS

Circuit Breakers

Approximately 15 percent of the circuit breaker population had a maintenance quality classified as "none." This compares with less than 1 percent of the population for transformers, motors, and generators.

It is of interest to note that data from Table 60, Part V also show that 15 percent of the circuit breaker population was less than one year old; this compares with less than 3 percent of the population for transformers, motors, and generators. This may possibly account for some of the listings of "none" under maintenance quality reported for failures of circuit breakers.

Motors

Motors with a maintenance quality of "fair" had a failure rate due to inadequate maintenance that was five times higher than motors with excellent maintenance quality.

Open Wire

Open wire with a maintenance quality of "fair" had a failure rate due to inadequate maintenance that was more than ten times higher than open wire with excellent maintenance quality.

DISCUSSION—MAINTENANCE QUALITY

From Table 79 it is possible to calculate for all equipment classes combined the ratio of the number of failures from inadequate maintenance to the number of failures from all other causes. This ratio versus maintenance quality is as follows: poor—0.49, fair—0.22, excellent—

0.13. This is a measure of how much improvement can be obtained by upgrading the maintenance quality from poor to fair to excellent. An excellent maintenance program has only 13 percent more failures added by inadequate maintenance, while a poor maintenance program has 49 percent more failures added by inadequate maintenance.

It is apparent from the data that excellent maintenance quality is very important on open wire and on motors.

It would also appear from the data in Table 65 that essentially everyone in the survey did excellent or fair maintenance on transformers, generators, and switchgear bus—bare. However, on circuit breakers 15 percent of the population had "none" and 4 percent had "poor" on maintenance quality. On motor starters 10 percent had "poor" on maintenance quality. Thus, it would appear that everyone did not maintain circuit breakers and motor starters as well as transformers, generators, and switchgear bus—bare.

One of the drawbacks to the results reported under maintenance quality was that there was no objective definition of "excellent," "fair," or "poor." There are no standards for maintenance quality, and thus this data must be considered to be individual judgment. However, data reported under "failure, months since maintained" does not have this same drawback; this data can be considered factual.

DISCUSSION—FAILURE, MONTHS SINCE MAINTAINED

The data in Table 80 show for all equipment classes combined that there is a close correlation between the percent of failures due to inadequate maintenance and the failure, months since maintained.

Failure, Months Since Maintained	Percent of Failures Due to Inadequate Maintenance
Less than 12 months ago	7.4
12–24 months ago	11.2
More than 24 months ago	36.7

Data from Tables 67 through 78 can also be used to calculate similar correlations for several equipment categories; however, in some cases the sample size is smaller than desirable for adequate statistical confidence.

COMMENTS—NORMAL MAINTENANCE CYCLE

A detailed analysis has not been made of the "maintenance, normal cycle" data in Tables 64 and 66. It is possible that some interesting conclusions could also be drawn from an analysis of this data.

REFERENCES

[1] W. H. Dickinson et al., "Report on reliability survey of industrial plants, part I: Reliability of electrical equipment," IEEE Trans. Ind. Appl., vol. 1A-10, pp. 213–235, Mar./Apr. 1974.
[2] W. H. Dickinson et al., "Report on reliability survey of industrial plants, part II: Cost of power outages, plant restart time, critical service loss duration time, and type of loads lost versus time of power outages," IEEE Trans. Ind. Appl., vol. 1A-10, pp. 236–241, Mar./Apr. 1974.
[3] W. H. Dickinson et al., "Report on reliability survey of industrial plants, part III: Causes and types of failures of electrical equipment, the methods of repair, and the urgency of repair," IEEE Trans. Ind. Appl., vol. 1A-10, pp. 242–249, Mar./Apr. 1974.

Appendix C

Cost of Electrical Interruptions in Commercial Buildings

By
Power Systems Reliability Subcommittee
Industrial and Commercial Power Systems Committee
IEEE Industry Applications Society

P. E. Gannon, Coordinating Author
A. D. Patton, *Chair*

C. E. Becker	M. D. Harris	D. W. McWilliams
M. F. Chamow	C. R. Heising	R. W. Parisian
W. H. Dickinson	R. T. Kulvicki	S. J. Wells

Industrial and Commercial Power Systems Technical Conference
Institute of Electrical and Electronics Engineers, Inc.
Toronto, Canada
May 5-8, 1975

Reprinted from IEEE Conference Record
75CHO947-1-1A, pp. 123-129

COST OF ELECTRICAL INTERRUPTIONS
IN COMMERCIAL BUILDINGS

by

Power Systems Reliability Subcommittee Report
Philip E. Gannon, Coordinating Author[1]

Abstract

An IEEE sponsored reliability survey to determine the cost of electrical interruptions in commercial buildings was completed in 1974. The survey form was a simplified version of forms used in 1972 reliability study of industrial plants. The survey included building types and locations, and length and cost of electrical service interruptions. The survey results reflect data from 48 companies covering 55 buildings in the United States. This information is useful in the design of electrical systems for commercial buildings.

Introduction

Knowledge of the cost of power outages, both for normal and critical services, is useful in the design of commercial building power systems, allowing cost-effective judgements to be made with respect to the installation of a second utility company service, an emergency generator, or possibly an uninterruptible power supply.

During 1974, the Reliability Subcommittee of the Industrial and Commercial Power Systems Committee completed a survey of the cost of electrical interruptions in commercial buildings in the United States. Included in this paper are the following results:

1 Cost of power outages to commercial buildings ($ per KWH of undelivered energy).

2 Cost of power outages to commercial buildings ($ per square foot/hr and $ per employee/hr).

3 Critical service loss duration time (length of time before an interruption causes a significant loss).

5 Miscellaneous items relative to provision of auxiliary generators, types of electrical service, and other physical data.

Survey Form

The survey form is shown in Appendix A (two pages). A simple multiple choice or single line fill-in form was utilized in an attempt to reduce the time of the responders, but still provide pertinent data for a meaningful analysis.

Response to Survey

A total of 48 companies reporting on 55 buildings responded to the survey with complete data. Incomplete data, omitting the critical outage cost information was received on 121 additional buildings. Unfortunately, this data was of no value in the present survey. Valid data was submitted almost equally for buildings located in the eastern, central, and western regions of the U.S.A.; with 43 percent of the buildings in downtown areas, 17 percent in urban areas, and 40 percent in suburban areas. Forty-six percent of the buildings were used 5 days per week; 39 percent, 6 days per week; and 15 percent, 7 days per week.

Survey Data Preparation

All of the returned survey forms were reviewed. Useable data was punched onto computer cards for use in data processing.

Survey Results -- Cost of Power Outages

Each respondent was asked to report on the cost of power outages as follows:

1 Dollars per failure -- 15-minute duration, one-hour duration, and greater than one-hour duration; total value of lost operation including wages, damages for delays, loss of computer time, and loss of retail sales minus cost of goods not sold was to be included.

2 Critical service loss duration time -- length of time before an interruption causes a significant loss.

3 Building maximum power demand, and usage, as well as area and number of employees.

The data made it possible to calculate the cost of power outages in terms of dollars per kilowatt-hours of undelivered energy at building peak load.

The average cost of power outages from the survey for the buildings surveyed is given in Table 1.

TABLE 1

AVERAGE COST OF POWER OUTAGES
FOR BUILDINGS IN THE UNITED STATES

All commercial buildings	$7.21/KWH not delivered
Office buildings only	$8.86/KWH not delivered

The average maximum demand was 3,095 KW for all commercial buildings reporting outage costs. The maximum demand for the office buildings was only 3,035 KW.

Additional details of the cost of power outages are given in Tables 2, 3, and 4. The tables present additional data including:

1 Outage costs for "office buildings" as a function of duration of outage for three time periods.

2 Effect of computers on outage costs.

3 Relationship of outage costs to: KWH not delivered, to cost per 1,000 square feet per hour of building affected, and to cost per employee per hour affected.

---
[1] Other members of Sub-Committee: A.D. Patton Chairman; C.R. Heising, Vice Chairman; C.E. Becker; M.F. Chamow; W.H. Dickinson; M.D. Harris; R.T. Kulvicki; D.W. McWilliams; R.W. Parisian; Stanley Wells

TABLE 2

OUTAGE COSTS FOR "OFFICE BUILDINGS"
AS A FUNCTION OF DURATION
(WITH AND WITHOUT COMPUTERS)

	Sample Size	Maximum	Minimum	Average
15-Minute Duration				
Cost/peak KW hr. not delivered	25	$ 22.22	$ 1.50	$ 7.54
Cost/1,000 sq. ft. of bldg./hr.	26	247.6	10.5	63.8
Cost/employee/hr.	26	52.0	3.0	16.0
1-Hour Duration				
Cost/peak KW hr. not delivered	29	$ 24.93	$ 0.64	$ 6.74
Cost/1,000 sq. ft. of bldg./hr.	32	125.00	5.24	53.12
Cost/employee/hr.	32	34.30	1.25	12.22
Duration 1 Hour				
Cost/peak KW hr. not delivered	13	$100.00	$ 0.16	$ 16.16
Cost/1,000 sq. ft. of bldg./hr.	14	320.00	1.05	68.06
Cost/employee/hr.	14	75.80	0.48	16.41

TABLE 3

OUTAGE COSTS FOR "OFFICE BUILDINGS"
AS A FUNCTION OF DURATION
(WITHOUT COMPUTERS)

	Sample Size	Maximum	Minimum	Average
15-Minute Duration				
Cost/peak KW hr. not delivered	11	$ 10.70	$ 1.50	$ 5.84
Cost/1,000 sq. ft. of bldg./hr.	11	107.4	10.54	49.54
Cost/employee/hr.	11	28.56	3.00	12.56
1-Hour Duration				
Cost/peak KW hr. not delivered	13	$ 13.33	$ 0.91	$ 5.30
Cost/1,000 sq. ft. of bldg./hr.	15	120.0	5.24	49.42
Cost/employee/hr.	15	28.57	1.25	10.64
Duration 1 Hour				
Cost/peak KW hr. not delivered	3	$100.00	$ 1.97	$ 36.66
Cost/1,000 sq. ft. of bldg./hr.	3	320.00	48.00	156.00
Cost/employee/hr.	3	50.00	4.00	27.52

TABLE 4

OUTAGE COSTS FOR "OFFICE BUILDINGS"
AS A FUNCTION OF DURATION
(WITH COMPUTERS)

	Sample Size	Maximum	Minimum	Average
15-Minute Duration				
Cost/peak KW hr. not not delivered	14	$ 22.22	$ 1.88	$ 8.89
Cost/1,000 sq. ft. of bldg./hr.	15	250.00	16.57	78.21
Cost/employee/hr.	15	52.00	4.00	18.53
1-Hour Duration				
Cost/peak KW hr. not delivered	16	$ 24.93	$ 1.88	$ 8.30
Cost/1,000 sq. ft. of bldg./hr	17	125.00	15.88	54.52
Cost/employee/hr.	17	34.30	4.00	13.62
Duration 1 Hour				
Cost/peak KW hr. not delivered	10	$ 67.66	$ 0.16	$ 9.81
Cost/1,000 sq. ft. of bldg./hr.	11	226.19	1.05	44.08
Cost/employee/hr.	11	75.82	0.48	12.70

TABLE 5

CRITICAL SERVICE LOSS DURATION TIME
FOR "ALL BUILDINGS"

	Service Loss Duration Time								
	1 Cycle	2 Cycles	8 Cycles	1 Sec.	3 Sec.	5 Min.	30 Min.	1 Hour	12 Hours
Percent of buildings with critical service loss duration less than or equal to the time indicated.	3%	6%	9%	15%	18%	36%	64%	79%	100%

TABLE 6

CRITICAL SERVICE LOSS DURATION TIME
FOR "OFFICE BUILDINGS"

	Service Loss Duration Time								
	1 Cycle	2 Cycles	8 Cycles	1 Sec.	3 Sec.	5 Min.	30 Min.	1 Hour	12 Hours
Percent of buildings with critical service loss duration less than or equal to the time indicated.	5%	10%	15%	25%	30%	50%	70%	75%	100%

TABLE 7

RELATIONSHIP OF AUXILIARY GENERATORS
AND SINGLE FEEDER SERVICE TO "ALL BUILDINGS"

	Number of Responses	Buildings with Auxiliary Generation	No Auxiliary Generation and Only Single Feeder
Buildings with computers	23	15	1
Buildings without computers	32	13	7
TOTAL	55	28	8

Survey Results -- Critical Service Loss Duration Time

The amount of time an electrical service can be interrupted before it causes significant losses is a question which our profession has not been able to suit-ably define. The results of the survey indicate that individual requirements for electrical energy are such that it is probably not possible to establish a general critical service loss duration time. The survey results are shown in Tables 5 and 6.

TABLE 8

TYPE OF ELECTRICAL SERVICE
TO "ALL BUILDINGS"

	Number of Responses	Type of Service			
		Single Feeder	Network	Multiple Feeder	Other
Buildings with computers	23	1	8	12	2
Buildings without computers	32	12	10	7	3
TOTAL	55	13	18	19	5

TABLE 9

PHYSICAL DATA -- "ALL BUILDINGS"

Item	Sample Size	Maximum	Minimum	Average
Area, sq. ft. x 10^3	54	2,085	3	400
Number of floors	55	52	1	12
Number of employees	51	7,000	12	1,384
Annual usage - Megawatt hours	52	101,349	210	11,973
Peak Kilowatt demand	52	17,250	95	3,095

TABLE 10

PHYSICAL DATA -- "OFFICE BUILDINGS"

Item	Sample Size	Maximum	Minimum	Average
Area, sq. ft. x 10^3	35	1,600	38	371
Number of floors	35	44	2	13
Number of employees	35	7,000	150	1,651
Annual usage - Megawatt hours	32	51,046	840	9,444
Peak Kilowatt demand	32	17,000	270	3,035

TABLE 11

AVERAGE OF PHYSICAL DATA
FOR "ALL BUILDINGS"
AND FOR "OFFICE BUILDINGS"

Item	All Buildings	Office Buildings
Megawatt hours/1,000 sq. ft. of buildings area/year	35.5	33.5
Megawatt hours/employee/year	20.2	7.5
Peak Kilowatt demand/1,000 sq. ft. of building area	11.3	11.5
Peak Kilowatt demand/employee	5.0	2.5
Employees/1,000 sq. ft. of building area	3.9	4.7

Thirty-six percent of "all buildings" reporting could be without electrical energy for 5 minutes before the lack of energy was considered to be critical, while 6 percent could be without energy for only 2 cycles and 3 percent for only one cycle before significant losses were incurred.

Fifty percent of the "office buildings" reporting could be without electrical energy for 5 minutes before the lack of energy was considered to be critical, while 10 percent could be without energy for only 2 cycles, and 5 percent for only one cycle before significant losses were incurred.

Precautionary measures taken to minimize critical outages in buildings where computers are installed are indicated in Table 7, where 65 percent (15 of 23) of the buildings reporting have auxiliary generating units. Only 4 percent (1 of 23) of the buildings reporting have no auxiliary generation and are served by a single feeder from the utility company. A like comparison is shown for buildings not having computers; in these instances, 41 percent of the buildings have auxiliary generation and 22 percent are served by single feeders from the utility company.

Table 8 shows the type of electrical service to all buildings reporting. Eighty-seven percent of the buildings with computers have network or multiple feeder service, while 53 percent of the buildings without computers have network or multiple feeder service.

Survey Results -- Demand and Usage Data

Each respondent was asked to report gross floor area, number of floors, number of employees, and electrical energy usage and demand. While not directly related to the subject of this paper, the data is of interest, and will perhaps allow the reader to make a better judgement of the validity of the data presented previously. The details are given in Tables 9, 10, and 11.

It is believed that the employee data for the "All Buildings" category may not be valid, since it appears that not all employees were reported for some multi-function buildings, the office/retail category in particular.

Conclusions and Discussion of Results

1 Cost of Power Outages (Tables 1, 2, 3, and 4)

   a There is a wide spread in the cost of power outages (KWH not delivered) in commercial buildings. Even within like types of buildings, with or without computers, there is a great difference in the costs assigned.

   b The cost per KWH not delivered increases greatly when the outage duration time exceeds one hour. An exception to this is buildings with computers.

      It is probable that for outages of less than one hour, employees may remain partially productive and the temperature of their environment remains tolerable. For longer outages, employees may have to be furloughed for the remainder of the day.

   c The cost of power interruptions for buildings with computers varies from $8.89/KWH average for outages of 15-minutes duration to $9.81/KWH for outages of greater than one hour. It is suspected that the small differential is due to the fact that a short duration as well as a long outage renders the computer inoperable, and the employees are either non-productive during this period or repairing possible damage caused by the outage.

   d A comparison of the average costs of outages for commercial buildings with that for industrial plants (Reference 1) is shown in Table 12. The data is interpreted to mean that short-term outages in industrial plants could be more costly than those in commercial buildings, while long-term outages are more costly in commercial buildings.

   e Additional information on the cost of power outages in Sweden, Norway, and the United States is contained in Reference 3.

2 Critical Service Loss Duration Time (Tables 5 and 6)

   a As would be expected, there is a wide spread in the critical time of a power interruption. This is probably due to the wide variations of type of work being accomplished, the type of equipment involved, and the general work environment. For example, a windowless building in which a sensitive computer operation is performed would be more rapidly affected than a window-wall building performing normal office functions.

   b It is suggested that a future survey attempt to define the reasons for the wide variances.

3 Demand and Usage Data (Tables 9, 10, and 11)

   a Of the "all building" data reported, the areas averaged 400,000 square feet, 12 floors in height, with an annual usage of almost 12,000 megawatt hours, and a demand of 3,095 KW. Minimum and maximum data were not available.

TABLE 12

COMPARISON OF AVERAGE COSTS OF POWER OUTAGES
IN COMMERCIAL BUILDINGS AND INDUSTRIAL PLANTS

Type	Cost
All commercial buildings	$7.21/KWH not delivered
Office buildings	$8.86/KWH not delivered
Industrial plants -- all	$1.89/KW interrupted + $2.68/KWH not delivered

The data for "office buildings" indicate average values within 10 percent of that for "all buildings," except for the number of employees, which is 16 percent greater.

   b The average electrical usage for all buildings and for office buildings only is nearly equal when placed on a per unit basis (33.5 KWH/Sq. Ft.) as is the peak demand (11.3 Watts/Sq. Ft. to 11.5 Watts/Sq. Ft.). The relationship of usage and demand to employees does not correlate for all buildings and office buildings only. As mentioned heretofore, the validity of employee data with regard to the Office/Retail category of buildings is questionable. On this basis, no attempt to draw conclusions has been made.

References

1 A.D. Patton, et al, "Report of Reliability Survey of Industrial Plants, Part 4 – Additional Detailed Tabulation of Some Data Previously Reported in the First Three Parts," IEEE I & CPS Conference Record, June 2-6, 1974.

2 W.H. Dickinson, et al, "Report of Reliability Survey of Industrial Plants, Part 2 – Cost of Power Outages, Plant Restart Time, Critical Service Loss Duration Time, and Type of Loads Lost vs. Time of Power Outages," IEEE I & CPS Conference Record, May 14-16, 1973.

3 R.B. Shipley, A.D. Patton, J.S. Denison, "Power Reliability Cost vs. Worth," IEEE Transactions on Power Apparatus and Systems, PAS-91, P. 2204-2212, September/October 1972.

SURVEY FORM ON COST OF ELECTRICAL INTERRUPTIONS IN COMMERCIAL BUILDINGS

**THE INSTITUTE OF
ELECTRICAL AND
ELECTRONICS
ENGINEERS, INC.**

INDUSTRY AND GENERAL APPLICATIONS GROUP

RELIABILITY SUBCOMMITTEE OF THE INDUSTRIAL & COMMERCIAL POWER SYSTEMS COMMITTEE

> *Electricity is an integral part of our every day life. If it isn't available -- what is its economic effect? Please help us to find out by filling out this form.*

Please address reply to:

A. D. Patton
Texas A & M University
Electric Power Institute
College Station, TX 77843

Date _____

1. **COMPANY NAME** (Fill in 3-letter abbreviation of name) _____

2. **BUILDING NO.** (Fill in sequence number 1, 2, 3, etc. for building(s) reported on) _____

3. **BUILDING TYPE** (Check type which best describes your building):

   ☐ Office   ☐ Office/Retail Sales   ☐ Office/Retail Sales/Apartment

   ☐ Retail Sales   ☐ Other (describe) _____

4. **BUILDING LOCATION** (Check applicable items):

   ☐ Downtown;   ☐ Urban;   ☐ Suburban;

   ☐ USA: Eastern;   ☐ USA: Central;   ☐ USA: Western

5. **BUILDING DATA - GENERAL**

   Gross Area, square feet _____

   Number of Floors _____

   Average Usage of Building: Hours/Day _____ Days/Week _____

   Estimated Number of Office Employees (if any) _____

   Estimated Annual Retail Sales (if any) _____

   Is Auxiliary or Emergency Generation Provided: ☐ Yes   ☐ No

6. <u>BUILDING ELECTRICAL USAGE DATA</u>

Electrical Energy Usage for 12-month Period   _____ KWH

Electrical Maximum Demand for this Period   _____ KW

Type of Service: ☐ Single Feeder;   ☐ Network;   ☐ Multiple Feeders With
                                                     Automatic Transfer

                  ☐ Other (Explain) _____

7. <u>COST OF A TOTAL INTERRUPTION OF ELECTRICAL SERVICE TO YOUR BUILDING</u>
<u>DURING PEAK PERIOD</u>: (Best Opinion - If no interruptions have
occurred, assume hypothetical instances)

a) 15-Minute Duration $ _____

b) 1-Hour Duration $ _____

c) _____ Hours Duration $ _____

Does a, b, or c include losses from
an "on-line" electronic computer?   ☐ Yes   ☐ No

For "Office Buildings" loss should include wages of all employees affected,
plus any other direct costs incurred including delays, and damage to equipment. This would include any losses from an "on-line" electronic computer.

For "Retail Sales" cost should include estimated loss of sales minus cost
of goods not sold, plus cost of any damage incurred.

8. <u>LENGTH OF INTERRUPTION OF ELECTRICAL SERVICE</u>

If there a definitive length of time before
an interruption causes a significant loss?   ☐ Yes   ☐ No

If "Yes", what is maximum time before
significant losses will be incurred?   _____ Hours _____ Minutes

Appendix D

Reliability of Electric Utility Supplies to Industrial Plants

By
Power Systems Reliability Subcommittee
Industrial and Commercial Power Systems Committee
IEEE Industry Applications Society

A. D. Patton, *Chair*

C. E. Becker   P. E. Gannon   D. W. McWilliams
M. F. Chamow   M. D. Harris   R. W. Parisian
W. H. Dickinson   Cr. R. Heising   S. J. Wells
                            R. T. Kulvicki

Industrial and Commercial Power Systems Technical Conference
Institute of Electrical and Electronics Engineers, Inc.
Toronto, Canada
May 5–8, 1975

Reprinted from IEEE Conference Record
75CHO947-1-1A, pp. 131–133

RELIABILITY OF ELECTRIC UTILITY
SUPPLIES TO INDUSTRIAL PLANTS

by
Power Systems Reliability Subcommittee
Industrial and Commercial Power Systems Committee
A. D. Patton, Coordinating Author[1]

ABSTRACT

The paper summarizes the results of a 1974 survey of the reliability of electric utility supplies to industrial plants. Results include the average rates of occurrence and durations of power interruptions as a function of type of electric utility supply. This information should help industrial plant operators choose the types of electric utility supplies best suited to their plants.

INTRODUCTION

The electric utility supply reliability survey reported here is a followup to the 1972 survey of the reliability of electrical equipment in industrial plants. [1,2] The 1972 survey showed that the electric utility supply is the most fallible "component" of an industrial plant system and therefore deserves careful consideration.

Certain of the data in the earlier survey were subject to possible error due to misinterpretation of the survey form. Hence, a prime objective of the present survey was to improve the accuracy of data on electric utility supplies. A second objective was to provide more detailed and definitive data on electric utility supply interruption rates and average durations as a function of the number of supply circuits, the type of switching scheme, and the voltage of the supply circuits. A third objective was to obtain data from a larger number of plants than in the 1972 survey thereby permitting interruption rates and average durations to be determined with greater precision. A total of 87 plants provided usable data, almost triple the number of plants providing data on electric utility supplies in the 1972 survey. Survey response broken down by industry is as follows: cement = 2, chemical = 14, metal = 4, petroleum = 30, pulp and paper = 1, rubber and plastics = 4, and other manufacturing = 32.

It should be emphasized that electric utility supply reliability is a function of a number of factors not directly identified in the data presented here. Included in these reliability-influencing factors are line exposure, weather and other environmental conditions, and utility operating and maintenance practice. Thus, the electric utility supply reliability data given in this paper represents average performance and should not be used in preference to specific data when this is available. Methods are available for computing the reliability performance of an electric utility supply when the reliability performance parameters of utility system components are known.[3]

SURVEY QUESTIONNAIRE

The survey questionnaire requested the following data for each electric utility supply.
1. Type of industry
2. Type of electric utility supply
   a. Number of utility circuits supplying the plant

[1] Members of the Power Systems Reliability Subcommittee are: A. D. Patton, chairman, C. E. Becker, M. F. Chamow, W. H. Dickinson, P. E. Gannon, M. D. Harris, C. R. Heising, R. T. Kulvicki, D. W. McWilliams, R. W. Parisian, and S. Wells.

   b. Mode of operation if more than one supply circuit: all circuit breakers normally closed, manual throw-over scheme, or automatic throw-over scheme
   c. Voltage of utility circuits supplying the plant
   d. Type of supply circuits: overhead or underground
   e. A sketch of the electric utility supply system
3. The period of time covered by the survey report. (Respondents were asked to limit their response to the period January 1, 1968 to the present.)
4. The number of interruptions to the plant due to loss of the electric utility supply during the time period of (3).
5. The duration of each electric utility supply interruption, an indication whether service was restored to the plant by a switching operation or by repair or replacement of failed equipment, and, if known, the equipment which failed causing the interruption.

SURVEY DATA SUMMARY AND DISCUSSION

Some respondents to the survey listed voltage dips which caused disruption of plant production as well as complete interruptions of electric utility service. Other respondents commented on production disruptions due to voltage dips without giving details. However, most respondents reported only on complete interruptions of service and this was the intent of the survey. The Subcommittee feels that the sensitivity to voltage dips is a rather unique characteristic of each plant and process and that average interruption rates including voltage dips would not be very meaningful. Therefore, all voltage dip events were removed from the survey data leaving only those interruptions due to complete loss of electric utility service. Hence, the interruption rates given in the summary tables reflect complete loss of electric utility service only. If a plant is sensitive to voltage dips, the rate of such events must be added to the reported interruption rates to obtain the total rate of production disruption due to utility supply troubles.

Almost all respondents indicated that utility supply circuits are overhead rather than underground. Hence, no effort was made to separate supplies with overhead and underground circuits. The data given in the summary tables essentially reflects overhead supply circuits due to the preponderance of such circuits in the survey response.

Preliminary analyses of utility supply interruption rates by industry category indicated no significant differences between industries. Further, there seems to be no good reason why utility supplies of the same type and voltage should differ between industries. Therefore, the data presented in the summary tables is not broken down by industry.

The survey response broken down by number of utility supply circuits, voltage of utility supply circuits, and mode of operation of multiple supply circuit utility supplies is given in Table I.

Table I
Number of Responding Plants
With Electric Utility Supplies
of Various Types

Number of Supply Circuits

1 circuit	- 20 plants
2 circuits	- 56 plants
3 or more circuits	- 11 plants

Supply Circuit Voltage

voltage $\leq$ 15 KV	- 22 plants
15 KV < voltage $\leq$ 35 KV	- 17 plants
voltage >35 KV	- 48 plants

Switching Scheme of Multiple Circuit Supplies

all breakers closed	- 45 plants
manual throwover	- ·9 plants
automatic throwover	- 13 plants

Table I shows that two-circuit supplies are the most common among the responding plants. A much smaller number of plants reported three or more supply circuits. All multiple-circuit supplies are combined in the data tables which follow because such supplies are expected to have similar interruption rates and because of the relatively small sample of supplies with three or more circuits. Responses have been broken into three voltage categories corresponding roughly to distribution voltages, subtransmission voltages, and transmission voltages. This was done because electric utility design and operating practice is rather different at these three function levels. Hence, it can be expected that utility supply reliability will be a function of the system level at which service is provided.

Table I indicates that about two-thirds of the responding plants having multiple circuit utility supplies operate with all circuit breakers closed. That is, service is supplied simultaneously over all supply circuits. Service may also be lost, however, by failures in the plant substation or by a widespread failure in the supplying utility's system. Plants having throwover schemes operate with a single circuit providing normal service. Thus, such plants suffer an interruption any time the normal supply circuit fails. The duration of interruption to such plants is usually limited to the time required ro reclose the normal supply circuit or to switch to the alternate supply circuit if the normal circuit is permanently faulted.

Table II summarizes interruption rate and average interruption duration data for single-circuit utility supplies broken down by voltage level. Interruption rates and average durations are given separately for interruptions reported terminated by utility switching operations and by repair or replacement of failed components. Also given are overall interruption rates and average durations.

Tables III and IV show interruption rates and average durations for multiple circuit utility supplies broken down by switching scheme and by voltage level. Table V shows interruption rates and average durations for multiple-circuit utility supplies which operate with all circuit breakers closed broken down by voltage levels. Similar breakdowns by voltage for throwover switching schemes were not possible due to lack of an adequate data base.

Interruption rates and average durations are given in Tables II through V for interruptions where service

is restored by: (a) some switching operation or sequence of switching operations in the electric utility system, and (b) repair or replacement of components which failed in the electric utility system. If service can be restored by some automatic or manual switching action in the electric utility system, whether remote or within the utility switchgear at the plant, interruptions are usually much shorter than if repair or replacement of failed components is required to restore service. The reason for providing data on both short-duration switching-terminated interruptions and on long-duration repair-terminated interruptions is because of possible differences in impact on plant operations.

It should be mentioned here that interruption rates and average durations computed from a small number of observed interruptions should be regarded as less accurate than those computed from a larger sample of observations. In particular, Reference [1] shows that interruption rates computed from an observed number of interruptions less than about 8 or 10 may well be in error by plus or minus 50 per cent or more due to random variations alone.

The data of Tables II through V show the expected trends.
(1) Utility supply interruption rates are lowest for multiple circuit supplies which operate with all circuit breakers closed and highest for single-circuit supplies. Tables II and III show that the interruption rate for single-circuit supplies is about six times that of multiple circuit supplies which operate with all circuit breakers closed. Interruption rates for multiple-circuit supplies which operate with a throwover scheme are comparable to those for single-circuit supplies, but throwover schemes have a smaller average interruption duration than single-circuit supplies.
(2) Interruption rates are highest for utility supply circuits operated at distribution voltages and lowest for circuits operated at transmission voltages.

Direct comparisons between interruption rates determined in this survey and in the 1972 survey are not possible in every case, but where possible show somewhat higher values in the present survey. Since the present survey is believed to be more accurate, has a larger data base, and is more up-to-date, the values presented here are to be preferred over those presented in 1972 survey.

REFERENCES

1. Reliability Subcommittee Report, "Report On Reliability Survey of Industrial Plants, Part I: Reliability of Electrical Equipment", IEEE Transactions on Industry Applications, pp. 213-235, March/April 1974.

2. Reliability Subcommittee Report," Report On Reliability Survey of Industrial Plants, Part III: Causes and Types of Failures of Electrical Equipment, the Methods of Repair, and the Urgency of Repair", Ibid., pp. 242-252, March/April 1974.

3. R. Billinton, R. J. Ringlee, and A. J. Wood, Power-System Reliability Calculations, The MIT Press, Cambridge, Mass., 1973.

Table II
Single Circuit Utility Supplies

Voltage Level	Unit-years of History	Number of Interruptions Reported*		Interruptions Per Year**			Average Interruption Duration, Minutes**		
		N_S	N_R	λ_S	λ_R	λ	r_S	r_R	r
v≤15KV	27.62	25	75	.905	2.715	3.621	3.5	165	125
15KV<v≤35KV	12.67	0	21	–	1.657	1.657	–	57	57
v>35KV	71.16	37	60	.527	.843	1.370	1.5	59	37
all	111.45	62	156	.556	1.400	1.956	2.3	110	79

Table III
Multiple Circuit Utility Supplies
All Voltage Levels

Switching Scheme	Unit-Years of History	Number of Interruptions Reported		Interruptions Per Year			Average Interruption Duration, Minutes		
		N_S	N_R	λ_S	λ_R	λ	r_S	r_R	r
all breakers closed	246.17	63	14	.255	.057	.312	8.5	130	31
man. throw-over	42.33	31	5	.732	.118	.850	8.1	84	19
auto. throw-over	64.36	66	11	1.025	.171	1.196	0.6	96	14
all	352.86	160	30	.453	.085	.538	5.2	110	22

Table IV
Multiple Circuit Utility Supplies
All Swiching Schemes

Voltage Level	Unit-Years of History	Number of Interruptions Reported		Interruptions Per Year			Average Interruption Duration, Minutes		
		N_S	N_R	λ_S	λ_R	λ	r_S	r_R	r
v≤15KV	81.31	52	12	.640	.148	.788	4.7	149	32
15KV<v≤35KV	78.00	39	5	.500	.064	.564	4.0	115	17
v>35KV	193.55	69	13	.357	.067	.424	6.1	184	34

Table V
Multiple Circuit Utility Supplies
All Circuit Breakers Closed

Voltage Level	Unit-Years History	Number of Interruptions Reported		Interruptions Per Year			Average Interruption Duration, Minutes		
		N_S	N_R	λ_S	λ_R	λ	r_S	r_R	r
v≤15KV	45.61	8	4	.175	.088	.263	0.7	335	112
15KV<v≤35KV	52.61	18	1	.342	.019	.361	7.0	120	13
v>35KV	147.95	37	9	.250	.061	.311	11.0	203	49

*N_S and N_R are, respectively, the number of service interruptions terminated by switch-ing and by repair or replacement.

**Interruption rates and average durations subscripted S and R are, respectively, rates and durations of interruptions terminated by switching and by repair or replacement. Un-subscripted rates and duration are overall values.

Appendix E

Report of Switchgear Bus Reliability Survey of Industrial Plants and Commercial Buildings

By
Power Systems Reliability Subcommittee
Power Systems Support Committee
Industrial Power Systems Department
IEEE Industry Applications Society

P. O. O'Donnell, *Coordinating Author*
P. E. Gannon, *Chair*

J. W. Aquilino	C. R. Heising	A. D. Patton
C. E. Becker	D. Kilpatrick	C. Singh
W. H. Dickinson	D. W. McWilliams	W. L. Stebbins
B. Douglas	R. N. Parisian	H. T. Wayne
I. Harley		S. J. Wells

**Industrial and Commercial Power Systems Technical Conference
Institute of Electrical and Electronics Engineers, Inc.
Cincinnati, Ohio
Jun. 5-8, 1978**

Published by
IEEE Transactions on Industry Applications
Mar./Apr. 1979, pp. 141-147

Report of Switchgear Bus Reliability Survey of Industrial Plants and Commercial Buildings

Power Systems Reliability Subcommittee
Power Systems Support Committee
Industrial Power Systems Department

PAT O'DONNELL, MEMBER, IEEE
COORDINATING AUTHOR[1]

Abstract—The Power Systems Reliability Subcommittee of the IEEE Industry Applications Society has been conducting surveys of the reliability of electrical equipment in industrial plants and commercial buildings. Switchgear bus was included in a previous survey published in 1973 and 1974 [1] and generated some controversy concerning bare and insulated bus. For this reason, and also for an ongoing effect to continually update the 1973 and 1974 survey [1], switchgear bus reliability has been investigated in a new survey in 1977, and the results are presented. Reference is made to a paper [2] given at the 1977 Industrial and Commercial Power Systems Technical Conference on reasons for conducting the new survey.

INTRODUCTION

CURRENT reliability data on failure rate of electrical equipment can provide a valuable tool for the power systems designer or planner. These data can also be a valuable tool for the manufacturer of the equipment concerned.

Many parameters were included in this new survey in an effort to uncover the most influencing factors on the reliability of bare bus and insulated bus and to allow any new obvious and significant applications considerations to be identified. The questionnaire submitted was condensed to a practical and useful form to obtain optimum response in as short of time period as possible.

Results of the survey are presented in tabular form, and discussion is included primarily where adequate response and population data were obtained. Many questions and uncertainties still exist, and the intent of the following presentation is to report the results of the survey with some discussion, but drawing of definite conclusions is left to the reader.

SURVEY FORM

The questionnaire form (Fig. 1) and cover letter used in the survey are included in the Appendix. Total populations data

Paper ISPD approved by the Power Systems Protection Committee of the IEEE Industry Applications Society for presentation at the 1978 Industrial and Commercial Power Systems Technical Conference, Cincinnati, OH, June 5-9. Manuscript released for publication October 25, 1978.

The author is with El Paso Natural Gas Company, El Paso, TX 79978.

[1] Other members of the subcommittee are Phillip E. Gannon (Chairman), J. W. Aquilino, Carl E. Becker, W. H. Dickinson, Bruce Douglas, Ian Harley, C. R. Heising, Don Kilpatrick, D. W. McWilliams, R. N. Parisian, A. D. Patton, Dr. Chanan Singh, Wayne L. Stebbins, Harold T. Wayne, and Stanley J. Wells.

categorize information into major areas of application. An area of primary concern is maintenance because of its obvious relation to failure rate. However, this is the most difficult datum to obtain in complete and uniform format for meaningful results. Responses in this survey did not permit these results to appear, partly due to the respondents' failure to submit information and partly due to the survey format.

Failed unit data were requested in the form shown in the second portion of the questionnaire. The major categories are causes of failure, types of failure, duration of failure, and failed components. This form is less extensive, but more specifically oriented for switchgear bus than in 1973 and 1974 survey [1].

SURVEY RESPONSE

Table I summarizes the survey response including number of buses, companies, and plants. In this survey, bus "unit-year" is defined as the product of the total number of switchgear connected circuit breakers and connected switches reported in a category times the total exposure time. In the previous survey, the unit-year did not include the number of connected switches; that is, only the connected circuit breakers were counted. Table II shows the 1973 and 1974 [1] survey and is included for comparison of responses. The total number of plants in the new survey response is considerably greater than in the 1973 and 1974 survey, but it is interesting to note that unit-year sample size is slightly less. Also some discrepancy appears in the total number of failures reported in Table I and those of some subcategories in tables to follow. This is due to all companies not responding to every category.

SURVEY RESULTS

Insulated and Bare Bus

A major controversy emerged in the results of the 1973 and 1974 survey [1] concerning bare and insulated switchgear bus. Insulated bus, 601-15 000 V, showed a higher failure rate than bare bus, above 600 V, but data were heavily influenced by the chemical industry. The new survey shows the opposite of this, as seen in Table I, with less chemical industry influence. Bare bus, above 600 V, shows a relatively high failure rate, but the sample size is not large, thus making this observation somewhat questionable. With more companies responding in the

Company Name and Plant: _____ _____

Industry Type: _____

Period Reported - From: Month _____ Year _____

To: Month _____ Year _____

Plant Climate: Temperature _____ Relative Humidity _____

Contamination Level and Type: _____

Total Population:

Bus No.	No. CB's & Sw's	Age of Bus (YRS)	Bare	Insulated	Outdoor	Indoor	Copper	Aluminum	L-L Voltage (KV)	Current (KA)	L-L Voltage (KV)	Ungrounded	Solid Ground	Imped. Ground	Maint Cycle (MO)	Extent of Maintenance
				Bus Type and Rating								System Application				Maintenance Data
1																
2																
3																
4																
5																
6																

Failed Unit Data:

Bus No.	Failure Primary Cause	Failure Contributing Cause	Short L-G	Short L-L	Open	Other	Last Maint. (MO)	Round Clock	Normal Hours	Schedule Later	Failure Duration (Hrs.)	Failed Component and Material
				Type of Failure					Restore Data			

Fig. 1. Switchgear bus reliability survey for metalclad and metal enclosed switchgear bus.

new survey but with less overall unit-year sample size, the failure rate for all bus shows to be slightly higher than in the previous survey. But on breaking this down further, bare bus failure rate is higher and insulated bus failure rate is lower in the new survey.

Table I shows the chemical industry data broken out since it is believed to be a major contributor in the controversy of the 1973 and 1974 survey [1]. In the new survey the chemical industry dominated the number of failures in each category, but did not dominate sample sizes. This supports the argument of some that bus utilized in the chemical industry should have a relatively high failure rate, especially in the use of bare bus.

Table I also shows median outage duration time after a failure of each category, in hours per failure. It is important to emphasize that these data are based on many factors, and without sufficient supplement from respondents concerning operating procedures, maintenance type, spare parts inventory, etc., the data relate to a very general or all-inclusive type of information.

Grounding Type

Survey results are shown in Tables III–V. Inadequate response and the general nature of the questionnaire format prohibit sufficient results for this category. It is believed that grounding type related to failures is important data, but data should be specific, for example, in types of failures in ungrounded systems and in impedance value of impedance grounded systems. This category may be pursued in greater detail in the next survey.

TABLE I
SWITCHGEAR BUS: INDOOR AND OUTDOOR

NUMBER OF COMPANIES	NUMBER OF PLANTS IN SAMPLE-SIZE	NUMBER OF BUSES	SAMPLE SIZE UNIT-YR	NUMBER OF FAILURES REPORTED	INDUSTRY	EQUIPMENT SUB-CLASS	FAILURE RATE FAILURE PER UNIT-YEAR	MEDIAN HOURS DOWNTIME PER FAILURE
39	56	444	51391	54	ALL	ALL	.001050	28
28	36	245	24855	28	ALL	INSULATED ABOVE 600V	.001129	28
25	35	199	26592	26	ALL	BARE (ALL VOLTAGES)	.000977	28
17	23	132	22420	18	ALL	BARE 0-600V	.000802	27
14	18	67	4172	8	ALL	BARE ABOVE 600V	.001917	36
14	19	92	7425	15	PETROLEUM CHEMICAL	INSULATED ABOVE 600V	.002020	40
11	13	135	7002	18	PETROLEUM CHEMICAL	BARE (ALL VOLTAGES)	.002570	28
10	11	83	4707	13	PETROLEUM CHEMICAL	BARE 0-600V	.002761	22
7	8	52	2295	5*	PETROLEUM CHEMICAL	BARE ABOVE 600V	*	48

* Small sample-size.

TABLE II
RESULTS OF PREVIOUS SURVEY PUBLISHED IN 1973 AND 1974 [1]
SWITCHGEAR BUS: INDOOR AND OUTDOOR

NUMBER OF PLANTS SAMPLE-SIZE	SAMPLE SIZE (UNIT-YEAR)	NUMBER OF FAILURES REPORTED	INDUSTRY	EQUIPMENT SUB-CLASS	FAILURE RATE FAILURES PER UNIT-YEAR	ACTUAL HOURS DOWNTIME/FAILURE			
						INDUSTRY AVERAGE	MINIMUM PLT. AVG.	MEDIAN PLT. AVG.	MAXIMUM PLT. AVG.
12	11740	20	ALL	INSULATED 601-15000V	0.001700	261	5	26.8	1613
12	32280	11	ALL	BARE 0-600V	0.000340	550	2	24	2520
5	20560	13	ALL	BARE >600V	0.000630	17.3	6.9	13	48
5	4003	15	PETROLEUM CHEMICAL	INSULATED 601-15000V	0.003750	340	18	26.8	1613
3	17270	10	PETROLEUM CHEMICAL	BARE >600V	0.000580	19.3	6.9	42	48

TABLE III
TYPE OF GROUNDING OVERALL, BUS INSULATED AND BUS BARE

	UNGROUNDED	SOLID-GROUND	IMPEDANCE-GROUND	NOT REPORTED	TOTAL
(Unit-Year) Sample-Size	20262	9787	17280	4062	51391
# FAILURE	17	12	23	2*	54
FAILURE RATE	.000839	.001226	.001331	-	.001050

* Small sample size.

TABLE IV
BUS INSULATED

	UNGROUNDED	SOLID-GROUND	IMPEDANCE-GROUND	NOT REPORTED	TOTAL
(Unit-Year) Sample-Size	4626	4274	14270	1685	24855
# FAILURE	7*	4*	16	1*	28
FAILURE RATE			.001121	-	.001126

* Small sample size.

TABLE V
BUS BARE

	UNGROUNDED	SOLID-GROUND	IMPEDANCE-GROUND	NOT REPORTED	TOTAL
(Unit-Year) Sample-Size	15636	5513	3010	2377	26536
# FAILURE	10	8	7*	1*	26
FAILURE RATE	.000640	.001451			.000980

* Small sample size.

TABLE VI
AVERAGE AGE OF SWITCHGEAR BUS

	ALL	INSULATED	BARE
AGE 1-10 yrs.	6526 unit-year	1899 unit-year	4627 unit-year
>10 yrs.	44596 unit-year	22887 unit-year	21709 unit-year

Age of Bus

Tables VI–VIII illustrate how failures of insulated and bare bus relate to age in this survey. An interesting observation here is that newer bus appears to experience a higher failure rate than older bus. This might be expected if one considers improper installation, new components failure rate, type of construction of new switchgear, etc. As discussed below under "causes" of failures, the logicality of this observation is not consistent.

As incoming data were analyzed, it became apparent that the period reported (it was assumed that the period reported was the period of best kept records) and the age of bus did not correlate as well as expected in every case, a fallacy in the questionnaire format perhaps. Note that the older bus sample size is much larger.

Indoor and Outdoor Bus

The results of this category are summarized in Tables IX–XI below. Table XI shows an overall result of outdoor bus failure rate versus indoor bus failure rate. Outdoor bus shows a higher failure rate than indoor bus, an observation not too surprising.

Failure Duration

Failure duration results are reported in Tables XII and XIII below and categorized into repair on a round-the-clock emergency basis and repair on a normal working hour basis. This adds more meaning to the data in Table I, but would be more meaningful if repair methods were known. Urgency of repair as shown in Table XIV reveals that most repairs were made on an emergency basis. The data of these tables compare very favorable with those of the previous survey.

Type of Maintenance

Response was disappointingly low in this category and results are presented in Tables XV and XVI. The tables show results of maintenance cycles and time since last maintenance in three groups: 1) less than 12 months, 2) 12-24 months, and 3) more than 24 months. This is a very important category regarding reliability, and hopefully the next survey will produce better results.

Causes of Failures

Primary and contributing causes of failures are summarized in Tables XVII and XVIII. As might be expected inadequate maintenance is a large contributor to failures. This does not necessarily follow from the observation above on age of bus. However, defective components are a large primary cause of failures, which is logical for new installations. Correlation between the two tables below is clearly evident from the contributing cause of exposure to contaminants and the primary cause of inadequate maintenance. Exposure to contaminants, which includes dust, moisture, and chemicals, also supports the data showing outside bus with a relatively high failure rate. Inadequate maintenance was reported as the single largest primary cause of failures in the 1973 and 1974 survey [1]. This prompted the effort to survey type of maintenance in the new survey.

TABLE VII
NUMBER OF FAILURES VERSUS AGE

	ALL	INSULATED	BARE
AGE 1-10 yrs.	15	5*	10
>10 yrs.	37	23	14

* Small sample size.

TABLE VIII
FAILURE RATE (FAILURE PER UNIT-YEAR)

	ALL	INSULATED	BARE
AGE 1-10 yrs.	.002298	*	.002161
>10 yrs.	.000829	.001005	.000645

* Small sample size.

TABLE IX
SWITCHGEAR BUS INSULATED

	OUTDOOR	INDOOR
Sample-Size Unit-Year	4275	20356
FAILURE	7*	19
FAILURE RATE	*	.000933

* Small sample size.

TABLE X
SWITCHGEAR BUS BARE

	OUTDOOR	INDOOR
Sample-Size Unit-Year	2750	22339
FAILURE	8	11
FAILURE RATE	.002909	.000492

TABLE XI
SWITCHGEAR BUS (OVERALL)

	OUTDOOR	INDOOR
Sample-Size Unit-Year	7825	42695
FAILURE	15	30
FAILURE RATE	.001917	.000703

TABLE XII
FAILURE DURATION: ROUND CLOCK VERSUS NORMAL HOUR
(HOURS DOWNTIME PER FAILURE)

FAILURE REPAIR URGENCY	BUS INSULATED		BUS BARE	
	MEDIAN	AVERAGE	MEDIAN	AVERAGE
ROUND CLOCK	24 hr.	87 hr.	32 hr.	39 hr.
NORMAL HOUR	240 hr.	430 hr.	24 hr.	154 hr.

TABLE XIII
FAILURE DURATION: ROUND CLOCK VERSUS NORMAL HOUR
(HOURS DOWNTOWN PER FAILURE)

	BUS INSULATED		BUS BARE	
	ROUND CLOCK	NORMAL HOUR	ROUND CLOCK	NORMAL HOUR
25 PERCENTILE	8 hr.	8 hr.	3 hr.	4 hr.
50 PERCENTILE	24 hr.	240 hr.	32 hr.	24 hr.
75 PERCENTILE	48 hr.	350 hr.	48 hr.	48 hr.

TABLE XIV
FAILURE REPAIR URGENCY

	ROUND CLOCK	NORMAL HOUR	SCHEDULE LATER
BUS INSULATED	64%	28%	8%
BUS BARE	53%	41%	6%

TABLE XV
NUMBER OF SWITCHGEAR BUS-INSULATED FAILURES VERSUS
MAINTENANCE CYCLE

	LESS THAN 12 MO.	12-24 MO.	MORE THAN 24 MO.
Sample-Size (Unit-Year)	3563	8812	7253
# FAILURE	2*	13	6*
FAILURE RATE	-	.001475	

* Small sample size.

TABLE XVI
NUMBER OF SWITCHGEAR BUS BARE FAILURES VERSUS MAINTENANCE CYCLE

	LESS THAN 12 MO.	12-24 MO.	MORE THAN 24 MO.
Sample-Size (Unit-Year)	980	10,455	6312
# FAILURE	2*	12	4*
FAILURE RATE	-	.001147	-

* Small sample size.

TABLE XVII
SUSPECTED PRIMARY CAUSE OF FAILURE

BUS INSULATED	BUS BARE	
26%	17%	1. Defective Component
4%	4%	2. Improper Application
7%	9%	3. Improper Handling
7%	13%	4. Improper Installation
19%	22%	5. Inadequate Maintenance
-	18%	6. Improper Operating Procedure
11%	4%	7. Outside Agency - Personnel
26%	-	8. Outside Agency - Other
-	13%	9. Overheating

TABLE XVIII
CONTRIBUTING CAUSE TO FAILURE

BUS INSULATED	BUS BARE	
6.6%	-	1. Thermocycling
3%	8%	2. Mechanical Structure Failure
6.6%	-	3. Mechanical Damage From Foreign Source
-	15%	4. Shorting By Tools or Metal Objects
3%	-	5. Shorting By Snakes, Birds, Rodents, etc.
10%	4%	6. Malfunction of Protective Device
	4%	7. Improper Setting of Protective Device
3%	-	8. Above Normal Ambient Temperature
3%	15%	9. Exposure to Chemical or Solvents
30%	15%	10. Exposure to Moisture
10%	19%	11. Exposure to Dust or Other Contaminants
6.6%	-	12. Exposure to Non-Electrical Fire or Burning
-	8%	13. Obstruction of Ventilation
10%	4%	14. Normal Deterioration from Age
3%	4%	15. Severe Weather Condition
-	4%	16. Testing Error

TABLE XIX
FAILURE TYPE

BUS INSULATED	BUS BARE	
57%	33%	1. Short L-G
40%	60%	2. Short L-L
-	7%	3. Open
3%	-	4. Other

Failure Type

The survey results on types of failures, shown in Table XIX, show a surprisingly high percentage of failures line-to-line.

GENERAL DISCUSSION

At this point it is well to note the confidence intervals of failure rate for bare and insulated bus. Table XX shows the limits for a 90 percent confidence interval. The table illustrates the statistical limits within which 90 percent of the failures could be expected to occur.

Lack of specific details limits the integrity of some data, and as previously indicated not all categories surveyed were reported in this paper, due primarily to small sample sizes and numbers of failures. As with most surveys, accurate data combined with large response are difficult to obtain since response definitely relates to simplicity in questionnaire format. Data of the effect of maintenance on failure rate are highly desirable for obvious reasons, and effort will be made to acquire this data in the future in a meaningful and usable form.

TABLE XX
CONFIDENCE INTERVALS FOR FAILURE RATE λ

FAILURE RATE (λ) FAILURE PER UNIT-YR	INSULATED BUS >600V	BARE BUS > 600V	BARE BUS ≤ 600V
λ L *	.000779	.000958	.000521
λ	.001129	.001917	.000802
λ U *	.001569	.003488	.001203
% DEVIATION - L	31%	50%	35%
% DEVIATION - U	39%	82%	50%

* Upper and lower limits of 90 percent confidence interval for λ

APPENDIX

A. D. Patton
Texas A & M University
Department of Electrical Engineering
College Station, Texas 77843

Dear Sir:

RE: Switchgear Bus Reliability Survey for Metalclad and Metal Enclosed Switchgear

The Reliability Subcommittee of the Industrial and Commercial Power Systems Committee requests your cooperation in a survey to determine the reliability of metal-clad and metal-enclosed switchgear bus in industrial plants. The survey is a follow-up to the general reliability survey of plant equipment in 1971 and is intended to provide more meaningful data on switchgear bus. Attached for your information is a report by the subcommittee on reasons for the survey.

The results of the survey will be published in an IEEE paper and are expected to be of value to system planners and designers in the reliability evaluation of alternatives. Individual responses will be held in confidence and only summaries published.

SURVEY INSTRUCTIONS

It is hoped that the survey form is reasonably self-explanatory. Nevertheless, a sample filled-out data sheet is attached for your guidance, and some brief instructions follow. We wish to emphasize that all requested data are important, but it is realized that some of the requested information may be unknown. In such cases, simply provide the information which is known and leave the other spaces blank. We also encourage you to provide explanatory comments on any of your data as you feel appropriate. If additional data sheets are needed, please duplicate the data sheet provided.

General Data

1) It is vitally important that the period reported be given.
2) The plant climate and contamination data should be your general estimates of the requested information.

Total Population Data

1) Using the total population data block, give requested data for all buses *in service during the period reported* whether or not failures have been experienced. (Note the period reported may not exceed the age of a bus. Use separate data sheets for newer busses.)
2) It is vitally important that the number of connected circuit breakers and switches be given for each bus.

Failed Unit Data

1) List each bus failure event separately.
2) Identify the bus in each failure event by specifying the bus number as assigned in the total population data block.
3) Specify failure cause and contributing cause, where known, using the code numbers on the attached sheet.
4) Specify months since bus was last maintained.
5) Check off urgency of restoration effort.
6) Specify time in hours from onset of failure until bus was restored to service.
7) Describe component which first failed, including component material.

Our schedule dictates that responses be received no later than April 1, 1977. Your participation in this project will be greatly appreciated.

Sincerely,

A. D. Patton
Chairman, Reliability Subcommittee

SURVEY QUESTIONNAIRE

Primary Cause of Failure:

1) defective component,
2) improper application,
3) improper handling,
4) improper installation,
5) inadequate maintenance.
6) improper operating procedures,
7) outside agency—personnel.
8) outside agency—other,
9) overheating.

Contributing Cause to Failure:

1) persistent overloading,
2) transient overvoltage,
3) overvoltage,
4) thermocycling,
5) mechanical structural failure,
6) mechanical damage from foreign source,
7) shorting by tools or metal objects,
8) shorting by snakes, birds, rodents, etc.,
9) malfunction of protective device,
10) improper setting of protective device,
11) above normal ambient temperature,
12) below normal ambient temperatures,
13) exposure to chemicals or solvents,
14) exposure to moisture,
15) exposure to dust or other contaminants,
16) exposure to non-electrical fire or burning,
17) obstruction of ventilation,
18) normal deterioration from age,
19) severe weather conditions,
20) loss or deficiency of cooling medium,
21) testing error.

Comments:

REFERENCES

[1] IEEE Committee Report, "Report on reliability survey of industrial plant," *IEEE Trans. Ind. Appl.*, Mar./Apr., July/Aug., and Sept./Oct., 1974. (Part 1—Reliability of electrical equipment; Part 3—Causes and types of failures of electrical equipment, the methods of repair, and the urgency of repair; Part 5—Plant climate, atmosphere and operating schedule, the average age of electrical equipment, percent production lost, and the method of restoring electrical service after a failure; Part 6—Maintenance quality of electrical equipment.)

[2] IEEE Committee Report, "Reasons for conducting a new reliability survey on switchgear bus-insulated and switchgear bus-bare," Industrial and Commercial Power System Tech. Conf., May 1977, Conf. Rec., p. 91–95.

Appendix F

Working Group Procedure for Conducting an Equipment Reliability Survey

By
Power Systems Reliability Subcommittee
Power Systems Technology Committee
Industrial Power Systems Department
IEEE Industry Applications Society

Procedure 1
Compiled Dec. 8, 1980; Approved May 4, 1981

WORKING GROUP PROCEDURE FOR
CONDUCTING AN EQUIPMENT RELIABILITY SURVEY

POWER SYSTEMS RELIABILITY SUBCOMMITTEE
POWER SYSTEMS TECHNOLOGY COMMITTEE
INDUSTRIAL POWER SYSTEMS DEPARTMENT
IEEE INDUSTRY APPLICATIONS SOCIETY

Scope: Conduct an equipment reliability survey of industrial plants and commercial buildings. Keep anonymous the names of those who submit data. Do not collect the equipment manufacturer's name. Publish an IEEE Working Group report. Collect data that may be included in future versions of IEEE Standard 493-1980, "IEEE Recommended Practice for the Design of Reliable Industrial and Commercial Power Systems". This will include failures, population and unit-years, outage duration time after failure, and other information that are considered important.

Review Approval: The final IEEE Working Group report must be approved before publication by the Chairman, Power Systems Reliability Subcommittee and anyone else that he delegates. Other members of the Power Systems Reliability Subcommittee may ask to review the IEEE Working Group report before the Chairman and/or his delegates give their approval, but they do not have a veto over what is published.

Steps: 1. The Power Systems Reliability Subcommittee (PSRS) will determine the equipment category to be surveyed.

2. The PSRS Chairman will appoint a Working Group Chairman. The Working Group Chairman (WGC) will select the members of the Working Group, subject to approval by the PSRS Chairman. Usually the WG will include a WGC from a previous survey who is familiar with conducting a reliability survey. It is expected that the WGC will do the most work, including survey preparation, data collection, data analysis, and will be the coordinating author of the final report and will present the report at an IEEE technical conference. The PSRS Chairman will compile a budget and submit it to IAS for approval.

3. The WGC will compile a schedule for steps 4 through 15.

4. The WGC will review previous reliability surveys (AIEE 1962 and IEEE 1973/1974, etc.) on this equipment category, if available, and will compile a report summarizing previous survey results and why the new survey is being conducted. This report will be used in the survey and will be sent out with the survey form to the prospective participants. In some cases in the past this report has become an IEEE paper at an IEEE conference (but this is not encouraged).

271

5. The WGC will compile a draft form for the survey and will send it
    to the members of the WG. In general the new survey will be a
    refinement of the previous version, geared to resolving questions
    raised by the past surveys. He will compile a second version,
    third version, etc. as necessary and develop a final form in-
    corporating comments received from Working Group members.

6. The WGC will ask all members of the PSRS: 1) if they wish to
    review the final form and, 2) if they wish to review the final
    WG report after the survey is completed. He will send copies
    to those who request it and should request comments back within
    twelve days.

7. The final form should be approved by the PSRS Chairman and those
    he has delegated. However, responses that take longer than two
    weeks may be considered "approval by default".

8. The WGC will have the material for the survey printed (cover letter
    on IEEE stationery, form & definitions, reasons for survey). He will
    obtain the mailing list from the Chairman, Mailing List Working Group.
    He will review the list and augment it if appropriate. The WGC will
    buy postage stamps and send the survey material out for the survey.
    A return envelope and postage will be included and a requested
    return date will also be included. The WGC will keep track of
    negative, moved, or deceased responses for feedback to the Mailing
    List Chairman.

9. A follow up letter will be sent out by the WG Chairman to all
    participants about 8 weeks later. This always brings in additional
    responses.

10. An oral pep talk (3 minutes long) should be given by WGC during a
    technical session at the I & CPS Conference (if the timing is
    convenient).

11. After the "cut off" date, the WGC will analyse and tabulate the
    results from the survey. (An attempt should be made to contact
    respondents for clarification of incomplete or inconsistent data).
    They will be sent to the WG members for comments and suggestions
    for additional analysis and for what should go into the WG report.

12. The WGC will compile a first draft WG report and will send a copy
    to the members for comments. A second draft, third draft will be
    compiled as needed. A final WG report will be compiled.

13. The final WG report will be sent to the PSRS Chairman and those he
    has delegated for approval. Fourteen days will be allowed for their
    review. The final WG report will also be sent to those PSRS members
    who have requested it in step 6, and comments should be requested back
    within twelve days.

14. The WG Chairman will have the approved final WG report typed on model paper for presentation at an IEEE technical conference. Only those who have contributed as Working Group members, or by commenting on the survey or report drafts should be listed as authors; the WGC will obtain written approval from each co-author to use their names on the report. Approval of the final report by those who request it to review should be adequate approval to use names. A copy of this paper should be sent to all members of the PSRS; written discussions should be invited back from them. Other solicitations for discussions are also encouraged as deemed appropriate by the WGC or the PSRS Chairman.

15. The WG Chairman should present the final WG report at the IEEE Conference. An alternate from the WG should be designated, by the WGC, to present the paper in his absence.

It is believed that the total time cycle for steps 1 through 15 is about two years.

Charles R. Heising
Charles R. Heising
Secretary
Power Systems Reliability Subcommittee

CRH:sk

THE **I**NSTITUTE OF
ELECTRICAL AND
ELECTRONICS
ENGINEERS, INC.

INDUSTRY APPLICATIONS SOCIETY

TYPICAL MAJOR MILESTONE SCHEDULE
for
EQUIPMENT RELIABILITY SURVEYS

YEAR 1:

1. May/June (I&CPS Conference) PSRSC Chairman appoints WG Chairman.

2. October (IAS Conference) WG Chairman presents first draft of survey form to WG.

3. November/December. WG Chairman finalizes survey form and obtains approval from PSRSC Chairman.

YEAR 2:

4. January/February. WG Chairman mails survey form to industries.

5. March/April. WG Chairman mails follow-up letter to industries.

6. May/June (I&CPS Conference) WG Chairman presents a pep talk to Conference, outlining reasons for survey.

7. August/September. WG Chairman evaluates data received; compiles first draft of report.

8. September/October (IAS Conference) WG Chairman reviews first draft of paper with members of WG and PSRSC.

9. November/December. WG Chairman prepares number of drafts required to satisfy need of WG.

YEAR 3:

10. January. WG Chairman sends final draft to PSRSC Chairman for approval.

11. February. WG Chairman prepares final manuscript and transmits for publication in I&CPS Conference record.

12. May/June (I&CPS Conference) WG Chairman presents results of survey at Conference.

Prepared by:

Philip E. Gannon, Chairman

Appendix G
Report of Transformer Reliability Survey—
Industrial Plants and Commercial Buildings

By
J. W. Aquilino
IEEE Transactions on Industry Applications
Sep./Oct. 1983, pp. 858–866

Report of Transformer Reliability Survey—Industrial Plants and Commercial Buildings

JAMES W. AQUILINO, MEMBER, IEEE

Abstract—The Power Systems Reliability Subcommittee of the IEEE Industry Applications Society has been conducting surveys of the reliability of electrical equipment in industrial and commercial power systems. A previous survey published in 1973 and 1974 [1] included data on the reliability of transformers. Some of the questions raised by the previous results, together with a general need for updated data, prompted a new survey which was conducted in 1979. The results of that survey are presented in this paper.

INTRODUCTION

ACCURATE reliability data on transformers, together with similar data on other types of electrical equipment, are necessary for evaluating power system reliability. Information of this type is often the only means of showing economic justification for spares, redundancy, or improved maintenance programs. The purpose of this 1979 transformer reliability survey of industrial plants and commercial buildings was to improve upon the results of the previous survey published in 1973-1974 [1] by answering some of the questions raised and eliminating some of the controversy created. The major reasons for conducting the new survey were outlined in a paper presented at the 1979 Industrial and Commercial Power Systems Technical Conference [2].

The most controversial items in the previous survey concerned the average outage duration time after a transformer failure in relation to the failure restoration method. Another item which raised questions was the comparatively high failure rate for rectifier transformers. The 1979 survey form was condensed considerably from the 1973-1974 version. Most of the items found to be of little significance in the past have been omitted. The remaining survey items are aimed at factors believed to have the most influence on the important transformer reliability and availability parameters.

Another major consideration in preparing the new survey form was simplicity. This was intended to enable the respondent to reply with minimal effort, thereby assuring maximum possible response. Obviously, the condensation could only be carried to a certain extent before the survey results would become so general that they would be of little practical value.

Results of the 1979 transformer survey are presented in this paper in tabular form. The discussion which follows under *Survey Results* attempts to expand upon some of the more

Paper IPSD 80-7, approved by the Power Systems Technologies Committee of the IEEE Industry Applications Society for presentation at the 1980 Industrial and Commercial Power Systems Conference, Houston, TX, May 12-15. Manuscript released for publication February 2, 1981.

The author was with Northrop Corporation, 100 Morse Street, Norwood, MA 02062. He is now with General Radio (GenRad), 170 Tracer Lane, Waltham, MA 02154.

significant survey data obtained. In any survey of this type there will undoubtedly be some new questions raised and also some old questions and controversies left unresolved. We feel, however, that this data will be of considerable value to system planners, designers, and users.

SURVEY FORM

The form used for the 1979 survey is shown in the Appendix. As mentioned before, the Total Population form was condensed to include data relating specifically to transformer reliability. Important influencing factors were rating, voltage, age, and maintenance. However, reporting the response to maintenance quality is difficult. The 1973-1974 survey asked the respondent to give his or her opinion of the maintenance quality as excellent, fair, poor, or none. It is very difficult to be completely objective in responding to this type of question. The new survey, therefore, asked for a brief description of the extent of maintenance performed, the idea being to enable the reader to judge for himself the benefits derived from a particular maintenance procedure. The failed unit data requested is basically the same as that in the previous survey. The most important categories here are the causes of failure, the restoration method, restoration urgency, duration of failure, and age at time of failure.

SURVEY RESPONSE

The response to the survey is summarized in Tables I and II. Responses were received from 25 different companies, and in many cases several locations within the companies were reported. Various types of industrial and commercial facilities are represented including chemical and petro-chemical plants, steel mills, paper mills, manufacturing plants, and hospitals, to name a few. Similar data from the 1973-1974 survey are shown in Table III for comparative purposes. A summarized comparison between the two survey results appears in Table IV. Direct comparisons cannot be made in some instances because of changes made in the sub-classes. For example, the new survey broke the ratings down into two groups, units 300-10 000 kVA and those greater than 10 000 kVA. The ratings in the previous survey were 300-750 kVA, 751-2 499 kVA, and 2 500 kVA and up.

One of the reasons for conducting this new survey was the need for reliability data on arc-furnace transformers. Unfortunately, the response to this category was very poor. The sample size reported was too small to obtain reliable results, therefore, the arc-furnace data were omitted. Hopefully, the response will improve in subsequent surveys. The response to the latest survey did improve over the 1973-

TABLE I
POWER TRANSFORMERS 1979 SURVEY

Type	Number of Units	Unit-Years	Number of Failures	Failure Rate Failures/ Unit-Year	Average Repair Time (Hours)	Average Replacement Time (Hours)
All liquid filled	1814	17 996	111	0.0062	356.1 N: 60 F[2]	85.1 N: 39 F[2]
Liquid 300–10 000 kVA	1750	17 410	102	0.0059	297.4 N: 56 F[2]	79.3 N: 37 F[2]
Liquid >10 000 kVA	64	586	9	0.0153	1178.5[1] N: 4 F[2]	192[1] N: 2 F[2]
Dry 300–10 000	159	1700	1[1]	0.0006[1]	6[1] N: 1 F[2]	– N: 0 F[2]

[1] Small sample size–less than eight failures.
[2] F is failures.

TABLE II
RECTIFIER TRANSFORMERS 1979 SURVEY

Type	Number of Units	Unit-Years	Number of Failures	Failure Rate Failures/ Unit-Year	Average Repair Time (Hours)	Average Replacement Time (Hours)
All liquid filled	85	841	16	0.0190	2316 N: 8 F[2]	41.4 N: 8 F[2]
Liquid 300–10 000 kVA	61	644	10	0.0153	1664[1] N: 3 F[2]	38.7[1] N: 7 F[2]
Liquid >10 000 kVA	24	197	6[1]	0.0303[1]	2707.2[1] N: 5 F[2]	60[1] N: 1 F[2]

[1] Small sample size–less than eight failures.
[2] F is failures.

TABLE III
ALL TRANSFORMERS[1]

Number of Plants in Sample Size	Sample Size Unit-Years	Number of Failures Reported	Industry		Failure Rate- Failures per Unit-Year	Industry Average	Minimum Plant Average	Median Plant Average	Maximum Plant Average
33	15,210	63	All..........	Liquid Filled - All.............	0.0041	529.0	2.0	219.	3744.
30	13,210	39	"	601-15,000 volts-All Sizes....	0.0030	174.	2.0	49.	840.
12	3,002	11	"	300-750 kVA.................	0.0037	61.0	4.5	10.7	336.
18	6,040	15	"	751-2,499 kVA...............	0.0025	217.	2.0	64.0	840.
11	4,036	13	"	2,500 kVA & up..............	0.0032	216.	24.0	60.0	403.
12	1,848	24	"	Above 15,000 volts..........	0.0130	1076.	12.8	1260.	3744.
16	4,937	18	"	Dry Type; 0-15,000 volts......	0.0036	153.	0.5	28.	720.
3	672	20	"	Rectifier; Above 600 volts......	0.0298	380.	24.0	80.	867.
14	8,598	43	Chemical.....	Liquid Filled - All.............	0.0050	338.	8.0	168.	1800.
12	6,838	24	"	601-15,000 volts-All Sizes....	0.0035	52.3	8.0	48.5	336.
7	3,274	10	"	300-750 kVA.................	0.0031	19.3	3.0	8.0	120.
9	1,601	19	"	Above 15,000 volts...........	0.0119	670.	12.8	708.	3600.
2	662	16	"	Rectifier; Above 600 volts......	0.0242	425.	80.0	474.	867.
3	2,512	14	Petroleum....	Liquid Filled - All.............	0.0056	843.	4.5	591.	1178.
3	2,334	10	"	601-15,000 volts-All Sizes....	0.0043	244.	4.5	204.	403.

[1] From IEEE Survey published in 1973-1974 [1].

TABLE IV
ALL TRANSFORMERS[1]

	Sample Size Unit-Years	Number of Failures	Type	Failure Rate Failures/ Unit-Year	Average Hours Downtime/ Failure
			Power-		
1979	17996	111	Liquid Filled	0.0062	249.3
Survey	1700	1[2]	Power-Dry	0.00062	6
	841	16	Rectifier	0.0190	1178.7
	15210	63	Liquid Filled	0.0041	529
1973/74	4937	18	Dry	0.0036	153
Survey	672	20	Rectifier	0.0298	380

[1] Comparison of 1979 and 1973-1974 surveys.
[2] Small sample size–less than eight failures.

TABLE V
FAILURE RATE VERSUS AGE

		Power Transformers			
Type	Age[1] (Yrs)	Number of Units	Sample Size Unit-Years	Number of Failures[2]	Failure Rate Failures/ Unit-Year
Liquid					
300-10 000 kVA	1-10	638	2625.5	19	0.0072
300-10 000 kVA	11-25	715	8846.5	47	0.0053
300-10 000 kVA	>25	397	5938.0	36	0.0060
Liquid					
>10 000 kVA	1-10	27	144.0	0[3]	–
>10 000 kVA	11-25	28	283.5	7[3]	0.0246[3]
>10 000 kVA	>25	9	158.0	2[3]	0.0126[3]

[1] Age is the age at end of reporting period.
[2] Relay or tap changer faults were not considered in calculations for failure rates or repair and replacement times.
[3] Small sample size–less than eight failures.

1974 survey as seen by comparing the total number of unit-years for both the power and rectifier transformers. Not too surprisingly, the largest sample size reported occurred among the power transformers 300-10 000 kVA which totaled 17410 unit-years.

SURVEY RESULTS

In Table IV it is clear that the results from the largest category, liquid filled power transformers, compared favorably between the 1973-1974 and 1979 surveys. This table also confirms the high failure rates for rectifier transformers. Before a further discussion on the results of the survey, in general, it would be worthwhile to note how the data compared with the controversial items in the previous survey.

The total number of hours (130 h) to replace a failed transformer with a spare appeared in Table 48 of the results of the 1973-1974 survey, under units 601-15 000 volts requiring a round-the-clock all out effort, and was felt by many to be too high. Units that were repaired showed an average outage time of 342 h. The new survey shows a considerable variation among power transformers depending upon size. The higher voltage units, reported in Table 49 of the results published in the 1973-1974 survey, showed an average repair time of 1842 h. This difference could be due to several factors, such as the transportation and han-

dling problems associated with the larger units and the greater likelihood of having spares for the smaller units on hand at the site.

The results of the new survey confirmed the long replacement time after a transformer failure. The much longer times needed to repair a failed transformer than to replace it with a spare were also confirmed. The new survey also confirmed the fact that the failure rates for rectifier transformers are much higher than those for the other transformer categories. This may be due to severe duties or the environments to which they are subjected.

AGE

Table V contains data broken down into three age groups. The failure rates for power transformers 300-10 000 kVA were approximately equal in all three age groups. The slightly higher failure rates for the units aged 1-10 years, and greater than 25 years, can probably be attributed to the infant mortality rate and units approaching end of life, respectively.

RESTORATION METHOD

Tables I and II also include data on restoration times versus restoration method. The data clearly indicate that the restoration of a unit to service by repair rather than replacement results in a much longer outage duration in all cases. This compares favorably with the previous survey which showed

279

TABLE VI
FAILURE INITIATING CAUSE

All Power Transformers

	No. of Failures¹	Percentage
Transient overvoltage disturbance (switching surges, arcing ground fault, etc...)	18	16.4
Overheating	3	2.7
Winding insulation breakdown	32	29.1
Insulating bushing breakdown	15	13.6
Other insulation breakdown	6	5.4
Mechanical breaking, cracking, loosening, abrading or deforming of static or structural parts	8	7.3
Mechanical burnout, friction or seizing of moving parts.	3	2.7
Mechanically caused damage from foreign source (digging, vehicular accident, etc.)	3	2.7
Shorting by tools or other metal objects	1	0.9
Shorting by birds, snakes, rodents, etc.	3	2.7
Malfunction of protective relay control device or auxiliary device	5	4.5
Improper operating procedure	4	3.6
Loose connection or termination	8	7.3
Others	1	0.9
Continuous overvoltage	0	0
Low voltage	0	0
Low frequency	0	0
	110	

¹ Failure initiating cause not specified for two failures.

TABLE VII
FAILURE CONTRIBUTING CAUSE

All Power Transformers

	No. of Failures¹	Percentage
Persistent overloading	1	1.1
Abnormal temperature	5	5.5
Exposure to aggressive chemicals, solvents, dusts, moisture or other contaminants	13	14.4
Normal deterioration from age	12	13.3
Severe wind, rain, snow, sleet or other weather conditions	4	4.4
Lack of protective device	2	2.2
Malfunction of protective device	7	7.8
Loss, deficiency, contamination, or degradation of oil or other cooling medium	9	10.0
Improper operating procedure or testing error	3	3.3
Inadequate maintenance	7	7.8
Others	27	30.0
Exposure to non-electrical fire or burning	0	0
Obstruction of ventilation by foreign object or material	0	0
Improper setting of protective device	0	0
Inadequate protective device	0	0
	90	

¹ Failure contributing cause not specified for 22 failures.

TABLE VIII
SUSPECTED FAILURE RESPONSIBILITY

All Power Transformers

	Number of Failures¹	Percentage
Manufacturer defective component or improper assembly	32	33.3
Transportation to site, improper handling	1	1
Application engineering, improper application	3	3.1
Inadequate installation and testing prior to start-up	6	6.3
Inadequate maintenance	25	26.0
Inadequate operating procedures	4	4.2
Outside agency–personnel	3	3.1
Outside agency–others	6	6.3
Others	16	16.7
	96	

¹ Suspected failure responsibility not specified for 16 failures.

repair times considerably longer than replacement times. Despite this fact, in most cases, a larger number of units was restored to service by repair. Results such as these show the obvious benefits in having spares at the site or readily available. The data may also help system planners and users determine the economic feasibility of purchasing spares. In computing the average repair and replacement times, those instances in which the repair or replacement was deferred were excluded to avoid distorting the averages. The averages shown represent only those cases where restoration was begun immediately.

FAILURE CAUSE

Tables VI–XI summarize the causes which initiate and contribute to the failure and the suspected failure responsibility for both power and rectifier transformers. Tables VI and IX show large percentages of failures initiated by some type of insulation breakdown or transient overvoltages. Table IX, however, shows a surprisingly large percentage of rectifier transformer failures initiated by mechanical causes.

Tables VII and X, which show the failure contributing causes, compare well with the 1973–1974 survey results. Normal deterioration from age contributed to a large number of both power and rectifier transformer failures. As in the past, Table VIII shows that respondents believed that manufacturer defects and inadequate maintenance were responsible for the greatest numbers of failures of power transformers. Table XI shows inadequate operating procedure was also a significant cause of failures of rectifier transformers.

MAINTENANCE CYCLE AND EXTENT OF MAINTENANCE

The large percentage of failures which resulted from inadequate maintenance shows the importance of accurate data on the extent and frequency of the maintenance performed. The latest survey attempted to obtain this data in a simple form. The response did not lend itself to reporting in tabular form. Maintenance information continues to be the most difficult to obtain in useful form, not only for transformers, but for all other equipment that have been surveyed as well. Hopefully in the future, we will be able to devise a method of obtaining this data and reporting it in a manner that will enable system users to establish effective preventive maintenance programs.

TYPE OF FAILURE

The 1979 survey limited the choices of failure type to "winding" and "other" (Tables XII and XIII). About half of the failures occurred on transformer windings.

All Rectifier Transformers		
	Number of Failures[1]	Percentage
Transient overvoltage disturbance (lightning, switching surges, arcing ground fault, etc.).	2	13.3
Overheating	1	6.6
Winding insulation breakdown	2	13.3
Insulation bushing breakdown	1	6.6
Other insulation breakdown	3	20
Mechanical breaking, cracking, loosening, abrading or deforming of static or structural parts	3	20
Mechanical burnout, friction or seizing of moving parts	2	13.3
Loose connection or termination	1	6.6
Continuous overvoltage	0	0
Mechanically caused damage from foreign source (digging, vehicular accident, etc.)	0	0
Shorting by tools or other metal objects	0	0
Shorting by birds, snakes, rodents, etc.	0	0
Malfunction of protective relay control device or auxiliary device	0	0
Low voltage	0	0
Low frequency	0	0
Improper operating procedure	0	0
Other	0	0
	15	

[1] Failure initiating cause not specified for 1 failure.

TABLE X
FAILURE CONTRIBUTING CAUSE

All Rectifier Transformers		
	No. of Failures[1]	Per- centage
Abnormal temperature	1	7.1
Exposure to aggressive chemicals, solvents dusts, moisture or other contaminants	1	7.1
Normal deterioration from age	4	28.6
Inadequate protective device	1	7.1
Loss, deficiency, contamination or degradation of oil or other cooling medium	3	21.4
Inadequate maintenance	3	21.4
Others	1	7.1
Persistent overloading	0	0
Exposure to non-electrical fire or burning	0	0
Obstruction of ventilation by foreign object or material	0	0
Severe wind, rain, snow, sleet or other weather conditions	0	0
Improper setting of protective device	0	0
Lack of protective device	0	0
Malfunction of protective device	0	0
Improper operating procedure or testing error	0	0
	14	

[1] Failure contributing cause not specified for two failures.

FAILURE CHARACTERISTIC

As would be expected, Tables XIV and XV show that about 3/4 of transformer failures resulted in removal from service by automatic protective devices, however, the percentage requiring manual removal was significant. Increasing use of transformer oil or gas analysis could be a factor here. This would enable detection of incipient faults in their early stages, allowing manual removal before a large scale failure occurs.

TABLE XI
SUSPECTED FAILURE RESPONSIBILITY

All Rectifier Transformers		
	Number of Failures	Percentage
Manufacturer–defective component or improper assembly	5	31.2
Application engineering-improper application	2	12.5
Inadequate maintenance	2	12.5
Inadequate operating procedures	5	31.2
Others	2	12.5
Transportation to site–improper handling	0	0
Inadequate installation and testing prior to startup	0	0
Outside agency–personnel	0	0
Outside agency–others	0	0
	16	

TABLE XII
TYPE OF FAILURE

Power Transformers		
Type of Failure	Number of Failures	Percentage
Winding	59	53
Others	53	47

TABLE XIII
TYPE OF FAILURE

Rectifier Transformers		
Type of Failure	Number of Failures	Percentage
Winding	8	50
Others	8	50

TABLE XIV
FAILURE CHARACTERISTICS

Power Transformers		
Failure Characteristic	Number of Failures	Percentage
Automatic removal by protective device	83	75
Partial failure reducing capacity	5	5
Manual removal	23	20

TABLE XV
FAILURE CHARACTERISTIC

Rectifier Transformers		
Failure Characteristic	Number of Failures	Percentage
Automatic removal by protective device	11	69
Partial failure reducing capacity	0	0
Manual removal	5	31

VOLTAGE

Table XVI shows the failure rate for liquid filled power transformers broken down by voltage rating. From Table XVI it is evident that the failure rates for 600-15 000 volt transformers are less than those for the higher voltage units in both

TABLE XVI
FAILURE RATE VERSUS VOLTAGE

		Power Transformers			
Type	Voltage (kV)	Number of Units	Sample Size Unit-Years	Number of Failures	Failure Rate Failures/Unit-Year
Liquid 300–10 000 kVA	.6–15	1626	15775	82	0.0052
Liquid 300–10 000 kVA	>15	124	1637	18	0.0110
Liquid >10 000 kVA	>15	52	490	9	0.0184

TABLE XVII
FAILURE RATE VERSUS VOLTAGE

		Rectifier Transformers			
Type	Voltage (kV)	Number of Units	Sample Size Unit-Years	Number of Failures	Failure Rate Failures/Unit-Year
All Liquid	.6–15	65	745	15	0.0201

size categories. The small sample sizes in several categories in Table XVII make it impossible to draw any definite conclusions on the effect of voltage on the failure rates of rectifier transformers.

CONCLUSION

The purpose of this survey was to update the results of the 1973–1974 survey and to clarify some of the issues raised by those results. In general, the data obtained in the latest survey confirm the previous results.

Only that data from which meaningful results could be obtained were included in this report. Obviously more information was requested in the survey than discussed in the previous sections. The remaining data were eliminated either because the sample sizes were too small, because analysis showed it to have little or no influence on transformer reliability, or because it could not be reported in a meaningful way.

APPENDIX

December 15, 1978

Subject: *Reliability Survey for Power, Rectifier, and Arc-Furnace Transformers*

Dear Sir:

The Power System Reliability Subcommittee of the Industrial and Commercial Power Systems Committee requests your cooperation in a survey to determine the reliability of power, rectifier, and arc-furnace transformers in industrial plants. This survey is part of a program to update the information obtained in our 1971 general reliability survey of plant equipment and to provide additional information on rectifier and arc-furnace transformers.

The results of this survey will be published in an IEEE paper. The information obtained is expected to be of value to system planners, designers, and users in the reliability evaluation of various alternatives. Individual responses will be held in confidence and only summaries published.

SURVEY INSTRUCTIONS

Definitions, brief instructions, and sample survey forms (Figs. 1-2) are provided for guidance. We wish to emphasize that all requested data is important, but it is also realized that some of the information requested may be unknown. In such cases, simply provide the information that is known, and leave the other spaces blank. If additional survey forms are needed, please duplicate the forms provided.

Definitions

1) *Failure*: A failure is any trouble with a power system component that causes any of the following to occur:
   a) partial or complete shutdown, or below standard plant operation,
   b) unacceptable performance of user's equipment,
   c) operation of the electrical protective relaying or emergency operation of the plant electrical system,
   d) de-energization of any electric circuit or equipment.

2) *Failure Duration:* Duration of a failure or repair time of a failed component is the clock hours from the time of the occurrence of the failure to the time when the component is restored to service, either by repair of the component or by substitution with a spare component. It includes time for diagnosing the trouble, locating the failed component, waiting for parts, repairing or replacing, and restoring the component to service. *It is not the time required to restore service to a load by putting alternate circuits into operation.*

Company Name and Plant: ______________________________

Location: ______________________________

Industry Type: ______________________________

Period Reported: From: Month ______________ Year ________

To: Month ______________ Year ________

TOTAL POPULATION: (List each transformer by number on a separate line.)

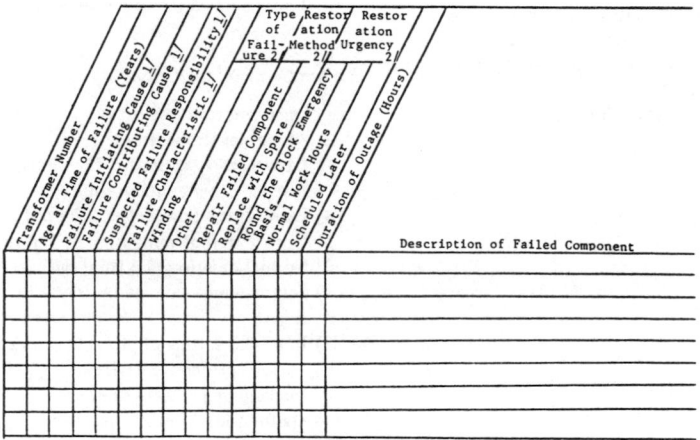

Transformer Number	Type 1/	Subclass Type 1/	Age at End of Period (Years)	Location 1/	Number of Phases	Primary Voltage Class (kV)	Rating 1/	Maintenance Cycle (Months)	Extent of Maintenance

1/ Use code from attached sheets.

Fig. 1. Reliability survey for power, rectifier, and arc-furnace transformers.

						Type of Failure 2/		Restoration Method 2/		Restoration Urgency 2/				
Transformer Number	Age at Time of Failure (Years)	Failure Initiating Cause 1/	Failure Contributing Cause 1/	Suspected Failure Responsibility 1/	Failure Characteristic 1/	Winding	Other	Repair Failed Component	Replace with Spare	Round the Clock Emergency Basis	Normal Work Hours	Scheduled Later	Duration of Outage (Hours)	Description of Failed Component

1/ Use code from attached sheets.
2/ Check applicable box.

Fig. 2. Failed Unit Data: Use transformer number from total population form.

General Data

1) It is vitally important that the period being reported be given.
2) Indicate the general type of industry involved at the plant being reported, such as auto, cement, chemical, metalworking, petroleum, pulp and paper, textile, etc.

Total Population Data

1) Using the Total Population data block, give the requested data for *all power, rectifier, and arc-furnace transformers in service during the period reported whether or not failures have been experienced.* Data should be reported on only those transformers used on a continuous basis. Transformers which are de-energized for substantial periods of time should not be included.
2) The age is the number of years from the time of installation to the end of the period reported under General Data.
3) Give a brief description of the extent of maintenance.

Failed Unit Data

1) List each failure separately.
2) Transformer Number for each failure is the Transformer Number used on the Total Population form.
3) Specify the age of the transformer at the time of failure.
4) Specify the failure initiating cause, contributing cause, and suspected failure responsibility using the code numbers on the attached sheets.
5) Check the restoration urgency.
6) Specify the time in hours from the onset of the failure until the transformer was restored to service.
7) Describe briefly the component that failed, including the component material.

Your participation in this survey will be greatly appreciated.

Sincerely,

J. W. Aquilino
Working Group Chairman

CODE NUMBERS TO BE USED WITH TOTAL POPULATION FORM

Transformer Type

1) Power
2) Rectifier
3) Arc-Furnace

Subclass Type

1) Liquid
2) Dry

Location

1) Indoor
2) Outdoor

Rating

1) 300–10 000 kVA
2) >10 000 kVA

CODE NUMBERS TO BE USED WITH FAILED UNIT DATA FORM

Failure Initiating Cause

1) Transient overvoltage disturbance (lightning, switching surges, arcing ground fault, etc.).
2) Continuous overvoltage.
3) Overheating.
4) Winding insulation breakdown.
5) Insulating bushing breakdown.
6) Other insulation breakdown.
7) Mechanical breaking, cracking, loosening, abrading, or deforming of static or structural parts.
8) Mechanical burnout, friction, or seizing of moving parts.
9) Mechanically caused damage from foreign source (digging, vehicular accident, etc.).
10) Shorting by tools or other metal objects.
11) Shorting by birds, snakes, rodents, etc.
12) Malfunction of protective relay control device or auxiliary device.
13) Low voltage.
14) Low frequency.
15) Improper operating procedure.
16) Loose connection or termination.
17) Others.

Failure Contributing Cause

1) Persistent overloading.
2) Abnormal temperature.
3) Exposure to aggressive chemicals, solvents, dusts, moisture, or other contaminants.
4) Exposure to nonelectrical fire or burning.
5) Obstruction of ventilation by foreign object or material.
6) Normal deterioration from age.
7) Severe wind, rain, snow, sleet, or other weather conditions.
8) Improper setting of protective device.
9) Lack of protective device.
10) Inadequate protective device.
11) Malfunction of protective device.
12) Loss, deficiency, contamination, or degradation of oil or other cooling medium.
13) Improper operating procedure or testing error.
14) Inadequate maintenance.
15) Others.

Suspected Failure Responsibility

1) Manufacturer-defective component or improper assembly.

2) Transportation to site-improper handling.
3) Application engineering-improper application.
4) Inadequate installation and testing prior to startup.
5) Inadequate maintenance.
6) Inadequate operating procedures.
7) Outside agency-personnel.
8) Outside agency-others.
9) Others.

Failure Characteristic

1) Automatic removal by protective device.
2) Partial failure reducing capacity.
3) Manual removal.

REFERENCES

[1] "Report on reliability survey of industrial plants," *IEEE Trans. Ind. Appl.*, Mar./Apr., July/Aug., and Sept./Oct., (Parts I–VI), 1974.
[2] "Reasons for conducting a new reliability survey on power, rectifier and arc-furnace transformers," in the *Ind. Comm. Power Syst. Tech. Conf.*, May 1979, pp. 70–75.

James W. Aquilino (M'76) was born in Arlington, MA, in 1949. He received the B.S.E.E. degree from Northeastern University, Boston, MA, in 1972.

After a brief period working in the field of special electronic instrumentation, he joined the Factory Mutual Engineering Association as a Loss Prevention Engineer specializing in electrical equipment and electric utility generating plants. He later joined the electrical section of the Factory Mutual Research Corporation, Boiler and Machinery Standards Department, where his duties included the preparation of standards for the protection and maintenance of electrical equipment. While employed at Factory Mutual, he represented the company as a member of the National Fire Protection Association on NFPA No. 75, Electronic Data Processing, on NEC Code Panels Nos. 11 and 23, and as an alternate on Panel No. 4. He then joined GenRad, Inc., in Concord, MA, as a Senior Facilities Engineer in the Corporate Plant Engineering Department where his responsibilities included plant electrical engineering, utilities monitoring, energy conservation, and preventive maintenance, among others. He was employed as a Plant Engineer at Northrop Corporation's Precision Products Division in Norwood, MA. He has since returned to GenRad.

Mr. Aquilino is a member of the IEEE Industry Applications Society Industrial Power Systems Department, Power Systems Support Committee, and the Power System Reliability Subcommittee. He served as the Chairman of the Working Group on Transformer Reliability. He is a Registered Professional Engineer in the State of Massachusetts.

Report of Large Motor Reliability Survey of Industrial and Commercial Installations
Parts I, II, and III

By
P. O'Donnell
IEEE Transactions on Industry Applications
Parts I & II, Jul./Aug. 1985, pp. 853–872; Part III, Jan./Feb. 1987, pp. 153–158

Motor Reliability Working Group
Power Systems Reliability Subcommittee
Power Systems Engineering Committee
Industrial and Commercial Power Systems Department
IEEE Industry Applications Society

Report of Large Motor Reliability Survey of Industrial and Commercial Installations, Part I

MOTOR RELIABILITY WORKING GROUP
POWER SYSTEMS RELIABILITY SUBCOMMITTEE
POWER SYSTEMS ENGINEERING COMMITTEE
INDUSTRIAL AND COMMERCIAL POWER SYSTEMS DEPARTMENT
IEEE INDUSTRY APPLICATIONS SOCIETY

Abstract—The Power Systems Reliability Subcommittee of the IEEE Industry Applications Society recently initiated a survey of the reliability of large motors in industrial and commercial installations in keeping with its commitment to support or update results of the survey published in 1973 and 1974. Moreover, the new survey has emphasized and expanded on one type of electrical equipment only. The previous survey results were heavily biased by one class of motors in the motor category and contained some results that appeared unreasonable and were considered questionable. The results of this new survey are presented here and intended to expand failure data to additional influencing categories and at the same time be oriented to the more common types in use today. A restriction to a lower limit in size also distinguishes the results to motors in relatively critical applications. A further explanation of the reasons for this survey and intended results is presented in a subcommittee report included for reference in the Appendix.

INTRODUCTION

THE RESULTS of the 1982 survey on the reliability of motors in industrial and commercial installations are summarized in Tables I–XIX. The data obtained allowed the various categories to be shown here which provide failure data on a more expanded and detailed basis, for the most part, than was presented in the 1973/1974 survey results. Also comparisons are made with the previous survey where results are of similar format.

To focus on motors that are of a critical nature, where reliability is most important, this survey differs from the other in that only motors larger than 200 hp are considered. In addition, to present data on motors most commonly manufactured and used today and to avoid distorted failure data from old motors that are expected to have high failure rates, this survey has limited the age of motors to no more than 15 years.

A brief discussion is included for each table identifying

Paper IPSD 83-12, approved by the Power Systems Technologies Committee of the IEEE Industry Applications Society for presentation at the 1984 Industrial and Commercial Power Systems Conference, Atlanta, GA, May 7–10, 1984. Manuscript released for publication May 7, 1984.
Members of IEEE Motor Reliability Working Group
   R. N. Bell is with E.I. du Pont de Nemours & Company, Engineering Department, Louviers Building, L5231, Wilmington, DE, 19898.
   D. W. McWilliams was with the Gates Rubber Company, 999 South Broadway, Denver, CO 80217. He is now deceased.
   P. O'Donnell, *Coordinating Author*, is with El Paso Natural Gas Company, Tex & Stanton Box 1492, El Paso, TX 79978.
   C. Singh is with Texas A&M University, Department of Electrical Engineering, College Station, TX 77843.
   S. J. Wells is with Union Carbide Corporation, P.O. Box 50, Hahnville, LA 70057.

significant points and results of the survey. The intent of this working group report is to present these results as updated experience in industry applications, and the drawing of definite conclusions is left to the reader.

SURVEY RESPONSE

The cover letter and questionaire form used in the survey are included in the Appendix. The form is specifically oriented to motors greater than 200 hp in size and no older than 15 years. As in other surveys succeeding the 1973 overall survey, this form is simplified into two sections: total population data and failure data.

Although the response was inadequate to identify a substantial number of industry types, the number of companies and plants identified was encouraging and the overall response was considered a success. Total population is less in this survey than in the 1973 survey, but this was anticipated due to the restriction on age and size. However, the total number of plants in the new survey is greater which adds credibility to the data as being representative of industry applications. The following list summarizes the magnitude of the response:

number of plants	75
number of companies	33
number of motors	1141
total population (unit years)	5085.0
total failures	360.

Some respondents did not submit data for every category evidenced by the comment "not specified" in the tables. Where response was insufficient to identify the motor and/or period reported the response was not used. As in previous survey reports, this report maintains the standard for credibility of failure rates by identifying categories that contain an insufficient number of failures to be representative.

SURVEY RESULTS

Summary

Table I summarizes the results in types of motors and voltage classes in similar fashion to the previous survey summary table. The previous data have been rearranged for comparison and presented here as Table II. In the new survey there was not enough response to separate the petroleum industry and chemical industry or to separate out other industry types and still show meaningful results.

TABLE I
OVERALL SUMMARY—LARGE MOTORS

Number of Plants in Sample Size	Sample Size (Unit Yr)	Number of Failures Reported	Industry	Equipment Subclass	Failure Rate (Failures/ Unit Yr)	Average Hours Downtime/ Failure	Median Hours Downtime/ Failure
75	5085.0	360	all	all	0.0708	69.3	16.0
				induction			
33	1080.3	89	all	0–1000 V	0.0824	42.5	12.0
52	2844.4	203	all	1001–5000 V	0.0714	75.1	12.0
5	78.1	2*	all	5001–15 000 V	*	*	*
1	13.5	—	all	not specified	—	—	—
				synchronous			
19	459.3	35	all	1001–5000 V	0.0762	78.9	16.0
2	29.5	3*	all	5001–15 000 V	*	*	*
				wound rotor			
5	137.0	10	all	0–1000 V	0.0730	*	*
9	251.1	8	all	1001–5000 V	0.0319	*	*
2	39.0	4*	all	5001–15 000 V	*	*	*
				direct current			
5	122.7	6*	all	0–1000 V	*	*	*
1	30.0	—	—	1001–5000 V	—	—	—
				induction			
11	484.3	39	petrochemical	0–1000 V	0.0805	88.3	40.0
28	1349.0	108	petrochemical	1001–5000 V	0.0801	109.4	48.0
2	10.3	1*	petrochemical	5001–15 000 V	*	—	—
				synchronous			
7	73.0	8	petrochemical	1001–5000 V	0.1096	72	16.0
				wound rotor			
2	20.8	4*	petrochemical	0–1000 V	*	—	—
3	17.6	3*	petrochemical	1001–5000 V	*	—	—

* Small sample size.

TABLE II
1973 OVERALL SUMMARY—MOTORS

Number of Plants in Sample Size	Sample Size (Unit Yr)	Number of Failures Reported	Industry	Equipment Subclass	Failure Rate (Failures/ Unit Yr)	Average Hours Downtime/ Failure	Median Hours Downtime/ Failure
—	42 463	561	all	all	0.0132	111.6	—
				induction			
17	19 610	213	all	0–600 V	0.0109	114.0	18.3
17	4229	171	all	601–15 000 V	0.0404	76.0	91.5
				synchronous			
2	13 790	10	all	0–600 V	0.0007	35.3	35.3
11	4276	136	all	601–15 000	0.0318	175.0	153.0
6	558	31	all	direct current	0.0556	37.5	16.2
				induction			
9	16 105	196	petrochemical	0–600 V	0.0122	123.4	—
10	3834	156	petrochemical	601–15 000 V	0.0407	74.3	—
				synchronous			
1	13 750	10	petrochemical	0–600 V	0.0007	35.3	35.3
6	4027	130	petrochemical	601–15 000 V	0.0323	175.8	—

Response was adequate in this survey to show an intermediate voltage class (1001–5000 V) not shown in the previous survey. Induction motors in the first two voltage classes show failures rates very nearly the same, with the lower voltage class slightly higher. Both are substantially higher than the earlier results (Table II).

The response for synchronous motors was dominated by the 1001–5000-V class, and again the new survey shows a failure rate twice that of the higher voltage rated synchronous motors in Table II. The new results show failure rates for synchronous and induction motors approximately equal for the same voltage class. The "petrochemical" industry shows a slightly higher failure rate for synchronous motors than for all industries.

The new survey obtained data on wound rotor induction motors with results showing a failure rate only slightly less than induction motors of the same lower voltage class. The next higher voltage class has a failure rate less than half that of synchronous and induction motors.

Although the sample size for dc motors was considered inadequate, this failure rate was the only one showing some consistency with the previous survey. The previous survey did not show a voltage class for dc motors.

Overall, the median hours downtime per failure was reported as less in the new survey than in the 1973 survey. Again the downtime reported was biased with unusually high periods and the average value for each class is consistently higher than the median value. The overall average and median downtime values calculated for all categories in this table include the downtime data omitted in the specific categories with "small sample size." Also, downtime for two failures was exceptionally and unusually high and therefore omitted from the results. One was reported as 960 h for an induction motor in the 0–1000-V class and replaced with a spare to restore service. The other was reported as 6570 h for an induction motor in the 1001–5000-V class and repaired during normal working hours.

Horsepower

Table III is presented to show a relationship of failure rate with size. The response gives a good comparison between the first two size categories with the failure rates calculating very nearly the same and also approximating those in Table I showing voltage classes. The third size category (5001–10 000 hp) shows a relatively high failure rate but calculated with a small population in sample size.

Speed

Failure rate is generally considered affected by speed, but Table IV shows somewhat unexpected results. The highest speed range, essentially 3600 r/min was included in this survey because of the increasing popularity in industry of two-pole motors. These results show the highest speed motors as most reliable and the lowest speed as least reliable.

Enclosure Type

This population type was added to expand on any notable effects on failure rate. Table V shows that open motors

TABLE III
HORSEPOWER VERSUS FAILURE RATE

	201–500 hp	501–5000 hp	5001–10 000 hp	> 10 000 hp	Not Specified
Sample size (unit-yr)	3185.6	1822.5	46.1	17.2	13.5
Number of failures	217	133	10	—	—
Failure rate (failures/unit-yr)	0.0681	0.0730	0.2169	—	—

TABLE IV
SPEED VERSUS FAILURE RATE

	0–720 r/min	721–1800 r/min	1801–3600 r/min	Not Specified
Sample size (unit yr)	657.1	3219.8	1194.6	13.5
Number of failures	66	232	62	—
Failure rate (failures/unit yr)	0.1004	0.0721	0.0519	—

experienced the highest failure rate among those with substantial sample size. Depending on the application this result might have been expected except the table below on causes does not support this result in the obvious causes of moisture and aggressive chemicals. It is suspected that more supporting data may be hidden in the relatively high response to causes reported as "other."

Environment

In Table VI the survey results show failure rate as affected by indoor and outdoor applications. It was expected that outdoor motors would show a higher failure rate than indoor motors, but the opposite was true. This follows from Table V which shows open type enclosures with the highest failure rate. One might conclude that when all environmentally related causes are combined as one, they support the results of Tables V and VI.

Duty Application

This population type breaks out continuous and intermittent application in Table VII. The total sample size was heavily dominated by continuous duty use with this category showing the highest failure rate at about twice that of intermittent duty. Some motors were reported as intermittent in a backup or standby role and operated only a small fraction of the period reported which may account partly for the large difference in failure rates.

Service Factor

Reliability versus service factor (SF) is an important consideration for those who must apply motors at varying load conditions that sometime exceed the normal nameplate rating of the motors. Table VIII shows a higher failure for 1.15-SF

TABLE V
ENCLOSURE TYPE VERSUS FAILURE RATE

	Open	Weather Protected	Totally Enclosed (TEFC, E.P., D.I.P)	Totally Enclosed (Open Pipe Vent)	Totally Enclosed (Water–Air)	Totally Enclosed (Air–Air)	Not Specified
Sample size (unit yr)	2597.6	569.5	1339.9	40.7	119.5	332.5	85.2
Number of failures	224	25	78	6*	6*	20	1*
Failure rate (failures/ unit yr)	0.0862	0.0439	0.0582	*	*	0.0602	*

*Small sample size.

TABLE VI
ENVIRONMENT VERSUS FAILURE RATE

	Indoor	Outdoor	Not Specified
Sample size (unit yr)	3359.9	1663.8	61.3
Number of failures	263	97	—
Failure rate (failures/unit yr)	0.0783	0.0583	—

TABLE VII
DUTY APPLICATION VERSUS FAILURE RATE

	Continuous	Intermittent	Not Specified
Sample size (unit yr)	4412.2	659.3	13.5
Number of failures	334	26	—
Failure rate (failures/unit yr)	0.0757	0.0394	—

TABLE VIII
SERVICE FACTOR VERSUS FAILURE RATE

	1.0SF	1.15SF	>1.15SF	Not Specified
Sample size (unit yr)	2557.9	2314.9	102.3	109.9
Number of failures	158	187	4*	11
Failure rate (failures/unit yr)	0.0618	0.0808	0.0391	0.1001

*Small sample size.

TABLE IX
AVERAGE NUMBER OF STARTS/DAY VERSUS FAILURE RATE

	<1	1–10	11–30	>30	Not Specified
Sample size (unit yr)	3654.8	1274.5	104.9	37.3	13.5
Number of failures	257	97	2*	4*	—
Failure rate (failure/unit yr)	0.0703	0.0761	0.0191	0.1072	—

*Small sample size.

TABLE X
POWER SUPPLY GROUNDING TYPE VERSUS FAILURE RATE

	Solid Ground	Impedence Ground	Ungrounded	Not Specified
Sample size (unit yr)	2287.7	1873.9	909.9	13.5
Number of failures	127	150	83	—
Failure rate (failures/unit yr)	0.0555	0.0800	0.0912	—

Average Number of Starts per Day

This population type was expected to provide data to show the effects of increasing severity in duty cycle, as related to starting, on failure rate. Surprisingly, the results (Table IX) show only a slight difference in failure rate between the first two categories. The response was disappointing in the last two categories, and no obvious trend in evident.

Power Supply Grounding Type

Much has been written about the effects of how the power supply system neutral is handled on reliability of electrical equipment and especially on motors. Table X shows results that support many generalizations and expected consequences of grounding types. The least failure rate is with solidly grounded power supplies, and the highest is with ungrounded power supplies. Commonly expected causes of failures in ungrounded systems include transient overvoltage and abnormal voltage levels, but the table on causes did not support this.

motors than for 1.0-SF motors. Under causes, overheating was reported as a significant failure initiator which raises the suspicion that exceeding temperature rises might be an application problem. These results do not show the effect of full service factor operation on field equipment of synchronous and dc motors or on secondary equipment of wound rotor motors. However, slip rings and brushes were not reported as obvious major problem areas as shown in Table XI.

TABLE XI
FAILED COMPONENT

Failed Component[a]	Number of Failures				
	Induction Motors	Synchronous Motors	Wound Rotor Motors	DC Motors	Total All Types
Bearing	152	2	10	2	166
Windings	75	16	6	—	97
Rotor	8	1	4	—	13
Shaft or CPLG	19	—	—	—	19
Brushes or slip ring	—	6	8	2	16
External device	10	7	1	—	18
Not specified	40	9	—	2	51

[a] Some respondents reported more than one failed component per motor failure.

However, insulation breakdown and deterioration from age might be interpreted as being affected by ungrounded systems.

Failed Component

Table XI shows which components failed most often for the four types of motors surveyed. Similar to the previous survey, bearings and windings were the predominate trouble areas. However, in this survey bearings by far led all other individual components in failures. In the previous survey windings failed most often. A significant number of failures occurred where the failed component was not specified in this survey.

Time Failure Discovered

The data in Table XII give an indication of when users discover most failures. Two-thirds of the failures were discovered during normal operation, and almost one third were discovered during testing or maintenance. Many feel that under a good maintenance program, most failures are discovered or prevented during testing or maintenance. Table XIV shows that about one-third of the total population reported excellent maintenance. The previous survey showed the same trend in when failures were discovered. The causes table lists major types that support the result of most failures being discovered during normal operation.

Causes of Failures

These results, shown in Table XIII are very close to those of the 1973 survey with some minor differences. The three most common failure initiators are mechanical breakage, overheating, and insulation breakdown. These causes, combined, are supportive of the previous survey results.

The major contributing cause reported is normal deterioration from age, as was also a major contributer in the other survey. Unlike the previous survey, high vibration and poor lubrication were also reported as significant causes which reinforce the problem areas of mechanical breakage and consequently bearing failures. Both surveys reported defective components and inadequate maintenance as major underlying causes.

Considering the combined contributing causes related to environmental conditions such as high ambient temperature, abnormal moisture, aggressive chemicals, and poor ventilation, the failure rates of open and indoor motors shown in

TABLE XII
TIME FAILURE DISCOVERED

	Number of Failures	Percent of Total
During normal operation	240	66.7
During routine maintenance or testing	101	28
Other	13	3.6
Not specified	6	1.7

TABLE XIII
CAUSES OF FAILURES

	Number of Failures	Percent
Failure Initiator		
1) Transient overvoltage	5	1.5
2) Overheating	45	13.2
3) Other insulation breakdown	42	12.3
4) Mechanical breakage	113	33.1
5) Electrical fault or malfunction	26	7.6
6) Stalled motor	3	0.9
7) Other	107	31.4
Failure Contributor		
1) Persistent overloading	14	4.2
2) High ambient temperature	10	3.0
3) Abnormal moisture	19	5.8
4) Abnormal voltage	5	1.5
5) Abnormal frequency	2	0.6
6) High vibration	51	15.5
7) Agressive chemicals	14	4.2
8) Poor lubrication	50	15.2
9) Poor ventilation or cooling	13	3.9
10) Normal deterioration from age	87	26.4
11) Other	65	19.7
Failure Underlying Cause		
1) Defective component	62	20.1
2) Poor installation/testing	40	12.9
3) Inadequate maintenance	66	21.4
4) Improper operation	11	3.6
5) Improper handling/shipping	2	0.6
6) Inadequate physical protection	19	6.1
7) Inadequate electrical protection	18	5.8
8) Personnel error	21	6.8
9) Outside agency other than personnel	12	3.9
10) Motor-driven equipment mismatch	15	4.9
11) Other	43	13.9

TABLE XIV
MAINTENANCE VERSUS FAILURE RATE

Maintenance Quality and Cycle	Sample Size (Unit Yr)	Number of Failures	Failure Rate (Failures/Unit Yr)	Median Hours Downtime/Failure	Average Hours Downtime/Failure
Excellent					
<12 mo	834.0	93	0.1115	8	53.6
12–24 mo	660.1	24	0.0364	24	40
>24 mo	285.5	9	0.0315	36	48
All	1779.6	126	0.0708	16	50.9
Fair*					
<12 mo	1776.8	155	0.0872	16	37.7
12–24 mo	967.7	39	0.0403	54	166.3
>24 mo	167.0	12	0.0719	165	264.4
Not Specified	4.0	1*	*	*	*
All	2915.5	207	0.0710	16	87.3
Poor[b]					
<12 mo	37.1	3*	*	*	*
12–24 mo	195.4	15	0.0563	96	83.6
>24 mo	6.0	1*	*	*	*
All	238.5	19	0.0797	72	70.7
None	123.3	7*	*	*	*
Not specified	28.0	1*	*	*	*

*Small sample size.
[a] 960 h downtime for one failure omitted.
[b] 6570 h downtime for one failure omitted.

Tables V and VI may not be abnormal. Additionally, this survey shows improper application as a significant problem area when the combined effects of poor installation/testing, physical and electrical protection, personnel error, and equipment mismatch are considered.

Maintenance Versus Failure Rate

Table XIV shows the results of failure rate compared to maintenance quality and maintenance cycle as reported in this survey. The previous survey results did not report maintenance cycle versus failure rate. However, Table XV has arranged available data to show quality versus failure rate. One notable difference can be seen in the maintenance cycle response in each quality category. The previous survey showed a trend in more frequent maintenance associated with higher quality. In the new survey response was greatest in the most frequent maintenance cycle in both the excellent and fair quality categories. So an obvious trend is not evident.

In both surveys, the largest response was to fair maintenance. However, the new survey had much more response to poor maintenance. Both had about the same division in response between fair and excellence qualities.

The most surprising result in the new data is the failure rate under reported excellent maintenance. Excellent maintenance with the most frequent cycle had the highest failure rate. Overall, in each quality category there is very little difference in failure rate.

The downtime listed in Table XIV does show an expected trend between the categories. The data suggest that the higher the quality and more frequent the cycle, the less severe the failure.

Description of Maintenance

Response was adequate to present a description of the methods of maintenance reported under the categories of

TABLE XV
1973 MAINTENANCE QUALITY VERSUS FAILURE RATE

Maintenance Quality and Cycle	Sample Size (Unit Yr)	Number of Failures	Failure Rate (Failures/Unit Yr)
Excellent			
<12 mo	14 650		
12–24 mo	1372		
>24 mo	1259		
All	17 281	77	0.0045
Fair			
<12 mo	121		
12–24 mo	21 930		
>24 mo	2958		
All	25 009	439	0.0175
Poor			
<12 mo	—		
12–24 mo	—		
>24 mo	74		
All	74	2*	0.0270*

*Small sample size.

quality and cycle. In Table XVI data are listed as percentages of the number of types of motor population reported (e.g., one plant reported six different types of motors with maintenance data listed for each type; these were counted as six population types for the purposes of this table). The differences and similarities between the various categories are quite obvious. The most commonly used method of maintenance under excellent and fair is "clean."

Failure Repair/Replace Urgency

Table XVII is intended to give some insight to the urgency reported for restoring motors to service and the resulting downtime of the failures. In these data the following two responses were considered unusual and exceptional and were omitted: downtime for one failure under "repair during normal working hours" was reported as 6570 h and downtime

TABLE XVI
DESCRIPTION OF MAINTENANCE REPORTED

Maintenance Description	Excellent				Fair				Poor			
	< 12 mo	12–24 mo	> 24 mo	All	< 12 mo	12–24 mo	> 24 mo	All	< 12 mo	12–24 mo	> 24 mo	All
1) Visual	12.5	2.3	—	6.5	24.7	43.1	41.7	32.6	—	31.2	—	33.3
2) Meggar	39.6	47.7	25.0	40.7	53.5	50.8	33.3	51.1	—	12.5	—	23.8
3) Clean	43.7	56.8	25.0	46.3	91.1	38.5	83.3	71.4	—	37.5	—	33.3
4) Lub. and/or filters	33.3	36.4	37.5	35.2	64.4	52.3	16.7	56.8	—	62.5	—	52.4
5) Vibration check	20.8	2.3	—	10.2	29.7	—	16.7	18.0	—	—	—	—
6) Bearing check	18.7	34.1	43.7	28.7	1.0	16.9	41.7	9.5	—	6.2	—	4.8
7) Reinsulate	4.2	—	18.7	4.6	—	3.1	33.3	3.4	—	6.2	—	4.8
8) Ampere or temperature check	4.2	—	—	1.9	3.0	13.8	8.3	7.3	—	12.5	—	9.5
9) Air gap check	2.1	20.5	—	9.3	8.9	—	—	5.1	—	—	—	—
10) Alignment	4.2	15.9	—	8.3	—	—	—	—	—	—	—	—
11) Change or check brushes	6.2	4.5	—	4.6	8.9	1.5	8.3	6.2	—	—	—	—
12) Overhaul	—	—	—	—	—	—	8.3	—	—	—	—	—
13) Paint	—	—	—	—	5.9	—	33.3	5.6	—	—	—	—
14) Check cooling system	—	—	—	—	3.0	—	—	1.7	—	—	—	—
15) Not specified	22.9	22.7	37.5	25.0	—	3.1	—	1.1	—	—	—	4.8
Number of Population Types	48	44	16	108	101	65	12	178	4	16	1	21

Percent of Population Types

TABLE XVII
REPAIR/REPLACE URGENCY VERSUS DOWNTIME

	Number of Failures	Average Hours Downtime/Failure	Median Hours Downtime/Failure
Normal working hours[a]	87	97.7	24.0
Round the clock	45	81.4	72.0
Replace with spare[b]	111	18.2	8.0
Low priority	4*	370.0*	400.0*
Not specified	6*	288.0*	240.0*
Total	251	69.3	16.0

*Small sample size.
[a] 6570 h for one failure omitted.
[b] 960 h for one failure omitted.

TABLE XVIII
1973 REPAIR/REPLACE URGENCY VERSUS DOWNTIME

	Number of Failures	Average Hours Downtime/Failure
Normal working hours	323	136.0
Working round the clock	54	110.3
Replace with spare	94	21.0
Low priority	7*	*
Total	478	108.5

*Small sample size.

for one failure under "replace with spare" was reported as 960 h. Data from the previous survey were rearranged and presented here as Table XVIII. Unlike the previous survey, median hours downtime per failure is included in the new data to reflect the influence of numerous long downtime periods reported.

In the first two categories the new survey shows obvious shorter average downtime per failure than the older survey, but the category on replace-with-spare is very close. An obvious uncertainty in the new results is evident in the median value for round-the-clock urgency. The downtime is higher than for less urgent repair. This suggests the possibility of some data being reported erroneously. Another interesting result is that half of the failures were reported as "replaced with spare" in the new survey. Only about one fifth of those of the old survey were in this category. This might be expected since the new survey covered only larger more critical

applications. The previous survey results presented no downtime data for the "low priority" category, and thus the total average in Table XVIII is calculated using only the data shown.

GENERAL DISCUSSION

It is the general consensus of the subcommittee sponsoring this activity that the new motor reliability data of this survey, contingent on reporting accuracy of the respondents, is more practical and useful for its intended purpose than the older survey data because of the restrictions on age and size. This survey also produced an attractive cross section of experience in the number of plants represented. One very obvious difference in the findings in this survey over the 1973 survey is the general trend of higher failure rates in the new data.

For obvious reasons, maintenance is expected to have a significant impact on failure rate and downtime. This paper, for the most part, presents results of responses to the population types as requested in the survey questionnaire.

TABLE XIX
90 PERCENT CONFIDENCE INTERVALS FOR FAILURE RATE

	Induction Motors	Synchronous Motors	Wound Rotor Motors	DC Motors	All
Lower limit	0.0659	0.0583	0.0350	0.0169	0.0644
Survey result	0.0732	0.0777	0.0515	0.0393	0.0708
Upper limit	0.0798	0.1026	0.0737	0.0699	0.0772
Percent deviation, L	10	25	32	57	9
Percent deviation, U	9	32	43	78	9

There are many possible combinations of categories, especially including those related to maintenance, that can be formulated from the responses. The questions and uncertainties stimulated by the results presented here warrant continued analysis and an additional report is planned to present this expanded analysis of the correlation between the various categorical results with particular emphasis on the effects of maintenance.

As an additional tool, Table XIX provides a measure of confidence in the use of the new data in this report. The table illustrates the statistical limits within which 90 percent of the failures could be expected to occur. The confidence limits are based on curves assuming a homogeneous population since it would be impractical to search out every variable affecting confidence levels and determine curves for each one.

APPENDIX

REASONS FOR CONDUCTING A NEW RELIABILITY SURVEY ON MOTORS

By: Power Systems Reliability Subcommittee,
Industrial and Commercial Power Systems Committee,
IEEE Industry Applications Society
September 1981

Charles R. Heising (*Chairman*)   Don W. McWilliams
James W. Aquilino   William T. Miles
Carl E. Becker   Joseph J. Moder
Richard N. Bell   John H. Moore
Thomas V. Booth   Pat O'Donnell
Williard H. Dickinson   A. D. Patton
Bruce Douglas   Chinan Singh
Phillip E. Gannon   Wayne L. Stebbins
Raymond E. Gibley   Howard P. Stickley
Ian Harley   Harold T. Wane
Thomas Key   Stanley J. Wells

The IEEE "Report on Reliability Survey of Industrial Plants, Part I: Reliability of Electrical Equipment" published in 1973 contained information on failure rates and downtime/failure for motors.

In their meeting on May 12, 1980, in Houston, TX, and in keeping with their commitment to update the previous survey, the Power Systems Reliability Subcommittee of the IEEE Industrial Power Systems Department is conducting a new survey on the reliability of motors.

Overall the main purpose of this reliability survey is to identify failure data and the effects of preventive maintenance

on important classes, types, and applications of motors, thus providing the designer and planner the valuable basic information needed to install a reliable and economic system.

The data in the previous reliability survey show that for motors rated 0–600 V the failure rate for induction motors is 15 times higher than synchronous motors. Since induction motors are normally considered more reliable than synchronous motors, it is presumed that the survey data were inadequate to cover enough applications to bring this out.

The data in the previous reliability survey shows that for induction motors 0–600 V (this category represents over 50 percent of the total motor population), the failure rate is 0.0109 (one unit failure per 92 unit years). This failure rate appears to be unreasonably low when compared with other equipment categories (i.e., motor starters = one failure per 72 unit years, steam turbine driven generators one failure per 32 unit years, transformers one failure per 244 unit years). Failure rate of this overall class of motors is obviously valuable information to users and manufacturers. This new survey will support or update this failure rate.

Motor designs, shop fabrication facilities, and manufacturing procedures for NEMA frame ac motors (ratings 1–200 hp) are significantly different from those for motors rated over 200 hp. In the previous motor reliability survey, the failure data for motors of all horsepower ratings were lumped together. The new motor reliability survey will collect failure data only on ac motors rated above 200 hp. Usually, motors rated above 200 hp are driving critical equipment. The reliability of these large motors is of prime importance to the industrial system design engineer. Recent user experience with reliability of the current generation of large ac motor designs (over 200 hp) indicates a trend toward a higher number of failures per unit time.

The previous survey data show that the industry average time to repair ac low-voltage motors (0–600 V) is 114 h compared to 76 h for medium-voltage ac motors (601–15 000 V). This information should be updated with a larger sample size of medium voltage motors.

The increased emphasis on minimizing capital investment in industrial facilities has resulted in a significant increase in the use of two-pole ac induction motors. Because of these relatively high speeds (3600 r/min), reliability of these two-pole motors is expected to be lower than the lower speed ac motors (four and six poles). The previous reliability study did not differentiate between 3600 r/min two-pole motors and the slower speed motors. The new motor reliability survey will collect separate reliability data on two-pole motors. Relative reliability data on two-pole motors and those with four or more poles will be useful to the industrial design engineer in evaluating the equipment cost savings inherent in two-pole (3600 r/min) operating speeds for motor and associated driven equipment.

The database for the previous reliability study (both unit years and number of units) represents something in the order of only a few hundredths of a percent of the total motor population.

The mailing list for the new survey will be expanded and edited to obtain failure data on a larger percentage of the total motor population.

COMPANY NAME AND PLANT:_____

INDUSTRY TYPE:_____

PERIOD REPORTED - FROM: MONTH_____ YEAR_____

TO: MONTH_____ YEAR_____

LOCATION:_____

CONTAMINATION LEVEL AND TYPE:_____

Fig. 1. Reliability survey for electric motors larger than 200 hp.

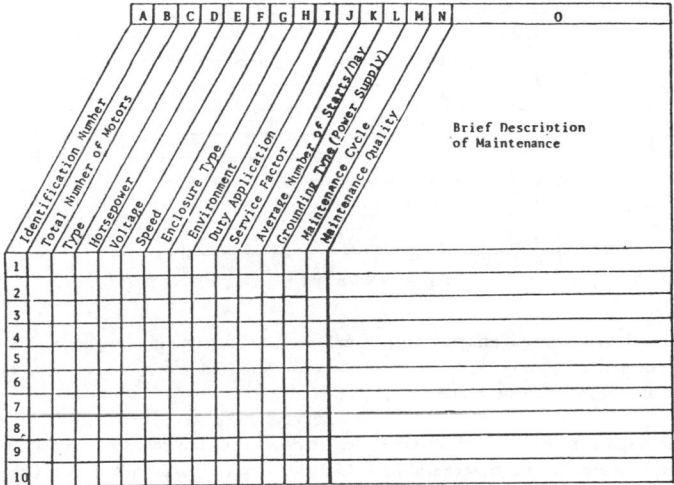

Fig. 2. Total population data.

COVER LETTER

Pat O'Donnell
El Paso Natural Gas Company
P.O. Box 1492
El Paso, TX 79978
(915) 541-2080

Dear Sir,

RE: Motor Reliability Survey for Motors Larger than 200 hp

The Reliability Subcommittee of the Industrial Power Systems Department requests your cooperation in a survey to determine the reliability of electric motors in industrial installations. As with previous surveys you may have seen, this survey is a followup to the general reliability survey of plant equipment in 1971 and is intended to provide more meaningful data on motors. Attached for your information is a report by the subcommittee on reasons for the survey.

The results of this survey will be published in an IEEE paper for value to system planners and designers in reliability evaluation of alternatives. Of course, individual responses will be held in strict confidence and only summaries published.

Survey Instructions

The survey form is reasonably self-explanatory, but a sample filled-out form is included for your guidance and some brief instructions follow. We emphasize that all requested data are important, but where some of these data are unknown, simply provide the known data and leave the other spaces blank. We also encourage any explanatory comments as you feel appropriate. If additional data sheets are needed, please duplicate those provided. *This survey is restricted to motors greater than 200 hp and no older than 15 years.*

General Data [Fig. 1]:

1) It is vitally important that the period reported be given.

2) Plant contamination level and type should be your best estimate.

Total Population Data [Fig. 2]:

1) Using the "total population" data block, give requested data for all motors greater than 200 hp and 15 years old or less, in service during the period reported *whether or not failures have occurred.* (Note: When the period reported exceeds the age of a motor, use separate data sheets for the new motors.)

2) Use the categories attached to the data block to describe the data.

3) When one data sheet is insufficient to list the total population of motors, use consecutive identification numbers in the first column of the data sheets (e.g., 1, 2, 3, etc., for first sheet; 11, 12, 13, etc., for the second sheet, and so on).

Failure Data [Fig. 3]:

1) List each motor failure event separately using the attached categories to describe the failure.

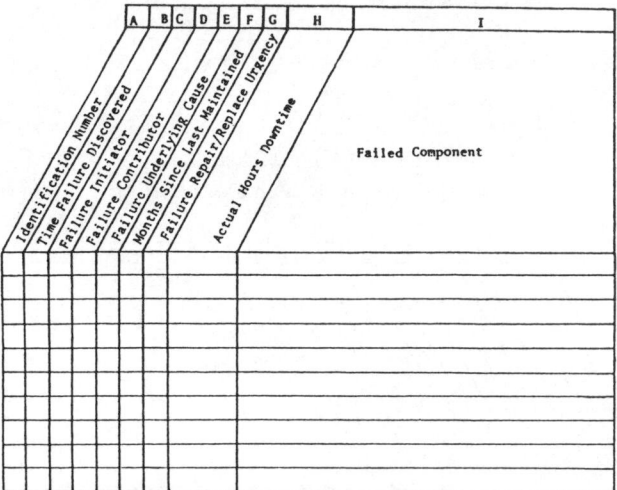

Fig. 3.   Failure data.

2) Identify each failure with the corresponding identification number in the "total population" data.

3) Under column I describe the component on the motor that failed.

Our schedule dictates that responses be received no later than April 15, 1982. Your participation in this project will be greatly appreciated.

Sincerely,

Pat O'Donnell
Chairman, Motor Reliability Survey Working Group

REFERENCES

[1] IEEE Committee Report, "Report on reliability survey of industrial plants," *IEEE Trans. Ind. Appl.*, Mar./Apr., July/Aug., and Sept./Oct. 1974.

[2] *IEEE Recommended Practice for Design of Reliable Industrial and Commercial Power Systems*, IEEE Standard 493, pts. I, III, IV, and VI.

Discussion

P. F. Albrecht (General Electric Company, Schenectady, NY), **E. L. Owen** (General Electric Company, Schenectady, NY), and **D. K. Sharma** (Electric Power Research Institute, Palo Alto, CA): This Working Group Report provides interesting and timely information which adds to a growing body of information about the reliability of electric motor drives. This information should be useful to owners, operators, and designers of motor equipments in their efforts to obtain improved motor reliability. The discussers welcome this additional information and support the objectives of the Working Group. We are hopeful that information of this type will become increasingly available as we feel it will assist all those involved in motor applications in obtaining increased reliability.

Surveys have been conducted by other groups seeking similar data for their industries. Under the sponsorship of the Electric Power Research Institute (EPRI), Palo Alto, CA, General Electric conducted an Industry Assessment Study (IAS) to evaluate the present reliability of powerhouse motors and to identify design and operational characteristics which, through advanced development, offer the potential of increased motor reliability [3]. Further work is presently underway to add data received after the closing date originally scheduled for the EPRI study. Analysis based on this additional data will be published at a later date.

We have compared the scope and results of this survey, as presented by the Working Group, with the results reported for the EPRI survey. Although the motor populations in the two studies are from different industries, we find many aspects of this Working Group Report which corroborates the findings of the EPRI study. The survey response achieved in the two studies are compared in Table XX.

In the EPRI study, it was found that failures subsequent to the first failure had a much different distribution than time to first failure. Therefore, the primary analysis was conducted in terms of time to first failure. Thus the failure rate from the EPRI study is not directly comparable with the Working Group results.

An important result of the EPRI study was to identify those motor components which are most subject to failure. This information was considered in setting priorities for development work to improve motor reliability. The type of motor involved in the EPRI survey was largely the squirrel-cage induction motor (approximately 97 percent of the "known" types were reported as cage induction motors), and the information about failure by component is most representative of this motor type. There are differences in the categories of

TABLE XX
SCOPE OF RELIABILITY STUDIES

Parameter		Working Group	EPRI Phase I
Working Group — Nomenclature — (EPRI)			
Number of companies	(Utilities)	33	56
Number of plants	(Units)	75	132
Number of motors	(Motors)	1141	4797
Total population (unit-years)	(Motor-years)	5085	24914*
Total failures		360	872
Failure rate (all motors)		0.0708	0.035*

*Based on first failure only.

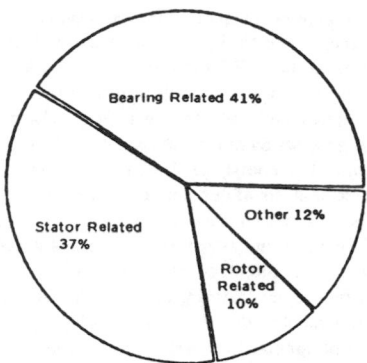

Fig. 4.   Percentage failure by component.

TABLE XXI
PERCENTAGE FAILURE BY COMPONENT

Bearing related	
Sleeve bearings	16
Antifriction bearings	8
Seals	6
Thrust bearing	5
Oil leakage	3
Other	3
Total	41
Stator related	
Ground insulation	23
Turn insulation	4
Bracing	3
Wedges	1
Frame	1
Core	1
Other	4
Total	37
Rotor related	
Cage	5
Shaft	2
Core	1
Other	2
Total	10

failed component as reported in the two studies, which makes a direct comparison of results very difficult.

However, both studies found that for squirrel-cage induction motors, bearing and stator winding related failures accounted for approximately three-fourths of all failures, while rotor related failures accounted for only ten percent of the failures. These results seem to corroborate each other and gives us greater confidence in our conclusions as to where emphasis should be placed. Fig. 4 and Table XXI show the percentage failure by component as reported by the EPRI study.

As a part of the EPRI study, additional analysis was performed to understand reliability issues better. We found that the most significant variable affecting motor failure rate was the plant (unit) where the motor was installed. For example, in the EPRI study a 90 percent confidence interval for failure rate of each of the 132 units was calculated. If all units had the same underlying failure rate, about 13 units would have a 90 percent confidence interval which does not include the failure rate for the entire population. However, in the EPRI study, 40 units had a 90-percent interval entirely below the population average, and 22 units were entirely above the population average.

We felt it was important to consider this unit variation when investigating other factors such as application or size effects. Was any such effect between respondents investigated in the Working Group survey? In particular, could the effect of horsepower noted in Table III of your report be *partly* due to the different companies represented in various size ranges.

Table III of the Working Group report suggests a tendency for the motor failure rate to increase with motor size. Booz, *et al.* also made an analysis based on motor size [4]. However, it was felt that horsepower per pole, rather than horsepower, better represented exposure to such failure mechanisms as

- fatigue resulting from differential expansion,
- high stress during operation,
- susceptibility to lateral vibration.

Would it be possible to analyze the Working Group data on the basis of horsepower per pole, similar to the EPRI analysis?

As a final comment, the detail of analysis must be commensurate with the size of the database. With the large database in the EPRI Phase II study, we hope to be able to investigate such factors as the effect of first failure on

subsequent failure rate. We again compliment the Working Group on a good survey and hope to see more of the same.

REFERENCES

[3] "Improved motors for utility applications," EPRI EL-2678, vol. 1, 1763-1, final rep., Oct. 1982.
[4] "Improved motors for utility applications, industry assessment study," EPRI EL-2678, vol. 2, 1763-1, final rep., Oct. 1982.

Pat O'Donnell (Coordinating Author): First, to address specific questions of the Discussion, we find the result of variation of reliability of motors in three different categories of units or groups very interesting and useful. However, the IEEE survey data do not lend themselves to this specific analysis. Our immediate response to this result is concern over the obvious cause or reason for this grouping to emerge. The IEEE data results attempted to classify industry types, which may follow a similar purpose, but the results related to maintenance more specifically categorize users in the IEEE report. We believe the IEEE and EPRI surveys are distinctly different in this respect but, as such, are complementary.

The IEEE survey collected data on a range of horsepower sizes and a range of speed ratings. We are not able to identify a fine resolution of horsepower per pole ratios but only general ranges. A quick analysis of our data for induction motors only allows the result shown in Table XXII.

The IEEE survey emphasized motor size and speed range separately with the intent of comparing these categories mutually and with others. Again, these results seem to be an excellent complement to the EPRI results, which diminish the significance of motor size in horsepower and speed as separate considerations. That is, a small high-speed motor might have the same horsepower/pole ratio as a large slow-speed motor.

We also are enthused about the added confidence in our data showing similarities in failed component trends. Bearing and winding failure trends were very similar in the two survey results. The IEEE survey did not collect detailed data to break down failed components into more subcategories of types, but data were collected on causes which helped determine *why* bearing and winding failures occurred. We are very interested in whether or not the difference in reliability between the "high" and "low" groups in the EPRI results supports the causes found in our survey results.

Finally, there is a significant difference in the basis of the two surveys that add, possibly, to some of the differences in results. The IEEE survey acquired data only on motors larger than 200 hp. The EPRI survey included sizes down to and including 100 hp. This surely accounts for some of the difference in total populations, but additionally, the IEEE data exclude standard NEMA frame size motors. It would be of interest to compare our results with EPRI results excluding motors 200 hp and smaller. This working group is enthused about the EPRI results, and we look forward to seeing further analysis of the data.

TABLE XXII
HORSEPOWER VERSUS SPEED
(INDUCTION MOTORS)

	Number of Failures	Unit Years	Failure Rate
0–720 r/min			
201–500 hp	7	137.92	0.0508
501–5000 hp	12	175.16	0.0685
5001–10 000 hp	—	—	—
>10 000 hp	—	—	—
721–1800 r/min			
201–500 hp	148	1922.43	0.0770
501–5000 hp	66	740.1	0.0892
5001–10 000 hp	1	2.83	0.3534
>10 000 hp	—	7.5	—
3600 r/min			
201–500 hp	42	655.75	0.0640
501–5000 hp	16	358.66	0.0446
5001–10 000 hp	—	—	—
>10 000 hp	—	—	—

Pat O'Donnell (S'64–M'68–SM'80) was born in El Paso, TX, in 1942. He received the B.S.E.E. degree from Texas Western College (now University of Texas at El Paso) in 1965.

After brief employment with Schlumberger Well Surveying Corporation, he joined El Paso Natural Gas Company in 1966 and is presently Principal Electrical Engineer in the main office Engineering Department in El Paso.

Mr. O'Donnell is currently active in the Industrial and Commercial Power Systems Department of the IEEE Industry Applications Society and currently serves as Secretary of the department. He is a member of and past Chairman of the Power System Technologies Committee and current Chairman of the Emergency and Standby Power Systems Subcommittee. He is also a member of the Power Systems Reliability Subcommittee, serving as Chairman of the Motor Reliability Working Group, and the Power Systems Analysis Subcommittee. Outside of IEEE, he is a member of the ASME Standards Committee on Ignition Systems for Industrial Engines. He is a Registered Professional Engineer in the States of Texas and New Mexico.

Report of Large Motor Reliability Survey of Industrial and Commercial Installations, Part II

MOTOR RELIABILITY WORKING GROUP
POWER SYSTEMS RELIABILITY SUBCOMMITTEE
POWER SYSTEMS ENGINEERING COMMITTEE
INDUSTRIAL AND COMMERCIAL POWER SYSTEMS DEPARTMENT
IEEE INDUSTRY APPLICATIONS SOCIETY

Abstract—In 1983 the initial results of an IEEE survey on large motors was published and presented at the 1983 I&CPS Conference. This was the first presentation of the results of a survey completed in 1982 of motors larger than 200 hp and no older than 15 years. The results presented here of the 1982 survey are to investigate the data further to address questions generated by the results of the earlier paper, to find additional correlations of the reliability criteria of some of the more interesting categories, and to bring out more results and categories available from the survey data. For information on the overall survey response and the general results of the surveyed categories, refer to the previous paper.

INTRODUCTION

THE SECOND set of results of the 1982 survey of the reliability of large motors in industrial and commercial installations is summarized in Tables I–XIII. Reference is occasionally made to the results presented in 1983 which will hereafter be called Part 1 [1].

In addition to new comparisons of categories to reveal more detailed analysis of the results of Part 1, these new results focus more on the effects of maintenance and especially more on the effects of causes. Of particular interest are the comparisons of reliability data for induction and synchronous motors, further analysis of service factor and speed, further analysis of bearing and winding failures, a closer look at the effect of inadequate maintenance on reliability, additional comparisons of indoor and outdoor applications, and additional grounding type comparisons.

Some comments about the data in the tables are in order to clarify some questions that may arise. Where no data are given, there was either no response or the number of failures (FLR's) and population were insufficient for meaningful

Paper IPSD 84-36, approved by the Power Systems Technologies Committee of the IEEE Industry Applications Society for presentation at the 1984 Industrial and Commercial Power Systems Conference, Atlanta, GA, May 7–10, 1984. Manuscript released for publication November 5, 1984.

Members of IEEE Motor Reliability Working Group
R. N. Bell is with E.I. du Pont de Nemours & Company, Engineering Department, Louviers Building, L5231, Wilmington, DE, 19898.
C. R. Heising is with Industrial Reliability Tech, 216 Farwood Road, Philadelphia, PA, 19151.
P. O'Donnell, *Coordinating Author,* is with El Paso Natural Gas Company, Tex & Stanton Box 1492, El Paso, TX 79978.
C. Singh is with Texas A&M University, Department of Electrical Engineering, College Station, TX 77843.
S. J. Wells is with Union Carbide Corporation, P.O. Box 50, Hahnville, LA 70057.

results. A footnote marks insufficient response where failures were reported, but the total was less than eight. This is in keeping with the standard of credibility previously established by the Power Systems Reliability Subcommittee. In preparation of this paper, a careful, closer look was taken and some of the minor errors in counting were corrected. Thus the total count in some areas will differ slightly from those of Part 1. However, the corrections are minor and no trends are affected. Also, as in the Part 1 results, downtime (DT) for two failures was omitted. One was 960 h for an induction motor, 0–1000 V and replaced-with-spare. The other was 6570 h for an induction motor, 1001–5000 V.

As with other survey results by this subcommittee, a brief discussion is included for each table emphasizing significant results, but there is no intent to draw definite conclusions. The tables are presented representing results from the data reported in the survey.

INDUCTION AND SYNCHRONOUS MOTORS

The results in Part 1 of the survey showed induction and synchronous motors with nearly equal failure rates. Some believe that synchronous motors, because of their complexity, should fail more than induction motors. Table I compares these types to various categories to identify any notable differences.

Two categories showed some deviation from the general results of Part 1. Where response was adequate in the first two classes, starts per day clearly affected synchronous motors more than induction motors. The induction motor failure rate changed very little, but the synchronous motor failure rate increased with an increase in starts per day. In the speed category it was the induction motors that showed some deviation from the trend of Part 1. One observation is the increase in failure rate with speed for the first two classes of speed. A second observation is the high failure rate for synchronous motors in the slowest speed class. So the two types of motors had opposite trends in failure rate with speed. The influence of synchronous motors on the slowest speed class is clearly evident where this class showed the highest failure rate in Part 1. For induction motors, the lowest failure rate was again in the highest speed class. The effects of speed are also evaluated in comparisons to horsepower, causes, and failed component.

TABLE I

	Starts/Day				Duty Application		Environment		Speed (r/min)			Grounding Type		
	1	1–10	11–30	>30	Contin-uous	Inter-mittent	Indoor	Outdoor	0–720	721–1800	3600	Solid	Imped-ance	Un-grounded
INDUCTION MOTORS														
Number of FLR's	234	58	—	—	274	20	203	91	19	216	59	101	123	70
Sample size (unit yr)	3215.8	756.0	88.4*	8.0*	3480.3	587.8	2485.9	1582.3	313.1	2817.9	1037.2	1909.6	1492.0	666.6
FLR rate (FLR's/unit yr)	0.0728	0.0767	—	—	0.0787	0.0340	0.0817	0.0575	0.0607	0.0766	0.0569	0.0529	0.0824	0.1050
Average hours DT/FLR	61.1	83.8	—	—	57.9	194.0	51.1	96.8	191.2	54.5	48.1	69.2	58.0	71.5
Median hours DT/FLR	12.0	18.0	—	—	12.0	54.0	8.0	48.0	72.0	8.0	36.0	36.0	10.0	8.0
Number of FLR's with no DT given	84	13	—	—	90	7	72	48	0	86	11	37	58	2
SYNCHRONOUS MOTORS														
Number of failures	13	23	2*	—	36	2*	38	—	27	10	1*	12	24	2*
Sample size (unit yr)	194.5	266.1	8.0	—	426.6	42.0	451.2	17.4*	254.9	200.9	12.7	251.7	200.3	16.5
FLR rate (FLR's/unit yr)	0.0668	0.0864	—	—	0.0844	—	0.0842	—	0.1059	0.0498	—	0.0477	0.1198	—
Average hours DT/FLR	97.5	68.4	—	—	58.4	—	74.2	—	33.1	139.1	—	166.0	39.8	—
Median hours DT/FLR	24.0	16.0	—	—	16.0	—	16.0	—	16.0	96.0	—	60.0	16.0	—
Number of FLR's with no DT given	2	1	—	—	16	—	3	—	0	3	—	2	1	—

*Small sample size.

TABLE II
MOTOR TYPE VERSUS SERVICE FACTOR

	Induction			Synchronous			Wound Rotor			Direct Current		
	1.0	1.15	>1.15	1.0	1.15	>1.15	1.0	1.15	>1.15	1.0	1.15	>1.15
Number of FLR's	127	165	2*	25	10	3*	10	12	—	6*	—	—
Sample size (unit yr)	2062.7	1943.0	62.5	274.2	152.8	41.5	160.7	246.4	—	94.2	30.0*	7.3*
FLR rate (FLR's/unit yr)	0.0616	0.0849	—	0.0912	0.0654	—	0.0622	0.0487	—	—	—	—
Average hours DT/FLR	54.4	75.0	—	81.2	63.4	—	52.3	192.2	—	—	—	—
Median hours DT/FLR	8.0	24.0	—	16.0	20.0	—	24.0	162.0	—	—	—	—
Number of FLR's with no DT given	28	71	—	0	3	—	3	6	—	—	—	—

*Small sample size.

SERVICE FACTOR

Another interesting result of this survey in Part 1 was that 1.15 service factor (SF) motors had a higher failure rate than 1.0-SF motors. Tables II–IV take a closer look at this category by comparison to other categories.

Table II compares service factor to the various types of motors surveyed. The results show that 1.15-SF induction motors failed more than 1.0-SF induction motors, but the opposite was true with synchronous and wound rotor induction motors. The lowest failure rate of all was in 1.15-SF wound rotor induction motors.

In Table III the service factor is evaluated in horsepower classes. Only the first two size classes had adequate response. As in the results of Part 1 failure rate increased with increase in service factor in the smallest size class. However, in the next larger size class the failure rate was approximately the same for 1.0 and 1.15 SF.

The next category broken out with service factor is voltage, shown in Table IV. The same trend evident in Part 1 is again evident here. The failure rate increased with increase in service factor for each voltage class where response was adequate. The service factor is evaluated further in Table VIII with comparisons to failed component and causes.

SPEED

Part 1 of the survey results showed a decrease in failure rate with increase in speed rating for all categories. Most expect that failure rate with speed is most affected by motor size. Table V is presented to show these categories from this survey. The results show the same trend as Part 1 except for a slight deviation in the smallest motor class. The 721–1800

302

TABLE III
HORSEPOWER VERSUS SERVICE FACTOR

	201–500 hp			501–5000 hp			5001–10 000 hp			10 000 hp		
	1.0	1.15	>1.15	1.0	1.15	>1.15	1.0	1.15	>1.15	1.0	1.15	>1.15
Number of failures	105	114	—	56	71	5*	7*	2*	—	—	—	—
Sample size (unit yr)	1758.0	1405.9	34.1*	777.4	961.4	77.2	39.2	4.8	—	17.2*	—	—
FLR rate (FLR's/unit yr)	0.0597	0.0811	—	0.0720	0.0739	—	—	—	—	—	—	—
Average hours DT/FLR	47.7	˙48.6	—	86.8	126.5	—	—	—	—	—	—	—
Median hours DT/FLR	8.0	12.0	—	16.0	50.0	—	—	—	—	—	—	—
Number of FLR's with no DT given	21	50	—	11	29	—	—	—	—	—	—	—

*Small sample size.

TABLE IV
VOLTAGE VERSUS SERVICE FACTOR

	0–1000 V			1001–5000 V			5001–15 000		
	1.0	1.15	>1.15	1.0	1.15	>1.15	1.0	1.15	>1.15
Number of FLR's	54	46	—	107	139	5*	8	1*	—
Sample size (unit yr)	745.5	509.0	7.3*	1725.4	1837.5	104.0	121	25.6	—
FLR rate (FLR's/unit yr)	0.0724	0.0904	—	0.0620	0.0756	—	0.0661	—	—
Average hours DT/FLR	38.8	88.3	—	75.3	75.9	—	22.7	—	—
Median hours DT/FLR	8.0	36.0	—	16.0	16.0	—	24.0	—	—
Number of FLR's with no DT given	6	18	—	24	61	—	2	—	—

*Small sample size.

TABLE V
HORSEPOWER VERSUS SPEED (r/min)

	201–500 hp			501–5000 hp			5001–10 000 hp			>10 000 hp		
	0–720	721–1800	8600	0–720	721–1800	3600	0–720	721–1800	3600	0–720	721–1800	3600
Number of FLR's	19	157	43	38	75	19	7*	2*	—	—	—	—
Sample size (unit yr)	277.3	2209.8	711.0	400.1	940.0	475.8	39.2	4.8	—	9.7*	7.5*	—
FLR rate (FLR's/unit yr)	0.0685	0.0710	0.0605	0.0950	0.0798	0.0399	—	—	—	—	—	—
Average Hours DT/FLR	156.2	35.7	39.9	99.4	109.2	116.1	—	—	—	—	—	—
Median Hours DT/FLR	70.0	8.0	36.0	16.0	24.0	52.0	—	—	—	—	—	—
Number of FLR's with no DT given	5	58	8	0	36	4	—	—	—	—	—	—

*Small sample size.

TABLE VI
ENCLOSURES—OUTDOOR

	Open	Weather Protected	Totally Enclosed (TEFC, E.P., D.I.P.)	Totally Enclosed (Open Pipe Vent)	Totally Enclosed (Water–Air)	Totally Enclosed (Air–Air)
Number of FLR's	18	17	49	2[a]	—	11
Sample size (unit yr)	111.1	379.0	1014.7	16.0	—	131.7
FLR rate (FLR's/unit yr)	0.1620	0.0449	0.0483	—	—	0.0835
Average hours DT/FLR	119.1	179.6	69.4	—	—	23.9
Median hours DT/FLR	48.0	80.0	48.0	—	—	12.0
Number of FLR's with no DT given	9	2	14	—	—	4
Failed component[b]						
Bearing	11	6	28	1	—	4
Winding	5	3	16	—	—	7
Rotor	1	1	2	—	—	—
Shaft or coupling	—	2	4	—	—	—
Brushes or slip rings	—	—	—	1	—	—
External dev.	—	3	—	—	—	—
Not specified	1	2	—	—	—	—

[a] Small sample size.
[b] Some respondents reported more than one failed component per failure.

TABLE VII
ENCLOSURES—INDOOR

	Open	Weather Protected	Totally Enclosed (TEFC, E.P., D.I.P.)	Totally Enclosed (Open Pipe Vent)	Totally Enclosed (Water–Air)	Totally Enclosed (Air–Air)
Number of FLR's	206	8	29	4[a]	6[a]	9
Sample size (unit yr)	2480.8	170.6	312.5	24.7	119.5	229.5
FLR rate (FLR's/unit yr)	0.0830	0.0469	0.0928	—	—	0.0392
Average hours DT/FLR	58.8	48.0	28.9	—	—	106.7
Median hours DT/FLR	16.0	16.0	10.0	—	—	8.0
Number of FLR's with no DT given	62	1	14	—	—	2
Failed component[b]						
Bearing	96	1	14	2	—	5
Winding	47	—	5	—	—	3
Rotor	3	—	2	—	—	—
Shaft or coupling	11	2	—	1	—	—
Brushes or slip rings	12	—	—	1	1	—
External dev.	6	4	—	—	4	—
Not specified	32	1	8	—	1	1

[a] Small sample size.
[b] Some respondents reported more than one failed component per failure.

r/min motors show a slightly higher failure rate than the 0–720 r/min motors. An interesting result is that the highest speed larger motors failed only approximately one-half the rate of the slowest speed smaller motors.

ENCLOSURES VERSUS ENVIRONMENT

Unexpected results of Part 1 were the relative failure rates of open and enclosed motors and the relative failure rates of indoor and outdoor motors. To evaluate these results further, the categories are combined in Tables VI and VII with failed components also included.

Table VI shows the highest failure rate with open type motors as would be expected since the environment is outdoor. In Table VII it was the second class of enclosed motors, which includes TEFC, explosion-proof (E.P.), and dust ignition proof (D.I.P.), with the highest failure rate. Combining all enclosed classes in each table shows very little difference in failure rate between indoor enclosed motors and outdoor enclosed motors.

TABLE VIII
SPEED AND SERVICE FACTOR VERSUS FAILED COMPONENT AND CAUSES[a]

| | Service Factor | | | Speed (r/min) | | |
	1.0	1.15	>1.15	0–720	721–1800	3600
Failed Component[b]						
Bearing	47.8	39.6	40	21.1	46.2	56.5
Winding	27.8	24.8	—	29.6	25.9	21.7
Rotor	2.8	5.0	—	8.5	2.0	5.8
Shaft or Coupling	6.7	6.4	—	5.6	6.9	5.8
Brushes or slip ring	7.2	1.5	—	15.5	2.0	—
External Device	0.6	6.4	60	8.5	2.8	5.8
Not Specified	7.2	16.3	—	11.3	14.2	4.3
Total FLR's	180	202	5	71	247	69
Failure initiator						
Transient Overvoltage	2.5	0.6	—	1.6	0.5	1.8
Overheating	15.2	11.2	20.0	8.1	13.5	17.9
Other Insulation Breakdown	12.7	12.8	—	12.9	14.4	5.4
Mechanical Breakage	36.7	30.2	20.0	16.1	36.0	41.1
Electrical Fault	10.1	3.9	60.0	12.9	6.8	5.4
Stalled Motor	1.3	0.6	—	3.2	—	1.8
Other	21.5	40.8	—	45.2	28.8	26.8
Total FLR's	158	179	5	62	222	56
Failure contributor						
Persistent Overloading	5.7	3.3	—	4.8	5.1	—
High-Ambient Temperature	5.7	1.1	—	1.6	3.3	3.8
Abnormal Moisture	7.1	4.9	—	4.8	6.5	3.8
Abnormal Voltage	2.1	1.1	—	1.6	0.9	3.8
Abnormal Frequency	—	1.1	—	—	4.7	1.9
High Vibration	14.2	16.8	—	14.5	14.9	18.9
Aggressive Chemicals	7.1	2.2	—	3.2	5.1	1.9
Poor Lubrication	19.9	10.9	40.0	9.7	14.4	24.5
Poor Ventilation or Cooling	2.1	4.9	—	8.1	2.8	1.9
Normal Deterioration/Age	17.0	33.2	60.0	25.8	28.8	18.9
Other	19.1	20.1	—	25.8	17.7	20.8
Total FLR's	141	184	5	62	215	53
Failure underlying cause						
Defective Component	12.9	25.6	60.0	19.4	19.6	23.1
Poor Installation/Testing	12.9	13.5	—	4.8	14.4	17.3
Inadequate Maintenance	22.4	20.5	20.0	16.1	25.8	11.5
Improper Operation	2.0	5.1	—	4.8	2.6	5.8
Improper Handling/Shipping	0.7	0.6	—	—	1.0	—
Inadequate Physical Protection	10.9	1.9	—	3.2	6.7	7.7
Inadequate Electrical Protection	9.5	3.2	—	4.8	6.7	5.8
Personnel Error	4.1	7.7	20.0	11.3	4.1	7.7
Outside Agency-Not Personnel	5.4	2.6	—	8.1	3.6	—
Motor-Driven Equipment Mismatch	4.1	5.8	—	8.1	4.6	1.9
Other	15.0	13.5	—	19.4	10.8	19.2
Total FLR's	147	156	5	62	194	52

[a] Number of failures in percent.
[b] Some respondents reported more than one failed component per failure.

The failed components followed the general overall trend with bearings and windings failing most, with bearings predominant. Only in the last enclosure class of outdoor motors was the trend between bearings and windings reversed.

FAILED COMPONENT AND CAUSES

Table VIII takes the speed analysis a step further by showing the failed components and causes of failure reported for the speed classes. With failed components distributed between the speed classes, the slowest speed motors show windings as the leading failed component and an increase in bearing failure percentages with increasing speed rating. Under causes an interesting result is the relative low percent blamed on inadequate maintenance for the highest speed rating. Also, deterioration from age was less for this class. This supports the low failure rate for high-speed motors.

Table VIII also breaks down service factor with failed component and causes. Bearings again led all components in failures with windings second. There seems to be no real outstanding difference in causes between 1.0 and 1.15 SF. However one difference that undoubtedly contributed to the failure rate of 1.15-SF motors is the contributing cause of

TABLE IX
CAUSES VERSUS VARIOUS CATEGORIES[a]

	Type		Solid	Grounding		Components	
	Induction	Synchronous	Solid	Impedence	Ungrounded	Bearings	Windings
Failure initiator							
Transient overvoltage	1.4	—	0.9	1.4	2.4	—	4.1
Overheating	14.7	—	14.0	11.7	14.5	12.4	21.4
Other insul. breakdown	11.9	21.1	16.7	11.0	9.6	1.9	36.7
Mechanical breakage	37.4	5.3	31.6	26.2	47.0	50.3	10.2
Electrical fault	5.8	23.7	8.8	4.8	10.8	3.7	11.2
Stalled motor	0.7	2.6	—	0.7	2.4	—	2.0
Other	28.1	47.4	28.1	44.1	13.3	31.7	14.3
Total FLR's	278	38	114	145	83	161	98
Failure contributor							
Persistent overload	4.9	2.7	4.5	4.4	3.7	1.4	6.5
High ambient temperature	3.4	—	3.6	0.7	6.1	.7	7.6
Abnormal moisture	6.7	2.7	8.0	4.4	4.9	2.7	18.5
Abnormal voltage	1.5	2.7	—	2.2	2.4	—	5.4
Abnormal frequency	0.7	—	0.9	0.7	—	—	1.1
High vibration	17.6	5.4	16.1	13.2	18.3	21.8	8.7
Aggressive chemicals	4.5	2.7	1.8	4.4	7.3	5.4	6.5
Poor lubrication	16.9	8.1	5.4	16.2	26.8	31.3	5.4
Poor ventilation or cooling	2.2	2.7	8.0	—	3.7	—	7.6
Normal deterioration/age	24.0	51.4	33.9	30.9	9.8	20.4	18.5
Other	17.6	21.6	17.9	22.8	17.1	16.3	14.1
Total FLR's	267	37	112	136	82	147	92
Failure underlying cause							
Defective component	20.3	22.2	23.5	14.5	24.4	17.8	10.9
Poor install/testing	15.9	—	7.8	12.9	19.5	14.5	10.9
Inadequate maintenance	22.8	11.1	25.5	18.5	20.7	27.6	19.6
Improper operation	3.3	2.8	3.9	4.0	2.4	2.0	6.5
Improper handling/shipping	.8	—	1.0	0.8	—	0.7	—
Inadequate physical protection	6.5	2.8	2.9	7.3	8.5	7.9	7.6
Inadequate electrical protection	5.3	11.1	6.9	6.5	4.9	2.6	15.2
Personnel error	5.7	5.6	3.9	6.5	8.5	7.2	5.4
Outside agency-not personnel	2.8	13.9	3.9	4.8	2.4	2.0	3.3
Motor-driven equip. mismatch	4.9	—	5.9	6.5	1.2	5.9	4.3
Other	11.8	30.6	14.7	17.7	7.3	11.8	16.3
Total FLR's	246	36	102	124	82	152	92

[a] Number of failures in percent.

normal deterioration from age which is about twice that for 1.0-SF motors.

Table IX is somewhat of a mix of some of the interesting categories brought out in other tables with emphasis on causes. Comparing induction and synchronous motors is difficult here because of the overwhelming response of induction motors. However, some of the results of other categories are supported. For instance, continuous duty induction motors had a higher failure rate than intermittent duty induction motors. Aside from the obvious influence of mechanical breakage, overheating and insulation breakdown are supportive. The contributing cause of normal deterioration from age is also evident.

The table correlates bearing and winding failures with causes rather well. Additionally, underlying causes show that both defective component and inadequate maintenance were reported as major factors in bearing and winding failures with inadequate maintenance the most significant. Failure initiators and contributors follow a reasonably logical trend.

The trend in failure rates for the categories of grounding do not appear supportive in this table if voltage related causes are expected to be obvious. This category exemplifies others where causes do not correlate well. It seems that in these results bearing and winding failures (especially bearing failures) and their related causes obscure some of the other cause reasoning.

MAINTENANCE

Tables X–XII attempt to delve further into the effects of maintenance on failure data. Table X reveals when the failed components were discovered. It gives some correlation to the effect of maintenance since one would expect a significant number of failures to be discovered during maintenance or testing under a good maintenance program. One observation for these data is that 56 percent of the bearing failures were discovered during normal operation. This is supported reasonably well by Table IX which shows inadequate maintenance as significant. Except for brushes and slip rings, all failed components show an obvious greater percentage of discovery during normal operation.

Tables XI and XII are presented to take a closer look at the underlying cause, inadequate maintenance, and associated failure data blamed on this cause. Again bearings by far led all other components in failures. Approximately 25 percent of all

TABLE X
FAILED COMPONENT VERSUS TIME DISCOVERED[a]

Failed Component[b]	Normal Operation	Time Discovered Maintenance or Test	Other
Bearing	36.6	60.6	50.0
Winding	33.1	8.3	28.6
Rotor	5.1	1.8	—
Shaft or coupling	5.8	8.3	14.3
Brushes or slip rings	3.1	7.3	—
External device	5.1	3.7	—
Not specified	11.3	10.1	7.1
Total FLR's	257	109	14

[a] Number of failures in percent.
[b] Some respondents reported more than one failed component per failure.

TABLE XI
INADEQUATE MAINTENANCE
FAILED COMPONENTS AND CAUSES[a]

Failed component[b]	
Bearing	59.6
Winding	25.4
Rotor	1.4
Shaft or coupling	—
Brushes or slip ring	8.5
External device	1.4
Other	4.2
Total FLR's	71
Failure initiator	
Transient overvoltage	—
Overheating	4.2
Other insulation breakdown	14.1
Mechanical breakage	52.1
Electrical fault	2.8
Stalled motor	—
Other	26.8
Total FLR's	71
Failure contributor	
Persistent overloading	—
High ambient temperature	4.2
Abnormal moisture	7.0
Abnormal voltage	—
Abnormal frequency	—
High vibration	4.2
Aggressive chemicals	9.9
Poor lubrication	43.7
Poor ventilation/cooling	1.4
Normal deterioration/age	18.3
Other	11.3
Total FLR's	71

[a] Number of failures in percent.
[b] Some respondents reported more than one failed component per failure.

TABLE XII
INADEQUATE MAINTENANCE FAILURE DATA

Number of FLR's	66
Sample size (unit yr)	603.6
FLR rate (FLR's/unit yr)	0.1093
Average hours DT/FLR	80.8
Median hours DT/FLR	9.0
Number of FLR's with no DT given	13

Maintenance quality and cycle	Number of FLR's (percent)
Excellent	
<12 mo	25.8
12–24 mo	—
>24 mo	—
Fair	
<12 mo	37.9
12–24 mo	7.6
>24 mo	3.0
Poor	
<12 mo	3.0
12–24 mo	12.1
>24 mo	—
Total FLR's	66

bearing failures were reported due to inadequate maintenance. Close to 44 percent of the brush and ship ring failures were reported due to this cause which does not follow well from Table X. The single largest contributor with this underlying cause is poor lubrication.

Table XII shows a definite higher failure rate for inadequate maintenance related failures than the Part 1 failure rates for maintenance quality. In Part 1 the failure rate results for excellent to poor maintenance ranged from 0.0708 to 0.0797, respectively.

Data for when failures were discovered versus maintenance quality are presented in Table XIII. It was expected that the fair and excellent categories would be significantly different in when failures were discovered, but the results show very little difference. The same table also includes months since last maintenance versus maintenance quality. The failures seem to follow the same trend as scheduled cycle reported with most occurring less than 12 mo since maintenance. This table is presented in the same format as [2, table 70]. Those results showed an obvious difference between fair and excellent maintenance overall. The trend in failures was to a certain degree increasing directly with months since maintenance and indirectly with maintenance quality. The new survey results here show a very different trend with most failures occurring where last maintenance was less than 12 mo prior to the failure.

GENERAL DISCUSSION

The additional comparisons and analyses made in this paper have supported results of Part 1 in some cases and in other cases have revealed results that were obscured in the general categorical tables of Part 1. Not all questions are answered here, and there are certainly many more categories and comparisons that can be made with the data of this survey. As examples, bearing and winding failures compared to starts per

TABLE XIII
MAINTENANCE QUALITY VERSUS TIME FAILURES DISCOVERED AND MONTHS SINCE MAINTENANCE[a]

Maintenance Quality	Normal Operation	Time Discovered Maintenance or Test	Other	Months Since Maintenance		
				< 12	12–24	> 24
Excellent	85	35	1	87	17	6
Fair	132	63	10	102	22	8
Poor	15	3	1	11	5	—
None	7	—	—	—	1	5
Total	239	101	12	200	45	19
Inadequate Maintenance Cause						
Excellent	5	12	—	17	—	—
Fair	22	8	2	16	1	1
Poor	8	1	1	4	1	—
None	7	—	—	—	1	5
Total	42	21	3	37	3	6

[a] Number of failures.

day and duty application could add meaning to the results. The Reliability Subcommittee is presently evaluating criteria that should be presented in a third set of results, Part 3. Interested readers should submit comments and suggestions on information they would like to see in Part 3. In the format presented in these results, bearing failures and their causes were very dominant and likely prevent other less significant correlations to be evident.

ACKNOWLEDGMENT

The Reliability Subcommittee acknowledges and expresses its appreciation to J. Rizo and W. H. Healy of El Paso Natural Gas Company for assistance in the computer program used for the analysis of the data for this presentation.

REFERENCES

[1] IEEE Committee Report, "Report of large motor reliability survey of industrial and commercial installations, Part I," in *1983 IEEE I&CPS Conference Record*; also *IEEE Trans. Ind. Appl.*, vol. IA-21, pp. 853–864, July/Aug. 1985.
[2] IEEE Committee Report, "Report on Reliability Survey of Industrial Plants, Part VI—Maintenance Quality of Electrical Equipment," in IEEE Standard 493, 1980.

Pat O'Donnell (S'64–M'68–SM'80), for a photograph and biography, please see page 864 of this issue.

Report of Large Motor Reliability Survey of Industrial and Commercial Installations: Part 3

MOTOR RELIABILITY WORKING GROUP
POWER SYSTEMS RELIABILITY SUBCOMMITTEE
POWER SYSTEMS ENGINEERING COMMITTEE
INDUSTRIAL & COMMERCIAL POWER SYSTEMS DEPARTMENT
IEEE INDUSTRY APPLICATIONS SOCIETY

Abstract—Results of a survey conducted in 1982 of the reliability of large motors have been presented and published in two parts [1], [2]. These results have generated numerous questions and comments and, consequently, the need to further analyze the data of the survey was recognized. Part 1 presents general results based on categories of motor types and applications specifically requested in the survey questionnaire. Part 2 combines various categories and addresses some questions resulting from Part 1. Part 3 of the survey results is presented here to address new questions and comments and to add more specific analyses of areas not yet explored. These results, along with Parts 1 and 2, provide the complete complement of analysis to date.

Introduction

THE THIRD part of the results of the 1982 survey of reliability of large motors is presented here and summarized in Tables I through VII. As with Part 2, these results focus on new comparisons of the data. The tables address some questions and comments received since presentation of Part 2 and provide additional analysis of causes. The order of the tables as presented is more or less random and there is no intent to portray a delibrate order.

As in Parts 1 and 2, where no data is given, there is insufficient response to the questionnaire. An asterisk represents failures reported but with insufficient number (less than eight) for credible results. Additionally it is again emphasized that the tables and corresponding discussions represent results of the survey and that there is no intent to draw definite conclusions. Finally, as in Parts 1 and 2, differences in total

Paper ICPSD 86-13, approved by the Power Systems Engineering Committee of the IEEE Industry Applications Society for presentation at the 1986 Industrial and Commercial Power Systems Technical Conference, Philadelphia, PA, September 8-11. Manuscript released for publication September 9, 1986.

Members of the IEEE Motor Reliability Working Group
C. Heising is with Industrial Reliability Technology, 216 Farwood Road, Philadelphia, PA 19151.
P. O'Donnell, *Coordinating Author*, is with the El Paso Natural Gas Company, P. O. Box 1492, El Paso, TX 79978.
C. Singh is with the Department of Electrical Engineering, Texas A&M University, College Station, TX 77843.
S. J. Wells is at 1743 Lake Oak Drive, Seabrook, TX 77586.
IEEE Log Number 8612073.

failures between the various categories of Part 3 reflect missing data from some survey responses.

Enclosure—Indoor and Outdoor

Tables I and II are presented to take a closer look at the causes of failures reported for various enclosures in both indoor and outdoor environments. As was evident in the previously published results, most indoor applications were "open" motors and most outdoor applications were totally enclosed fan-cooled (TEFC), explosion-proof or dust ignition-proof motors.

For the outdoor motors with the above enclosures, Table I shows that the major failure initiators are well supported by the failure contributors. The main underlying causes point to defective components and inadequate maintenance. For indoor open motors in Table II, failure initiators and failure contributors again match, but inadequate maintenance was by far the single largest underlying cause.

Comparison of indoor and outdoor environments also reveals certain opposite trends relative to causes of all failures (Part 1, Table 13). For instance, the following causes show opposite trends between indoor and outdoor applications when their respective percentages of total are compared to the same for all applications of Part 1, Table 13: mechanical breakage, electrical fault or malfunction, abnormal moisture, poor lubrication, inadequate electrical protection, inadequate maintenance, and personnel error. An example will make this more clear. For outdoor motors, mechanical breakage is 26/90 or 28.9 percent of the total number of failures for "failure initiator," while for all applications 113/341 is 33.1 percent of the number of failures for "failure initiator." Indoor motors show 85/240, or 35.4 percent versus 33.1 percent.

High Vibration Cause

Tables III and IV present additional results to Parts 1 and 2 for failures blamed on vibration. Table III shows 48 failures blamed on vibration where data are also available on failure initiator and underlying cause. As would be expected, most failures were initiated by mechanical breakage. It is interesting that most underlying causes were reported as defective component and poor installation or testing. Only three failures list inadequate maintenance as a contributing cause. For

TABLE I
ENCLOSURES—OUTDOOR
(No. of Failures)

Causes	Open	Weather-protected	Totally Enclosed TEFC, Exp., D.I.	Totally Enclosed Open Pipe Vent	Totally Enclosed Water–Air	Totally Enclosed Air–Air	Total	All Applications (Part I)
Failure Initiator								
Transient overvoltage	1	—	1	—	—	—	2	5
Overheating	2	4	9	—	—	—	15	45
Other insulation breakdown	4	1	10	1	—	—	16	42
Mechanical breakage	6	7	11	1	—	1	26	113
Electrical fault/malfunction	1	—	4	—	—	4	9	26
Stalled motor	—	—	1	—	—	—	1	3
Other	4	6	5	—	—	6	21	107
Failure Contributor								
Persistent overload	—	2	—	—	—	—	2	14
High ambient temperature	—	1	1	—	—	—	2	10
Abnormal moisture	2	2	5	—	—	—	9	19
Abnormal voltage	2	—	—	—	—	—	2	5
Abnormal frequency	—	—	—	—	—	—	—	2
High vibration	1	3	6	—	—	1	11	51
Aggressive chemicals	1	—	1	1	—	3	6	14
Poor lubrication	—	2	3	—	—	1	6	50
Poor ventilation/cooling	—	—	—	—	—	4	4	13
Normal deterioration/age	2	2	7	1	—	2	14	87
Other	2	5	9	—	—	—	16	65
Failure Underlying Cause								
Defective component	3	4	9	—	—	2	18	62
Poor installation/testing	2	3	4	—	—	—	9	40
Inadequate maintenance	3	2	7	1	—	—	13	66
Improper operation	—	—	—	—	—	1	1	11
Improper handling	1	—	—	—	—	—	1	2
Inadequate physical protection	4	2	—	—	—	—	6	19
Inadequate electrical protection	2	2	3	1	—	—	8	18
Personnel error	—	—	1	—	—	1	2	21
Outside agency-not pers.	—	—	—	—	—	2	2	12
Motor–load mismatch	—	1	1	—	—	3	5	15
Other	—	3	9	—	—	2	14	43

convenience, the total of 51 failures blamed on high vibration (Part 1) is also shown.

Table IV compares vibration failure causes to size. Only two size ranges have sufficient response to allow meaningful results. The table shows that the percent of vibration failures to total failures increases slightly with size.

STARTS/DAY VERSUS CONTINUOUS DUTY APPLICATION

The results in Table V attempt to further evaluate the effects of starting on failures. Only continuous duty applications are considered, to avoid confusion over trying to distinguish between various degrees of intermittent duty. The first two voltage classes of induction motors, in which most of the survey data were collected, are emphasized. Also, very little data were collected for the categories of more than ten starts per day.

As can be seen from the table, overall there is very little difference in failure rates between less-than-one and one-to-ten starts per day, and very little difference between the two voltage classes. There does, however, seem to be a trend in longer downtimes for the one-to-ten starts per day category, suggesting that failures were more severe.

DOWNTIME VERSUS REPAIR URGENCY AND TIME DISCOVERED

Downtime is expected to be affected by the urgency with which repairs are made and also by when failures are discovered, which would seem to affect the severity of failures. Table VI compares downtime with these categories to get a different view than Parts 1 and 2 provide. Overall the trend in number of failures decreases as downtime increases. There are some obvious deviations from this trend at the range of 51–100 h downtime per failure. Also this trend is obscure under the repair urgency "round-the-clock." It is interesting that for this category there are practically as many failures in the higher downtime ranges as in the lower downtime ranges. Another somewhat unexpected result is that there is no obvious difference in the distribution of failures between the categories under the heading "time discovered." However, the results show that failures corrected by "replace with spare" are predominantly in the least downtime range, as would be expected.

HORSEPOWER VERSUS SPEED: INDUCTION MOTORS

A recent motor reliability survey [3] sponsored by the Electric Power Research Institute (EPRI) and conducted by the

TABLE II
ENCLOSURES—INDOOR
(No. of Failures)

Causes	Open	Weather-protected	Totally Enclosed TEFC, Exp., D.I.	Totally Enclosed Open Pipe Vent.	Totally Enclosed Water–Air	Totally Enclosed Air–Air	Total
Failure Initiator							
Transient overvoltage	3	—	—	—	—	—	3
Overheating	25	—	3	—	1	1	30
Other insulation breakdown	22	—	3	—	1	1	27
Mechanical breakage	68	1	11	2	—	3	85
Electrical fault/malfunction	—	5	1	—	—	3	9
Stalled motor	—	—	—	—	—	—	—
Other	68	1	10	2	4	1	86
Failure Contributor							
Persistent overload	—	—	3	—	—	1	4
High ambient temperature	—	—	3	—	—	1	4
Abnormal moisture	10	—	—	—	—	—	10
Abnormal voltage	—	—	—	—	—	—	—
Abnormal frequency	—	—	—	—	—	—	—
High vibration	35	1	1	—	—	2	39
Aggressive chemicals	—	1	—	1	—	—	2
Poor lubrication	38	—	3	—	1	2	44
Poor ventilation/cooling	—	1	—	2	1	—	4
Normal deterioration/age	38	3	14	1	—	3	59
Other	38	1	5	—	4	—	48
Failure Underlying Cause							
Defective component	27	4	6	1	5	—	43
Poor installation/testing	28	—	1	—	—	1	30
Inadequate maintenance	41	1	8	1	—	2	53
Improper operation	—	—	1	—	—	1	2
Improper handling	—	—	—	—	—	—	—
Inadequate physical protection	10	—	1	1	—	1	13
Inadequate electrical protection	—	1	—	—	—	2	3
Personnel error	16	—	—	—	—	1	17
Outside agency—not pers.	7	—	1	1	1	—	10
Motor-load mismatch	9	—	1	—	—	—	10
Other	23	1	4	—	—	1	29

TABLE III
VIBRATION FAILURES
(No. of Failures)

	Transient overvoltage	0
	Overheating	6
	Other insulation breakdown	2
Failure Initiator	Mechanical breakage	23
	Electrical fault/malfunction	3
	Stalled motor	1
	Other	13
	Defective component	14
	Poor installation/test.	15
	Inadequate maintanance	3
	Improper operation	0
	Improper handling/shipping	1
Failure Underlying Cause	Inadequate physical protection	3
	Inadequate electrical protection	0
	Personnel error	4
	Outside agency—not pers.	0
	Motor-load mismatch	3
	Other	5
Total Vibration Failures (From Part 1)		51

TABLE IV
VIBRATION FAILURES VERSUS SIZE

Motor Size	No. of Vibration Failures	Total No. Of Failures— All Causes	Percent
201–500 hp	27	218	12.4
501–5000 hp	22	131	16.8
5001–10 000 hp	1	9	*
< 10 000 hp	—	—	—

* Small sample size.

TABLE V
STARTS PER DAY VERSUS CONTINUOUS DUTY

	No. of Starts Per Day	No. of Flrs	Total Population U-Yrs	Flr Rate	Avg. Hrs D.T./Flr	Med Hrs D.T./Flr
All Motors	<1	241	3111.6	0.0775	48.7	12
	1–10	90	1178.1	0.0764	90.8	16
0–1000 V	All motors					
	<1	71	854.5	0.0831	36.1	8
	1–10	22	244.5	0.0900	111.1	48
	Individual Motors					
	<1	68	768.7	0.0885	37.2	8
	1–10	13	148.4	0.0876	50.7	36
1000–5000 V	All motors					
	<1	163	2185.0	0.0746	55.7	12
	1–10	66	859.1	0.0768	83.6	16
	Individual Motors					
	<10	152	1876.9	0.0810	54.7	12
	1–10	38	497.0	0.0765	102.6	16

TABLE VI
DOWNTIME VERSUS REPAIR URGENCY AND TIME DISCOVERED
(No. of Flrs)

Downtime Per Flr. (Hours)	Repair Urgency				Time Discovered		
	Normal Working Hours	Round the Clock	Replace with Spare	Low Priority	During Normal Operation	During Maintenance or Test	Other
1–12	14	2	89	—	66	35	4
13–24	32	13	9	—	35	20	—
25–50	10	6	2	—	12	6	—
51–100	13	11	2	—	20	6	—
101–150	6	6	—	—	12	—	—
151–200	4	4	1	1	5	4	2
201–350	3	3	1	3	7	3	—
<350	5	—	1	2	8	1	—

TABLE VII
HORSEPOWER VERSUS SPEED
INDUCTION MOTORS

	No. of Failures	Unit Years	Failure Rate
0–720 r/min			
201–500 hp	7	137.92	0.0508
501–5 000 hp	12	175.16	0.0685
5001–10 000 hp	—	—	—
>10 000 hp	—	—	—
721–1800 r/min			
201–500 hp	148	1922.43	0.0770
501–5000 hp	66	740.1	0.0892
5001–10 000 hp	1	2.83	
>10 000 hp	—	7.5	—
3600 r/min			
201–500 hp	42	655.75	0.0640
501–5 000 hp	16	358.66	0.0446
5001–10 000 hp	—	—	—
>10 000 hp	—	—	—

ª Small sample size.

General Electric Company focused on electric utility powerhouse motors. Several interesting correlations between the EPRI survey and the IEEE survey emerged. In a Discussion [4] of Part 1 of the IEEE results by participants in the EPRI survey it was noted that hp per pole had been analyzed in past studies as affecting failure rate. The data in the IEEE survey did not allow this specific analysis. Table VII, presented here, is a more general representation of this subject, showing ranges of speed and of size. Induction motors are the most common type in use and consequently most survey data were collected for this type. Table VII has been limited to induction motors. It should be noted that this table was also published in the Closure to the Discussion referenced in the aforementioned. Similar results were published in Part 2, Table 5, but included all types of motors surveyed.

The highest failure rate appears in the middle speed range and at 501–5000 hp. One might observe that within the first two speed ranges, as hp per pole increases (assuming that, specifically, 720 r/min and 1800 R/min are predominant in these speed ranges) so also does failure rate. However, the highest speed range reverses this trend. Aside from this observation there is not a significant difference in failure rates between the different horsepower ranges within the first two speed ranges. Table 5 of Part 2, which included all motor types surveyed, showed similar trends.

GENERAL DISCUSSION

The results of Part 3 have presented several new aspects of the data. Most are a result of questions and comments received concerning Parts 1 and 2, but in some cases the data did not allow exact analysis. In some cases trends are evident and in some cases they are not. Some of the results expected or at least anticipated, for example, were that most failures occurred with lower downtime per failure, high vibration resulted in mechanical breakage, and longer downtime per failure occurred with induction motors starting more than once per day. Some of the interesting results were the opposite trends in causes of failures between indoor and outdoor applications and vibration causes being blamed mostly on defective component and poor installation or testing.

Overall, Part 3 has added credibility to some previously published results and has reinforced some areas of causes that are otherwise normally speculated.

REFERENCES

[1] Pat O'Donnell, Coordinating Author, "Report of large motor reliability survey of industrial and commercial installations—Part 1," IEEE Committee Report, in IEEE Trans. Ind. Appl., vol. IA-21, no. 4, pp. 853–864, July/Aug. 1985.

[2] Pat O'Donnell, Coordinating Author, "Report of large motor reliability survey of industrial and commercial installations—Part 2," IEEE Committee Report, in IEEE Trans. Ind. Appl., vol. IA-21, no. 4, pp. 865–872, July/Aug. 1985.

[3] P. F. Albrecht, J. C. Appiarius, R. M. McCoy, E. L. Owen, and D. K. Sharma, "Assessment of the reliability of motors in utility applications—Updated," IEEE Trans. Energy Conversion, vol. EC-1, no. 1, pp. 39–46, March 1986.

[4] E. L. Owen, Discussion of "Report of large motor reliability survey of industrial and commercial installations—Part 1," IEEE Trans. Ind. Appl., vol. IA-21, no. 4, pp. 863–864, July/Aug. 1985.

DISCUSSION

Richard Bloss *(Independent Consultant, #5462 Banbury Drive, Cleveland, OH 44139, formerly with Booz, Allen & Hamilton):* I applaud the IEEE Motor Reliability Working Group for their efforts to build a better understanding of the factors that influence large motor reliability. I would like to add that my remarks here are my own and not those of the Electric Power Research Institute, the General Electric Company, the prime contractor for the EPRI study, or of Booz, Allen & Hamilton, the subcontractor for the survey phase.

There are certain differences in the focus of the two studies that are important to understand. The EPRI study was looking at power generation plant applications. The IEEE was looking at a much broader commercial and industrial application base. To capitalize on the commonality of applications, the EPRI study focused on possible effects of applications as well as basic motor failure modes. The EPRI study permits conclusions to be drawn across similar applications.

The General Electric Company representatives may have already drawn what may be the most significant conclusion to the EPRI study in earlier remarks they made relating to the first part of the IEEE study. That conclusion is that the most significant variable in motor reliability in the EPRI study was "who was the owner." My personal analysis of the findings leads me to draw the conclusion that those utilities which had developed their own motor specifications over and beyond the industry standards had the best reliability history.

As a result of the larger sample in the EPRI study and the greater focus on a limited range of applications, more conclusions can be drawn relating to applications. As an example, in the EPRI study a problem was identified relating to the failure of Weatherproof II enclosures to protect motors in outdoor installations in coastal regions affected by severe weather. In another case, a pattern of motor misapplication in purchased subsystems was identified. Data from a number of owners of a particular subsystem served to pinpoint the use of motors designed for horizontal use, with adequate axial thrust capacity, in vertical applications. The subsystem supplier had failed to understand the problem of lubrication of the bearings. Owners who had researched the problem of bearing failure were installing their own redesigned lube system while others who were unaware of the root cause were continuing to repair the same bearing failure over and over.

It does appear from the EPRI study that customer-generated specifications can impart a favorable impact on motor reliability. The IEEE may want to pursue, in conjunction with the EPRI and others, a further study of what specific factors in customer-generated motor specifications have this positive effect on motor reliability.

The payoff is clear. In the EPRI study the average cost per year of motor failures was identified as $300 000 per power generating unit. The "best" owners had much lower motor failure costs, approaching *zero* cost. The average unit had just 40 motors. The average cost per motor per year for failures was about $7500, *plus* the cost to repair the motor!

I feel the IEEE Working Group must enlist the help of major customers of large motors to develop improved specifications that will reduce motor failures.

Manuscript released for publication October 9, 1986.

C. R. Heising and Pat O'Donnell: The Discussion by Mr. Bloss presents some additional views and comparisons of the EPRI and IEEE surveys of the reliability of large motors.

A notable difference in the published results from the IEEE survey is the omission of conclusions except for some obvious conclusions from the data. This omission is deliberate and may possibly lead to a false impression that the IEEE results are not conducive to definite conclusions. We believe the results present facts as accurate as can possibly be obtained in a survey conducted by mail. The IEEE survey was successful in obtaining data covering causes of failures, and in some cases this was related to pertinent design factors.

A major difference in the surveys by the EPRI and the IEEE is the population base of each. The EPRI results, based on a large population base, appear to be more complete and contain more detail in some specific areas such as the failed part and the application of the motor. The IEEE survey results are based upon a lesser population, but are more complete on the causes of the failures and the effect of maintenance. The cause data included failure initiating cause, failure contributing cause, and failure responsibility.

Mr. Bloss's comments about the effect that customer-generated specifications can have on improving the reliability of motors are very pertinent. He suggests that the IEEE may want to pursue this subject further and identify some of the most pertinent factors that could be specified in order to improve the reliability of motors. The IEEE–IAS Power Systems Reliability Subcommittee will consider this matter further.

Accurate and well-engineered specifications are certainly found desirable by most users and manufacturers. The inability to provide such specifications may often be caused by insufficient experience and expertise, and this may lead to poor reliability. The IEEE survey results are intended to aid this cause by revealing what is actually happening in the industry, thus allowing improved standards and specifications. These results reveal existing reliability with existing specifications. Mr. Bloss reports from his experience on the EPRI study that good specifications can coincide with good reliability.

The data from the IEEE motor reliability survey will be included in the next revision to IEEE Standard No. 493 (Gold Book), "Recommended Practice for Design of Reliable Industrial and Commercial Power Systems." This recommended practice standard and its future revisions contain much of the data collected in the IEEE equipment reliability surveys of industrial and commercial installations.

Manuscript released for publication October 9, 1986.

Pat O'Donnell (S'64–M'68–SM'80) was born in El Paso, TX in 1942 and received the B.S.E.E. degree from Texas Western College (now University of Texas at El Paso) in 1965.

After brief employment with the Schlumberger Well Surveying Corporation, he joined the El Paso Natural Gas Company in 1966 and is presently a Consultant Electrical Engineer in the main office, Engineering Department, in El Paso.

He is presently active in the Industrial and Commercial Power Systems Department of the IEEE Industry Applications Society and serves as Secretary of the department. He is a Past Chairman of the Power System Technologies Committee and within the Power Systems Engineering Committee he is presently Chairman of the Orange Book (IEEE Std. 446) Working Group in the Emergency and Standby Power Systems Subcommittee, Chairman of the Motor Reliability Working Group in the Power Systems Reliability Subcommittee, and a member of the Power Systems Analysis Subcommittee. He is a Registered Professional Engineer in the States of Texas and New Mexico.

Appendix I
Reliability Study of Cable, Terminations, and
Splices by Electric Utilities in the Northwest

By
W. F. Braun
IEEE Transactions on Industry Applications
Jan./Feb. 1987, pp. 144–153

Reliability Study of Cable, Terminations, and Splices by Electric Utilities in the Northwest

WILLIAM F. BRAUN, MEMBER, IEEE

Abstract—The results for cable, terminations, and splice reliability are summarized from a reliability report prepared annually by the Northwest Electric Light and Power Association (NELPA). Failure rates are given for primary cable, secondary cable, plug-in elbow connectors, primary splices and loadbreak junctions, pole top terminators, and secondary connections. Pertinent factors that affect the failure rates are identified.

INTRODUCTION

FOR THE PAST 18 years the Northwest Underground Distribution Committee[1] of the Northwest Electric Light and Power Association (NELPA) has prepared an annual report titled "URD equipment and materials reliability in the Northwest" [1].

Of particular interest to the IEEE Power Systems Reliability Subcommittee on Industrial and Commercial Power Systems is the portion of the report pertaining to cables, terminations, splices, and connections, since similar equipment is often used on industrial or commercial power systems.

The data in the NELPA report appears to be more complete and represents a much larger sample size than the data from the IEEE reliability survey of industrial plants [2] that was published in 1973–1974 and incorporated into the present ANSI/IEEE Standard No. 493-1980 [3]. The standard is being revised and updated in 1986. This paper will sumarize the NELPA report with the intent of using it as a source for the 1986 revision of ANSI/IEEE Standard No. 493.

BACKGROUND

NELPA companies serve most of the Northwest areas of the United States. Because the geographical makeup of this area consists of some very wet areas, some very dry areas, some very hot areas, and some very cold areas, the data from the report should be valuable for evaluating URD equipment for use around the country, particularly for such items as corrosion resistance and insulation failure.

NELPA consists of the following member companies.

Paper ICPSD 86-14, approved by the Power Systems Engineering Committee of the IEEE Industry Applications Society for presentation at the 1986 Industrial and Commercial Power Systems Technical Conference, Cleveland, OH, May 5-8. Manuscript released for publication September 9, 1986.

W. F. Braun is with the Owens-Corning Fiberglas Corporation, Fiberglas Tower SG14, Toledo, OH 43659.

IEEE Log Number 8612031.

[1] Kenneth W. Prier of Portland General Electric Company was Chairman of the Northwest Underground Distribution Committee in 1984, when Report No. 17 was issued. Richard M. Snell of Montana Power Company is the present Chairman.

TABLE I
CABLE FAILURE RATES—ALL VOLTAGE CLASSES
(Failures per 100 Conductor Miles)

1969	0.67	1977	0.98
1970	1.11	1978	1.47
1971	0.73	1979	1.82
1972	0.91	1980	1.68
1973	1.00	1981	2.55
1974	1.03	1982	2.51
1975	0.89	1983	2.27
1976	1.10		

Idaho Power Company
Montana Power Company
Pacific Power & Light Company
Portland General Electric Company
Puget Sound Power & Light Company
Utah Power & Light Company
Washington Water Power Company

FAILURE DATA REPORTING

The report is only concerned with natural failures of equipment or materials. All failures caused by abnormal external means, such as through dig-ins or damage prior to installation, are not intended to be included in the data. In the cases where the cause of a failure could not be determined, the cause of the failure is assumed and reported in that way.

The member utilities are continuously improving their efforts to accumulate basic data. However, there are still problems with field people not reporting the material failures. All failure rates in this report should be considered on the low side.

PRIMARY CABLE

Table I lists the failure record for all voltage classes (15 kV, 25 kV, and 35 kV) and insulation types of primary cable used on the systems.

In general the failure record is excellent, although high molecular weight polyethylene (HMWPE) insulated cable is failing at a much greater rate than crosslinked polyethylene (XLPE). (See Tables IIIA and IIIB for complete data.)

The failure rates for the last few years are high because one utility has just started reporting failures and they have been having problems with 175-mil 15-kV cable.

A comparison was also made between 15-kV HMWPE and 15-kV XLPE cable for the last ten years. (See Table II.)

TABLE II
15-kV CLASS CABLE
(Failures per 100 Conductor Miles of Cable)

Failure Year	175-mil HMWPE	220-mil HMWPE	175-mil XLPE	220-mil XLPE
1973	0.90	1.40	0.27	0.0
1971	0.41	1.41	0.53	0.56
1975	0.72	1.51	0.33	0.0
1976	1.19	1.28	0.47	1.72
1977	1.25	1.04	0.39	0.0
1978	2.07	0.69	1.06	0.08
1979	2.67	0.90	0.68	0.12
1980	3.42	0.65	0.03	0.0
1981	5.38	0.95	0.10	0.05
1982	4.77	1.69	0.07	0.0
1983	4.40	1.73	0.53	0.0

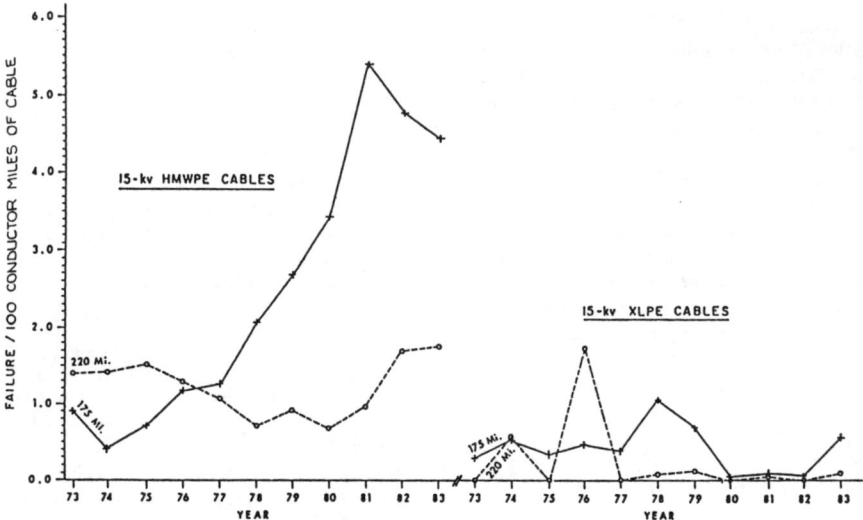

Fig. 1. Failure rates of 15-kV URD cables.

HMWPE cable seems to be failing at a much higher rate than XLPE cable. These failures seem to be related to treeing problems, which break down the insulation material. (See Fig. 1 for a plot of the data.)

Some member utilities have started to purchase tree-retardant insulation material. The usage has been limited and no failures have been reported.

In addition, 175-mil thickness insulation seems to have a much higher failure rate than 220-mil insulation. This is apparently due to the larger electrical stress capability of 220-mil insulation.

SECONDARY CABLES

The failure rate for secondary low-voltage cables (600 V and below) has remained fairly constant for the last two years. The failure rates since 1969 are as follows.

1969	1.5	failures per 100 conductor miles
1970	1.25	failures per 100 conductor miles
1971	0.74	failures per 100 conductor miles
1972	0.62	failures per 100 conductor miles
1973	0.35	failures per 100 conductor miles
1974	0.50	failures per 100 conductor miles
1975	0.39	failures per 100 conductor miles
1976	0.53	failures per 100 conductor miles
1977	0.73	failures per 100 conductor miles
1978	0.71	failures per 100 conductor miles
1979	0.73	failures per 100 conductor miles
1980	0.48	failures per 100 conductor miles
1981	0.80	failures per 100 conductor miles
1982	0.70	failures per 100 conductor miles
1983	0.78	failures per 100 conductor miles

Failures of this cable seem to be related mostly to mechanical-type damage occurring during or after installation. Corrosion problems due to moisture do not seem to be a problem. (See Table IV for complete data.)

PLUG-IN ELBOW CONNECTORS

The failure rate for 15-kV, 25-kV, and 35-kV loadbreak elbows of 0.41 failures per 1000 units (unit defined as one

TABLE IIIA
PRIMARY CABLE—15 kV

	Type	Company	Miles[a] Installed	Failures This Year	Failures to Date	Average Life Before Failure	Neutral Corrosion
(A)	HMWPE 175 mil, 15 kV	A	18	0	23	Unknown	
		B	1170	34	118	7–8 yrs	
		C	1388	97	412		
		E	3161	86	545	13 yrs	8
		G	1085	83	529	12 yrs	10
	Total		6822	300	1627		
(B)	HMWPE 175 mil, 15 kV tree retardant	C	504	0	0		
		E	1106	0	0		
	Total		1610	0	0		
(C)	HMWPE 220 mil, 15 kV	A	1	0	0	12 yrs	
		C	74	16	86	Unknown	
		D	2373	31	255	Not reported	
		F	488	3	6	20 yrs	
	Total		2896	50	351		
(D)	HMW-Poly 220 mil, 15 kV tree retardent	D	60	0	0		
(E)	XLP 175 mil, 15 kV	A	1960	21	117	Unknown	
		B	150	0	0		
		C	615	0	0		
		E	1519	0	14		
		G	60	1	3		
	Total		4154	22	134		
(F)	XLP 175 mil, 15 kV tree retardant	E	318	0	0		
(G)	XLP 175 mil, 15 kV tree retardant with insulated jacket	E	90	0	0		
(H)	XLP 175 mil jacket, 15 kV	G	149	0	0		
	Total		149	0	0		
(I)	XLP 220 mil, 15 kV	C	1	0	0		
		D	2417	0	10	Not reported	
		F	b	b	b		
	Total		2418	0	10		
(J)	Butyl-neoprene, 15 kV	A	1	1	1		
		D	10	0	1		
		E	79	2	15	21 yrs	
	Total		90	3	17		
(K)	EPR 175 mil, 15 kV	A	3	0	4		
GRAND TOTAL			18 610	375	2143		
LAST YEAR'S GRAND TOTAL			17 751	376	1768		

[a] Conductor miles (not circuit miles).
[b] Accurate data not available.
(Data from NUDC Report No. 17, October 8, 1984.)

TABLE IIIB
PRIMARY CABLE—25 AND 35 kV

	Type	Company	Miles[a] Installed	Failures This Year	Failures to Date	Average Life Before Failure	Neutral Corrosion
(A)	HMWP 260 mil, 25 kV	G	125	0	8		
(B)	HMWP 260 mil, 25 kV	B	80	8	15	8–10 yrs	
(C)	HMWP 260 mil, 25 kV	C	930	75	432		
(D)	HMWP 295 mil, 25 kV	C	108	26	67		
(E)	HMWP 280 mil, 25 kV	A	2	0	55	11 yrs	
	Total		1245	109	577		
(F)	HMWP 260 mil, 25 kV tree retardant	C	348	0	0		
(G)	XLP 260 mil, 25 kV	B	11	0	0		
(H)	XLP 260 mil, 25 kV	C	409	0	4		
(I)	XLP 295 mil, 25 kV	F	b	b	b		
	Total		420	0	4		
(J)	XLP 260 mil, 25 kV with jacket	B	326	0	0		
(K)	EP 295 mil, 25 kV	A	5	0	2		
(L)	HMWP 345 mil, 35 kV	C	74	0	22		
(M)	HMWP 345 mil, 35 kV tree retardant	C	31	0	0		
(N)	HMWP 345 mil, 35 kV tree retardant	E	98	0	0		
	Total		129	0	0		
(O)	XLP 280 mil, 35 kV	A	10	0	2		
(P)	XLP 345 mil, 35 kV	A	102	1	2	2 yrs	
(Q)	XLP 345 mil, 35 kV	C	34	0	0		
(R)	XLPE 345 mil, 35 kV	E	29	0	0		
(S)	XLPE 345 mil, 35 kV	G	5	0	0		
	Total		180	1	4		
(T)	XLP 345 mil, 35 kV tree retardant	E	48	0	0		
(U)	XLP 345 mil, 35 kV tree retardant with insulated jacket	E	17	0	0		
	GRAND TOTAL		2792	110	609		
	LAST YEAR'S GRAND TOTAL		2251	127	498		

[a] Conductor miles (*not* circuit miles).
[b] Accurate data not available.
(Data is from NUDC Report No. 17, October 8, 1984).

TABLE IV
LOW-VOLTAGE CABLE

Type Insulation	Company	Thickness	Miles[a] Installed	Failures This Year	Failures to Date	Average Life Before Failure	Neutral Corrosion
(A) Poly	C		0	0	0		
	D	(Sodium)	12	0	4[c]	$2\frac{1}{2}$ yrs	
	E		0	0	0		
Total			12	0	4		
(B) XLPE	A	70–110 mil	2983	165	543	Unknown	
	B	70–110 mil	1300	1	21		
	C	60–110 mil	11 295	87	914	Unknown[b]	All neutral
	D	80–95 mil	8128	5	89		
	E	Min. IPCEA	5520	5	47	(Failures[b] starting with 1976 data)	
	G	60–110 mil	2096	Unknown	Unknown		
Total			31 322	263	1614		
(C) Abrasion-resistant	E		9	0	0		
XLPE	G		0.2	0	0		
Total			9.2	0	0		
(D) Abrasion-resistant	A		46	0	0		
HMWP	E		697	0	0		
Total			743	0	0		
(E) PVC	A	80 mil	10	0	1		
	C		0	0	0		
	D		7.1	0	0		
	E		0	0	0		
Total			17.1	0	1		
(F) Rubber neoprene	C		195	0	0		
	D	65 & 65 mil	1162	0	11		
	E	5/64 in	99	0	3		
Total			1456	0	14		
GRAND TOTAL			33 550.1	263	1633		
LAST YEAR'S GRAND TOTAL			31 978.9	225	1370		

[a] Insulated conductor miles (*not* circuit miles).
[b] Some of these failures could have been rubber neoprene, but no record is available as to which type cable failed. More failures are being reported due to a computer-managed reporting system.
[c] Cable insulation failed due to mechanical stress placed on it by the design of the connector. (A stainless-steel hose clamp around the insulation makes a quick fix.)
(Data is from NUDC Report No. 17, October 8, 1984.)

single-phase terminator) has been fairly constant over the last four years. Many of the recent problems were due either to molding problems of one manufacturer, cross-threading of connectors, or bad compression joints. These failures include units that were improperly installed, which is a significant number. The failures also include units that were replaced during maintenance due to problems such as visible tracking overheating, etc. (See Table V for complete data.)

PRIMARY CONNECTIONS

This is the fifth year for the study of primary cable splices and primary load break functions. Since the data is relatively new the results should be used carefully.

The failure rate for 15-kV splices was 2.1 failures per 1000 units (unit defined as one single-phase splice), compared to last year's 2.4 failures; the failure rate for 15-kV primary junctions was 1.0 failures per 1000 units, compared to last year's 1.3 failures. Splice failures over the last couple of years have been due mostly to the molding problems experienced by one manufacturer and to improper installation by line crews. (See Table VI for completed data.)

POLE TOP TERMINATORS

The outstanding performer for all voltage classes (15 kV, 25 kV, and 35 kV) is the molded rubber terminator. The failure rate is 0.06 per 1000 units (unit defined as one single-phase

TABLE V
PLUG IN PRIMARY TERMINATORS (ELBOWS)

	Type	Company	Total Number on System	Failures This Year	Failures to Date	Average Life Before Failure
(A)	Non-LB rubber, 15 kV	A	65	0	0	
		C	1628	0	3	
		D	2302	0	41	
		E	6440	0	51	
		G	200	0	0	Unknown
	Total		10 635	0	100	
(B)	Non-LP rubber, 600 A–15 kV	A	707	0	0	
		C	0	0	0	
		D	322	0	2	
		E	1972	0	9	
		G	25	0	0	
	Total		3026	0	11	
(C)	Non-LB rubber, 600 A–25 kV	C	33	0	0	
		E	0	0	0	
		F	a	a	a	
	Total		33	0	0	
(D)	Non-LB rubber, 600 A–35 kV	E	30	0	0	
	Total		30	0	0	
(E)	Non-LB metal	A	40	0	0	
		C	15	0	1	
	Total		55	0	1	
(F)	LP rubber 15 kV	A	31 138	18	168	
		B	21 514	32	71	
		C	40 997	27	231	
		D	76 525	21	180	
		E	160 506	44	370	
		F	a	a	a	
		G	24 179	2	15	
	Total		354 859	144	1035	
(G)	LB rubber, 25 kV	B	797	4	19	
		C	24 311	5	27	
		E	0	0	0	
		F	a	a	a	
		G	565	0	4	
	Total		25 673	9	50	
(H)	LB rubber, 35 kV	A	2465	0	2	1 yr
		C	2293	5	19	
		E	730	0	0	
	Total		5488	5	21	
GRAND TOTAL			399 799	158	1218	
LAST YEAR'S GRAND TOTAL			371 119	160	1060	

a Accurate data not available.
(Data is from NUDC Report No. 17, October 8, 1984.)

TABLE VI
PRIMARY CONNECTIONS

Type	Company	Total Number on System	Failures This Year	Failures to Date	Average Life Before Failure
Primary splices—15 kV,	A	13 454	48	443	10 yrs
molded rubber	B	7484	68	149	
	C	17 624	6	15	
	D	21 481	4	35	
	E	18 990	65	389	
	G	11 813	3	25	
Total		90 846	194	1056	
Primary splices—25 kV,	B	309	3	5	
molded rubber	C	11 332	1	3	
	G	520	0	4	
Total		12 161	4	12	
Primary splices—35 kV,	A	711	1	6	
molded rubber	C	1057	0	16	
	E	437	0	0	
	G	32	0	0	
Total		2237	1	22	
GRAND TOTAL		105 244	199	1090	
LAST YEAR'S GRAND TOTAL		94 281	203	891	
Primary loadbreak junctions	A	3474	13	138	4–8 years
(lateral taps)—15 kV	B	3224	18	30	
	C	7321	3	9	
	D	8742	4	31	
	E	30 555	19	213	
	G	2103	0	2	
Total		55 419	57	423	
Primary loadbreak junctions	B	42	1	4	
(lateral taps)—25 kV	C	3587	3	7	
	G	16	0	0	
Total		3645	4	11	
Primary loadbreak junctions	C	306	1	2	
(lateral taps)—35 kV	E	261	0	0	
Total		567	1	2	
GRAND TOTAL		59 631	62	436	
LAST YEAR'S GRAND TOTAL		55 195	71	374	

Note: Data on taped primary splices has been discontinued due to lack of data.
(Data from NUDC Report No. 17, October 8, 1984.)

terminator). The porcelain elastomeric type has a rate of 0.43 per 1000 units. Overall the record for these devices is excellent. (See Table VII for complete data.)

SECONDARY CONNECTIONS

This is the sixth year of evaluating the different types of secondary connections 600 V and below made by the member utilities. This data should be used carefully due to the

difficulty in tabulating failures from previous years. The section on taped–insulated connections has been discontinued since the data is not dependable. Even though the data is new, the numbers on heat-shrink connections appear to be particularly interesting due to the failure rate of 0.002 per 1000 units (unit defined as one single-phase connection) for 1983 on 513 280 units installed. This compares with the failure rate of 0.00 per 1000 units for 1982. The failure rate for the molded

TABLE VII
POLE TOP TERMINATORS

	Type	Company	Total Number on System	Failures This Year	Failures to Date	Average Life Before Failure
(A)	Porcelain compound	A	192	0	7	8 yrs
		C	125	0	3	
		D	115	0	0	
		E	75	0	3	17 months
	Total		507	0	13	
(B)	Porcelain epoxy	E	125	0	3	
(C)	Porcelain elastomer—15 kV	B	Unknown	1	10	
		C	1631	2	20	
		D	2732	0	2	
		E	25 522	11	230	Unknown
	Total		37 126	16	271	
(D)	Porcelain elastomer—25 kV	C	1320	2	12	
	Total		1320	2	12	
(E)	Porcelain elastomer—35 kV	A	137	0	0	
		E	448	0	0	
	Total		585	0	0	
(F)	Porcelain elastomeric compound 35 kV	C	18	0	1	
		A	37	0	0	
	Total		55	0	1	
(G)	Molded rubber—15 kV	A	1840	0	1	
		B	14 359	2	15	
		C	17 861	2	34	Unknown
		D	32 576	2	8	2 yrs
		F	a	a	a	
		G	10 045	0	12	
	Total		76 681	6	70	
(H)	Molded rubber—25 kV	B	600	0	0	
		C	12 064	0	7	
		F	a	a	a	
		G	369	0	2	
	Total		13 033	0	9	
(I)	Molded rubber—35 kV	A	1071	0	0	
		C	972	0	4	
		G	40	0	0	
			a	a	a	
	Total		2083	0	4	
(J)	Taped	A	2^r	0	1	
		C	0	0	0	
		D	200	0	21	
	Total		229	0	22	
(K)	Scotch 83A3	A	227	0	23	
		F				
	Total		227	0	23	
(L)	Heat shrink—15 kV	E	9397	0	4	
		C	7	0	0	
	Total		9404	0	4	

TABLE VII
(*Continued*)

	Type	Company	Total Number on System	Failures This Year	Failures to Date	Average Life Before Failure
(M)	Heat shrink—25 kV	C	80	0	1	
	Total		80	0	1	
(N)	Heat shrink—35 kV	E	363	0	0	
		C	21	1	3	
	Total		384	1	3	
	GRAND TOTAL		141 839	25	436	
	LAST YEAR'S GRAND TOTAL		133 271	21	411	

a Accurate data not available.
(Data from NUDC Report No. 17, October 8, 1984.)

TABLE VIII
SECONDARY CONNECTIONS

	Type	Company	Total Number on System	Failures This Year	Failures to Date	Average Life Before Failure
(A)	Molded rubber/	A	53 162	5	36	
	plastic insulated	B	—	—	—	
	connections	C	1093	0	2	
		D	292 399	6	242	
		E	155 284	2	171	
		F	*a*	*a*	*a*	
		G	44 245	0	10	
	Total		546 183	13	461	
(B)	Heat shrink	A	47 987	1	11	1 yr
	connections	B	—	—	—	
		C	147 184	Unknown	Unknown	
		D	53 100	0	1	
		E	265 009	0	6	
		F	*a*	*a*	*a*	
	Total		513 280	1	18	
	GRAND TOTAL		1 059 463	14	479	
	LAST YEAR'S GRAND TOTAL		965 121	13	465	

a Accurate data not available.
(Data from NUDC Report No. 17, October 8, 1984.)

rubber–plastic units is 0.02 failures per 1000 units for 1983, which is the same as for 1982. (See Table VIII for data.)

REFERENCES

[1] Northwest Underground Distribution Committee of the Northwest Electric Light and Power Association (NELPA), "URD equipment and materials reliability in the Northwest," no. 17, October 8, 1984.
[2] IEEE Committee Report, "Report on reliability survey of industrial plants," published in six parts, *IEEE Trans. Ind. Appl.*, pp. 213-252, 456-476, 681, Mar./Apr., July/Aug., Sept./Oct. 1974. (Included as Appendices in [3].)
[3] ANSI/IEEE Standard No. 493-1980, "IEEE Recommended Practice for the Design of Reliable Industrial & Commercial Power Systems."

William F. Braun (M'76) received the B.S.E.E. degree from the University of Toledo in 1970.
He worked at the Ohio Power Company, Newark, OH, as an Engineer in the Transmission and Distribution Department from 1971 to 1973. Since 1973 he has been employed at the Owens-Corning Fiberglas Corporation, Toledo, OH. Presently he is working as a Senior Electrical Engineer in the Power Systems Section of Corporate Electrical Engineering of Owens-Corning.
Mr. Braun is a member of IEEE/IAS Power Systems Reliability Subcommittee.

Appendix J
Summary of CIGRE 13-06 Working Group
Worldwide Reliability Data, Maintenance Cost
Data, and Studies on the Worth of Improved Reliability
of High-Voltage Circuit Breakers

By
C. R. Heising
IEEE Industrial and Commercial
Power Systems Technical Conference
Cleveland, Ohio, May 5 – 8, 1986

Reprinted from IEEE Conference Record
86CH2279-8, pp. 93 – 101

SUMMARY OF CIGRE 13-06 WORKING GROUP WORLD WIDE RELIABILITY DATA,
MAINTENANCE COST DATA, AND STUDIES ON THE WORTH OF IMPROVED RELIABILITY
OF HIGH VOLTAGE CIRCUIT BREAKERS

by

Charles R. Heising, U. S. Representative, CIGRE 13-06 Working Group
Industrial Reliability Tech
216 Farwood Road
Philadelphia, PA   19151

ABSTRACT

A summary is given of the most significant
reliability data and maintenance cost data
from the CIGRE 13-06 Working Group world wide
survey of high voltage circuit breakers above
63 kV for the years 1974-77. A summary is
given of their studies on the worth of
improved reliability and the worth of reduced
maintenance costs for high voltage circuit
breakers applied on power transmission and
distribution systems and in power generating
stations. A brief description is given of the
objectives, scope, membership and future plans
of CIGRE 13-06 Working Group. A brief summary
is given of some of the highlights from their
studies on the reliability of high voltage
circuit breakers during the period 1971-85.

INTRODUCTION

CIGRE 13-06 Working Group has carried out
world wide reliability studies on high voltage
circuit breakers during the fifteen year
period 1971 through 1985. This work is
reported in three CIGRE 13-06 committee final
reports [1][2][3]. Eighteen countries have
participated in these studies, and eight
additional countries submitted data for the
world wide reliability survey [1].

CIGRE (Conference Internationale des Grand
Réseaux Electriques à haute tension) is one of
the technical arms of the IEC (International
Electrotechnical Commission). The IEC is
responsible for international standards and
IEC Subcommittee 17A on High Voltage Switchgear and Controlgear has set up Working Group
11 to write new test requirements to improve
the reliability of high voltage circuit
breakers (above 1000 volts). The three CIGRE
13-06 committee final reports supply back up
information for proposed changes in the IEC
standards; and these reports also include the
results from the studies by WG 11 of IEC 17A.

CIGRE 13-06 was a very active Working Group
and the work is believed to be a significant
advance in the mechanical reliability of high
voltage circuit breakers. Strong leadership
was provided by France, Japan, and Italy
because research work had been conducted by
national electric utilities or pooled utility
industry research groups on: 1. reliability
data collection and analysis, 2. improved
methods of testing, and 3. improved methods
of maintenance.

OBJECTIVES OF CIGRE 13-06 WORKING GROUP

The CIGRE 13-06 WG studies had very ambitious
objectives and were part of the following four
part program on high voltage circuit breakers:

1. Collect world wide reliability data and
    maintenance cost data in order to
    determine what are the facts. Draw
    conclusions from these facts.

2. Make studies to improve reliability
    covering:
    a. New testing methods,
    b. Improved maintenance methods.

3. Make studies on the worth of improved
    reliability and reduced maintenance
    costs.

4. Follow up with a world wide corrective
    action program:
    a. Upgrade IEC 17A Standards with new
        test requirements,
    b. Add maintenance guidelines in IEC 17A
        Standards,
    c. Develop a new IEC 17A Standard for
        reliability data collection; this
        should include the necessary
        reliability definitions.

These objectives were established in 1971 and
the first three steps are now completed, and
steps #1 and #3 above are briefly summarized
in this paper. Steps #2 and #4 are briefly
summarized in another paper [4].

SCOPE AND MEMBERSHIP OF CIGRE 13-06 WG

The original scope only covered high voltage
circuit breakers above 63 kV. In 1976 the
scope was expanded downward to above 1000
volts because IEC 17A has only one standard to
cover all high voltage circuit breakers above
1000 volts. But, the world wide reliability
survey for the years 1974-77 only collected
data on circuit breakers above 63 kV.

The Working Group was also given the task of
establishing some relevant definitions
concerning the reliability of circuit breakers.

Eighteen countries each had one representative
on the Working Group plus the Chairman
from France for a total of nineteen. This
included 4 manufacturers, 13 electric
utilities, 1 independent test lab, and 1
consultant. Five of the original members were
active for the entire 15 years, and three
others were active for at least 13 years; this
gave continuity to the work. The countries
and their representatives are listed at the
end of this paper.

NEED FOR COLLABORATION BETWEEN USER AND MANUFACTURER

One of the significant problems identified in 1971 by the CIGRE 13-06 WG was the need for close collaboration between user and manufacturer in identifying the causes of failures and any necessary corrective action that should be taken. This collaboration includes the feedback of field reliability information from user to manufacturer in order to enable the manufacturer to analyse the failure and make improvements in the design and manufacture of circuit breakers. It also enables the manufacturer to communicate recommended maintenance improvements to the user.

The CIGRE 13-06 WG world wide reliability survey attempted to solve this problem by recommending that the failure reports: 1. be filled out by the electric utility, 2. sent to the relevant manufacturer who should make comments with a minimum of delay, and 3. finally the utility should send the failure report to CIGRE. Many countries followed this recommended approach. Some countries did not. It was the hope of CIGRE 13-06 WG that this handling of the failure reports would improve the communication between user and manufacturer and would result in a more accurate identification of the true failure cause. This would also improve the credibility of the CIGRE 13-06 data, particularly from the manufacturer's viewpoint.

The importance of credibility can be better understood from an example that exists in the United States on circuit breaker reliability data collection; this is discussed in [5]. The data collected by the Edison Electric Institute, Electrical Systems & Equipment Committee for 1973 did not not have the manufacturer's input; and 94% of the failures and defects were attributed to the manufacturer. In contrast to this, the CIGRE world wide reliability survey data attributed to the manufacturer only 45% of the major failures and 52% of the minor failures plus defects.

RELIABILITY DATA FROM CIGRE 13-06 WG WORLD WIDE RELIABILITY SURVEY

A total of 102 electric utilities from 22 countries located in 6 continents submitted data on 20,000 circuit breakers above 63 kV. Data were collected for the years 1974-77 on circuit breakers installed after January 1, 1964. This resulted in a total of 78,000 breakers-years of service during the four year period. This was a pioneering effort that required the development of: 1. reliability and maintenance definitions, and 2. survey questionaire. Countries submitting data were: Australia, Belgium, Brazil, Canada, Czechoslovakia, Denmark, Ireland, Finland, France, Federal Republic of Germany, Greece, Italy, Japan, Morocco, Netherlands, New Zealand, Norway, Portugal, Spain, Sweden, United Kingdom, and Yugoslavia. The United States did not submit data; and if it had participated, this would have added another 11,000 breakers.

The results from the world wide reliability survey are published in "Electra" [1]. A summary is published in a U. S. paper [6]; this includes the questionaire and some of the problems encountered in the survey.

Table 1 shows the failure rate and outage duration time data. Table 4 shows the origin of the failures and Table 5 shows the causes of the failures. The failure modes for major failures are shown in Table 6, and Table 7 shows the correlation between origin/cause and the failure modes of major failures. Table 8 shows a statistical distribution of the yearly average number of operating-cycles.

DATA NEEDED FOR STUDIES OF THE WORTH OF BETTER CIRCUIT BREAKER RELIABILITY AND THE WORTH OF REDUCED MAINTENANCE COSTS

One of the major objectives of the CIGRE 13-06 WG was to make studies on the worth of better circuit breaker reliability and the worth of reduced maintenance costs. These studies were needed in order to have some guidelines of how much cost could be permitted to be added to the circuit breaker if new requirements were to be added in the IEC standards.

Studies on the worth of better circuit breaker reliability require circuit breaker reliability data that can be used in system reliability studies of a power system. There was a lack of credible data on this subject. Thus a decision was made in 1971 to ask for the necessary data in the CIGRE questionaire. Data obtained from the world wide survey are shown in Table 2. This is the best data ever collected on this subject and fulfills a need that has existed in several foreign countries as well as in the United States.

Table 3 shows the data on the cost of scheduled servicing of circuit breakers that was obtained from the world wide survey. The cost data for both labor effort and spare parts consumed have been expressed in manhours. This has the effect of being an international currency. Data were also collected on the cost of non-scheduled servicing of circuit breakers after failures; these costs were much lower than the costs shown in Table 3 for scheduled servicing.

RELIABILITY AND MAINTENANCE DEFINITIONS

The most important of the CIGRE 13-06 circuit breaker reliability and maintenance definitions are given after Table 3. Additional definitions used are given in [1][6]. In early 1973 the Electrical Systems & Equipment Committee of the Edison Electric Institute in the U. S. rewrote their reliability definitions so as to be consistent with CIGRE 13-06. Thus world wide reliability definitions now exist for high voltage circuit breakers.

STUDIES ON THE WORTH OF BETTER CIRCUIT BREAKER RELIABILITY AND THE WORTH OF REDUCED MAINTENANCE COSTS

During 1970-77, seven life cycle cost studies were made by electric utilities in various foreign countries and the results published at international technical conferences.

These studies evaluated the worth of better circuit breaker reliability and the worth of reduced circuit breaker maintenance costs on both power transmission and distribution systems and in power generating stations. The methods used in these studies are summarized in [7]. They used data on the average costs of power outages to consumers of electricity, and through system reliability studies related this back to the failure rate of the circuit breaker. This then enabled a worth to be calculated for an improved failure rate of the circuit breaker. Three general conclusions can be made from these studies:

1. A 50% reduction in the failure rate is worth between 4 to 10% on the breaker initial price

2. The present value of breaker maintenance on distribution system feeders can have a capitalized value as high as 25% of the breaker initial price.

3. The savings from #1 and #2 should be added together to get the total savings.

OBSERVATIONS FROM CIGRE 13-06 WORLD WIDE RELIABILITY SURVEY DATA

1. 70.3% of the major failures and 85.6% of the minor failures and defects have a mechanical origin. 19.1% of the major failures and 11.7% of the minor failures and defects have an elecrical auxiliary and control circuit origin. These percentages are approximately the same for all voltages.

2. 45.3% of the major failures and 52.5% of the minor failures and defects are caused by "design or manufacture." These values are approximately the same for all voltages. A separate study found that one half resulted from design and one half from manufacture.

3. 9.3% of the major failures and 10.7% of the minor failures and defects are caused by "incorrect erection." These values are approximately the same for all voltages.

4. 8.1% of the major failures and 4.5% of the minor failures and defects are caused by "incorrect maintenance." These values are approximately the same for all voltages.

5. The major failure rate increases significantly as the voltage is increased. The major failure rate at 500 kV and above is 26 times higher than for circuit breakers between 63 kV and 100 kV.

6. The calculated "Mean Operating-Cycles Between Major Failures" (MOCBF) based upon commands to the circuit breaker is equal to $1/\lambda_c$ from Table 2. This is another way to measure the reliability of a circuit breaker. The MOCBF decreases with increasing voltage, ranging from 13,230 for breakers 63-100 kV to 440 for breakers 500 kV and above.

7. 47.8% of all major failures concern the failure modes "does not close on command" or "does not open on command."

8. 48.9% of the "mechanical origin" major failures concern the failure modes "does not close on command" and "does not open on command."

9. 68.7% of the "auxiliary circuit origin" major failures concern the failure modes "does not open on command" or "does not close on command."

10. 51.4% of the major failures caused by "design or manufacture" concern the failure modes "does not close on command" or "does not open on command."

11. 43.5% of the major failures caused by "incorrect erection" concern the failure modes "does not close on command" or "does not open on command."

12. 44.3% of the major failures caused by "incorrect maintenance" concern the failure modes "does not close on command" or "does not open on command."

13. A high voltage circuit breaker sees an average of 26.5 operating-cycles per year; this gives an average life of 660 operating-cycles during 25 years of service. 95% of the circuit breakers will see less than 2,000 operating-cycles during 25 years of service.

14. A special study was made by countries with a cold climate on 1,000 circuit breakers exposed for a relatively long time to low temperature. 10% of the failures plus defects were attributed to the low temperature stress.

HIGHLIGHTS FROM SOME OF THE CONCLUSIONS FROM THE CIGRE 13-06 WG RELIABILITY STUDIES

Conclusions from the CIGRE 13-06 WG reliability studies are summarized in [4]. Some of the highlights are listed below:

1. Reliability improvements in high voltage circuit breakers are needed. The need is greater at the highest voltages, particularly at 300 kV and above.

2. There is a need to improve the mechanical reliability and also the reliability of the electrical auxiliary and control circuit.

3. A greater number of operating-cycles during testing is needed to improve the reliability. Proposals by reliability experts for changes in the IEC standards include more operating-cycles testing during design, manufacture, and erection.

4. Improvements in maintenance are needed. The number of failures caused by incorrect maintenance needs to be reduced. Better communication is needed between users and manufacturers on maintenance. Maintenance procedures need to rationalized so that less frequent dismantling of the circuit breaker can be achieved and also result in improved maintenance and lower costs.

5. Based upon the CIGRE 13-06 WG studies the type test requirement in the IEC 17A standards was increased in 1982 from 1,000 to 2,000 "no load" mechanical operating-cycles. Some experts have proposed that this be increased to 10,000; this is very controversial and produces emotional discussions between users and manufacturers. France, Japan, and Italy now require passing a 10,000 mechanical operating-cycles type test in their standards; this has the effect of being a reliability demonstration test during the type test. The Japanese have stated that 10,000 was chosen to represent the equivalent of 5 samples for 2,000 operating-cycles each, and a test of 5 samples with no failures demonstrates 60% reliability with 90% confidence.

6. Based upon CIGRE 13-06 WG studies several proposals by WG 11 have been accepted by IEC 17A S/C for international standards. This has included new type tests involving low temperature, high temperature, humidity, and icing. In addition new tests at the erection site have been added and are called "comissioning tests".

7. Several proposals by WG 11 that were based upon CIGRE 13-06 WG studies were rejected in 1983 for international standards by IEC 17A S/C. This has included the following proposed new type tests: 10,000 mechanical operating-cycles, a sudden temperature variation test, a water penetration test, and a low and high temperature soak test with some mechanical operating-cycles at the temperature extremes.

8. Proposals by manufacturers and other reliability experts for additional tests during production to improve the quality control include:

   a. A "no load" mechanical operating-cycle run-in test on every breaker as part of the routine test prior to shipment from the factory. Proposals on this range from 100 to 500 operating-cycles.

   b. A "no load" mechanical operating-cycle endurance test on one circuit breaker (or module) at periodic intervals at the ambient air temperature of the test room. Proposals vary widely on this for both the time interval and the number of operating-cycles. This test is usually a rerun of a type test or a reliability demonstration test (called reliability test for compliance success ratio in IEC Publication 605-5).

   c. The repetition of the mechanical and climatic type tests at periodic intervals.

Procedures "b" and "c" could be called "sampling tests" and should also be used after each design change, after a substantial change in production process, or if field problems are reported which have not been observed during the initial reliability testing.

FUTURE PLANS FOR CIGRE 13-06 WG

CIGRE 13-06 WG was disbanded at the end of 1985 because the original objectives had been achieved. A new Working Group on the Reliability of High Voltage Circuit Breakers has been set up under CIGRE Study Committee No. 13. The scope will probably include:

1. Conduct a new world wide reliability survey of high voltage circuit breakers. Devise a simpler questionaire than was used in the 1974-77 survey. Consider having this questionaire become an IEC standard.

2. Examine the laws of wear and ageing due to the operation of switchgear under mechanical and electrical stresses in order to improve the present tests and to make them more representative of the service conditions.

3. Make studies of diagnostic techniques in service in order to bring about preventive maintenance that avoids major failures.

4. Make further studies on the subject of mechanical close/open testing during the design phase covering reliability growth testing followed by a type test or a reliability demonstration test.

5. Study the problems of quality control of circuit breakers during manufacturing. This study should include defining better tests covering: routine tests, periodic sampling tests, run-in tests, reliability growth tests.

6. Make studies on the worth of tests leading to improvements in reliability.

FUTURE PLANS OF WG 11 OF IEC 17A

In May 1985 IEC Subcommittee 17A authorized WG 11 to write a plan of work to:

1. Develop a card for failure data collection.

2. Develop a maintenance document based upon documents developed by IEC T/C No. 56 (Reliability and Maintainability) also considering diagnostic and monitoring techniques.

IEC 17A did not authorize further work at this time on mechanical reliability tests or on tests to assure reliability; such as, type tests, routine tests, periodic sampling tests, and tests for the verification of maintenance.

TABLE 1 – FAILURE RATES AND OUTAGE DURATION DATA FOR
HIGH VOLTAGE CIRCUIT BREAKERS ABOVE 63 kV****

----------MAJOR FAILURE RATES----------						----------MINOR FAILURE* RATES----------				
Sample Size Breaker Years	Number of Major Failures	λ_M Major Failures per Breaker Year	Hours Downtime per Failure Average	*** Median	VOLTAGE kV	Sample Size Breaker Years	Number of Failures*	Failures* per Breaker Year	Hours Downtime per Failure Average	*** Median
77,892	1,231**	.0158	81.6	12.0	All Voltages	46,272	1,641	.0355	30.0	6.0
33,877	138	.0041	29.3	5.0	63 ≤ V <100	24,716	409	.0165	16.5	5.0
26,743	437	.0163	94.4	12.0	100 ≤ V <200	13,915	581	.0417	19.9	5.0
9,939	257	.0258	58.5	11.0	200 ≤ V <300	5,614	359	.0639	27.7	6.0
6,224	283	.0455	83.8	11.0	300 ≤ V <500	1,682	275	.1635	73.1	8.0
1,109	116	.1045	142.0	27.0	500 ≤ V	345	17	.0493	58.2	9.0

CODE
* Minor failures plus defects
** 45 of the 1,231 major failures had a fire and/or explosion
*** Downtime includes: time required to get to site, analyse the failure, obtain spare parts, repair and return circuit breaker to service. Deliberate delays have been excluded.
**** From CIGRE 13-06 World Wide Reliability Survey for 1974-77; see Ref. 1

Table 2 – RELIABILITY DATA ON HIGH VOLTAGE CIRCUIT BREAKERS ABOVE
63 Kv THAT CAN BE USED IN SYSTEM RELIABILITY STUDIES****

1. Major Failures per Open Command $\lambda_{c1} + \lambda_{c2}$
2. Major Failures per Close Command $\lambda_{c3} + \lambda_{c4}$

3. Major Failures per Operating Cycle** $\lambda_c = \lambda_{c1} + \lambda_{c2} + \lambda_{c3} + \lambda_{c4}$

4. Average Number of Operating-Cycles** per Year C
5. Major Failures per Breaker-Year During Commands to Open or Close $C \cdot \lambda_c$
6. Major Failures per Breaker-Year Occuring Without A Command to Open or Close λ_s
7. Total Major Failures per Breaker-Year $\lambda_M = \lambda_s + C \cdot \lambda_c$

λ_{c1} Does Not Open On Command	λ_{c2} Does Not Break the Current	λ_{c3} Does Not Close On Command	λ_{c4} Does Not Make the Current	λ_c Major Failures per 10,000 Operating Cycles**	C Average Number of Operating Cycles** per Year	VOLTAGE kV	$C \cdot \lambda_c$ Major Failures per Breaker Year	λ_s Major Failures per Breaker Year	λ_M Total Major Failures per Breaker Year
FREQUENCY PER 10,000 COMMANDS									
0.84	0.11	2.01	0.10	3.06	26.5	All Voltages	.0081	.0077***	.0158
0.166	0.018*	0.562	0.010*	0.756	24.7	63 ≤ V <100	.0019	.0022	.0041
0.81	0.12*	2.60	0.05*	3.58	23.8	100 ≤ V <200	.0085	.0078	.0163
1.42	0.07*	2.54	0.32*	4.35	32.0	200 ≤ V <300	.0139	.0119	.0258
3.16	0.64*	5.39	0.24*	9.43	25.0	300 ≤ V <500	.0236	.0219	.0455
9.75*	0.00*	12.98*	0.00*	22.73*	26.8	500 ≤ V	.0609	.0436	.1045

CODE
* Small sample size in failure mode data - less than 8 failures
** An operating-cycle is one open command and one close command
*** Approximately 10.7% of these major failures are "breakdown across open pole" and another 3.5% are "closes without command"
**** Calculated from CIGRE 13-06 World Wide Reliability Survey Data for 1974-77; see Tables 10.1a, 17.1a, and 8.1 in Ref. 1

TABLE 3 - COST OF SCHEDULED SERVICING of HIGH VOLTAGE
CIRCUIT BREAKERS ABOVE 63 kV*** (Includes
Ordinary Servicing and Detailed Servicing)

Interval Between Scheduled Servicing Average Median YEARS		VOLTAGE kV	Labor Effort Average Median		Spare Parts** Consumed Average Median	
			MANHOURS PER BREAKER PER YEAR			
2.1	3.0	All Voltages	38.6	82.0	67.6	15.0
2.3	3.0	63 ≤ V < 100	19.6	17.5	55.0	5.0
2.0	2.5	100 ≤ V < 200	34.0	30.0	38.2	12.0
2.0	3.0	200 ≤ V < 300	47.4	44.0	87.5	20.0
1.4	2.0	300 ≤ V < 500	48.5	50.0	72.7	38.0
2.0	3.0	500 ≤ V	91.4*	114.0*	477.2*	50.0*

CODE
  * Small sample size - less than 8 data points
 ** Each country converted the cost of spare parts consumed
    into equivalent manhours using their labor rate. This
    resulted in manhours being used as an international
    currency for both labor effort and spare parts consumed.
*** From CIGRE 13-06 World Wide Reliability Survey Data for
    1974-77; see Tables 8.2, 8.3, and 8.4 in Ref. 1.

DEFINITIONS

MAJOR FAILURE - Complete failure of a circuit
breaker which causes the lack of one or more
of its fundamental functions. A major failure
results in an immediate change in the system
operating conditions; for example, the backup
protective equipment is required to remove the
fault, or results in mandatory removal from
service for non-scheduled maintenance. The
following are considered major failures:
- Does not close on command
- Does not open on command
- Closes without command
- Opens without command
- Does not make the current
- Does not break the current
- Fails to carry the current
- Internal breakdown to earth
- External breakdown to earth
- Internal breakdown between poles
- External breakdown between poles
- Internal breakdown across open pole
- External breakdown across open pole
- Other failures necessitating intervention
  within 30 minutes.

MINOR FAILURE - Failure of a circuit breaker
other than a major failure. This includes a
failure of a component or subassembly which
does not cause a major failure of the circuit
breaker.

FAILURE - Lack of performance by an item of
its required performance or functions.

DEFECT - Imperfection in the state of an item
which can result in one or more failures of
the item itself or of another item under the
specified service conditions or maintenance
conditions.

  SERVICING - Action which may lead to
dismantling for examination and/or replacement
of parts on a circuit breaker out of service.

This includes Ordinary Servicing, Detail
Servicing, and Special Servicing.

ORDINARY SERVICING - Servicing scheduled
according to given operational conditions
which would include a check of the operati
of the principal control devices, the
measurement of the characteristics of
insulation and arc-extinguishing media,
cleaning, washing, lubricating, tightening
adjusting, replacing worn parts in accorda
with given instructions, and the measureme
of the operation chacteristics such as
lock-out pressures, operating time, insula
of auxiliary circuits, etc.

DETAILED SERVICING - Scheduled servicing i
accordance with the given instructions
necessitated by long service, large number
operations, etc. It will include a more
detailed examination of all the parts than
that carried out during the Ordinary
Servicing.

SPECIAL SERVICING - Unscheduled servicing
necessitated by a circuit breaker failure
defect or because it has been subjected to
excessive stresses.

MAINTENANCE - Every operation of inspectio
and servicing.

INSPECTION - Visual periodical examination
the principal features of the circuit brea
in service without any dismantling. This
examination is generally directed toward
pressures and/or levels of fluids, tightne
positions of relays, pollution of insulati
parts, but actions such as lubricating,
cleaning, washing, etc., which can be carr
out with the circuit breaker in service, a
included. The observations resulting from
inspections can be grounds for carrying ou
servicing.

TABLE 4 - ORIGIN OF THE FAILURE

	Major failures						Minor failures **					
	All Voltages	63 ≤ V <100	100 ≤ V <200	200 ≤ V <300	300 ≤ V <500	500 ≤ V	All Voltages	63 ≤ V <100	100 ≤ V <200	200 ≤ V <300	300 ≤ V <500	500 ≤ V
	%	%	%	%	%	%	%	%	%	%	%	%
- Mechanical	70.3	70.2	62.6	75.1	77.1	76.9	85.6	88.2	84.3	83.6	86.7	90.0
- Electrical (main circuit)	10.6	10.7	14.7	7.9	6.2	23.1	2.7	2.4	0.9	4.1	5.4	0
- Electrical (auxiliary circuit)	19.1	19.1	22.7	17.0	16.7	0	11.7	9.4	14.8	12.3	7.9	10.0
Number of answers	775	138	273	177	177	10	1602	408	561	351	266	16

** Minor failures plus defects

TABLE 5 - CAUSE OF THE FAILURE

	Major Failures						Minor failures**					
	All Voltages	63 ≤ V <200	100 ≤ V <200	200 ≤ V <300	300 ≤ V <500	500 ≤ V	All Voltages	63 ≤ V <100	100 ≤ V <200	200 ≤ V <300	300 ≤ V <500	500 ≤ V
	%	%	%	%	%	%	%	%	%	%	%	%
- Design or manufacture	45.3	52.0	39.6	50.7	42.9	58.3	52.5	46.4	54.9	48.6	62.3	45.5
- Transport or storage	0	0	0	0	0	0	0.3	0	0	0.5	1.2	0
- Incorrect erection	9.3	9.5	7.0	10.6	11.1	16.7	10.7	14.2	9.1	7.8	11.7	36.4
- Inadequate instruction for operation and maintenance	0.7	0	1.1	0	1.3	0	0.3	0	0	0.5	1.2	0
- Failure to follow operating instructions	1.2	0	1.4	1.8	1.3	0	0.2	0	0.3	0.5	0	0
- Incorrect maintenance	8.1	9.5	8.1	5.3	10.2	0	4.5	5.9	3.9	5.0	3.1	0
- Stresses beyond those specified	4.8	2.8	7.0	3.5	4.0	8.3	0.7	0.4	0.8	0.9	0.6	0
- Other external causes (e.g. animals)	2.3	1.7	2.8	1.3	3.1	0	1.7	2.0	1.7	0.9	2.5	0
- Unknown origin	28.3	24.5	33.0	26.8	26.1	16.7	29.1	31.1	29.3	35.8	17.4	18.1
Number of answers	751	135	268	170	169	9	1604	408	568	350	261	17

** Minor failures plus defects

TABLE 6 - FAILURE MODES OF MAJOR FAILURES

	All Voltages	63 ≤ V < 100	100 ≤ V < 200	200 ≤ V < 300	300 ≤ V < 500	500 ≤ V
	%	%	%	%	%	%
- Does not close on command	33.7	34.1	38.0	31.4	29.6	33.3
- Does not open on command	14.1	10.1	11.7	17.5	17.4	25.0
- Closes without command	1.7	2.8	0.7	2.7	1.3	0
- Open without command	5.2	4.5	4.4	7.2	4.3	16.7
- Does not make the current	1.6	0.6	0.7	4.0	1.3	0
- Does not break the current	1.9	1.1	1.8	0.9	3.5	0
- Fails to carry current	2.5	1.1	2.9	1.3	4.3	0
- Breakdown to earth (internal)	1.3	1.7	1.5	1.3	0.9	0
- Breakdown to earth (external)	1.3	1.7	0.7	1.3	1.7	0
- Breakdown between poles (internal)	0	0	0	0	0	0
- Breakdown between poles (external)	0.5	0.6	0.7	0	0	8.3
- Breakdown across open pole (internal)	4.0	2.7	6.2	3.6	1.7	8.3
- Breakdown across open pole (external)	1.2	0	2.2	0.4	1.3	0
- Other failure necessitating intervention within 30 minutes	31.0	39.0	28.5	28.4	32.7	8.4
Number of answers	773	138	274	172	179	10

TABLE 7 - CORRELATION OF THE FAILURE MODES OF MAJOR FAILURES WITH THE ORIGIN/CAUSE

	Origin			Cause			
	Mechanical	Electrical (Main Circuit)	Electrical (Aux. Circuits)	Design or Manufacture	Incorrect Erection	Incorrect Maintenance	Unknown Origin
	%	%	%	%	%	%	%
- Does not close on command	35.2	1.2	46.9	36.6	26.1	27.9	38.2
- Does not open on command	13.7	2.5	21.8	14.8	17.4	16.4	11.5
- Closes without command	2.2	0	0.7	2.4	2.9	0	1.4
- Open without command	5.6	1.2	5.4	5.9	8.7	3.3	4.8
- Does not make the current	1.5	3.7	0.7	1.8	2.9	0	1
- Does not break the current	1.3	9.9	0	0.9	1.5	4.9	2.9
- Fails to carry current	2.4	7.4	0	1.8	5.8	11.5	0
- Breakdown to earth (internal)	0.2	9.9	0.7	1.2	0	1.6	1.9
- Breakdown to earth (external)	0.6	7.4	0	0.3	0	0	1.4
- Breakdown between poles (internal)	0	0	0	0	0	0	0
- Breakdown between poles (external)	0	4.9	0	0	0	0	0.5
- Breakdown across open pole (internal)	0.9	32.2	0	2.1	0	8.2	2.4
- Breakdown across open pole (external)	0.4	8.6	0	0.6	0	0	1
- Other failure necessitating intervention within 30 minutes	36	11.1	23.8	31.6	34.7	26.2	33
Number of answers	539	81	147	338	69	61	209

336

TABLE 8 - ESTIMATED AVERAGE NUMBER OF OPERATING-CYCLES PER YEAR

	ALL VOLT.	63≤V<100	100≤V<200	200≤V<300	300≤V<500	500≤V
NUMBER OF ANSWERS	422	75	151	116	72	8
AVERAGE	26.5	24.7	23.8	32.0	25.0	26.8
10% PERCENTILE	3.3	3.3	3.1	3.4	5.0	5.0
25% PERCENTILE	6.3	4.0	6.5	8.7	7.3	5.0
MEDIAN	13.1	8.2	12.0	14.7	15.7	24.9
75% PERCENTILE	28.8	29.4	24.0	31.1	38.6	26.6
90% PERCENTILE	53.1	40.4	36.8	71.1	67.0	29.8
95% PERCENTILE	78.0	55.4	54.7	138.7	73.0	29.8
MAXIMUM	548.6	491.2	548.6	381.6	85.4	107.5

REFERENCES

1. G Mazza, R. Michaca, "The First
International Enquiry on Circuit Breaker
Failures and Defects in Service," "Electra"
No. 79, December 1981, Paris, France,
pp 21-91 (in French and English).

2. R. Michaca, C. R. Heising, G. Koppl,
"Summary of CIGRE Working Group 13-06 Studies
on the Test and Control Methods Intended to
Assure the Reliability of High Voltage Circuit
Breakers," "Electra" No. 102, October 1985,
pp 133-175 (in French and English).

3. J. Beierer, R. Kearsley, J. Verdon,
"Maintenance of Modern High Voltage Circuit
Breakers," "Electra" No. 102, October 1985,
pp 119-131 (in French and English).

4. C. R. Heising, "Summary of the CIGRE 13-06
Working Group World Wide Studies on Testing
and Maintenance to Improve the Reliability of
High Voltage Circuit Breakers," Thirteenth
Annual Reliability - Availability -
Maintainability Conference for the Electric
Power Industry," IEEE/American Society for
Quality Control, Syracuse, June 2-5, 1986.

5. C. R. Heising, "Comparison of Results from
Several Power Circuit Breaker Reliability
Surveys in the United States and the CIGRE
13-06 World Wide Survey," Tenth Annual
Engineering Conference on Reliability -
Availability - Maintainability for the
Electric Power Industry, IEEE/ASQC, Montreal,
May 25-27, 1983.

6. R. Michaca, G. Mazza, C. R. Heising, C. J.
Essel, "Summary of CIGRE 13-06 Working Group
World Wide Reliability Survey of Power Circuit
Breakers Above 63 kV for 1974 Thru 1977,"
Ninth Annual Reliability Engineering
Conference for the Electric Power Industry,
IEEE/ASQC, Hershey, PA, June 16-18, 1982.

7. C. R. Heising, "Reliability and
Maintenance Data Needed for High Voltage
Circuit Breakers When Making Life Cycle Cost
Studies," Sixth Annual Reliability Engineering
Conference for the Electric Power Industry,
IEEE/ASQC, Miami, Florida, April 19-20, 1979.

CIGRE 13-06 WORKING GROUP
MEMBERS AND COUNTRY (1971-1985)

1985 Members	Country	Past Members
R. Michaca--	Chairman (France)	--------
J. Beierer---	Federal Republic	E. Pflaum,
	of Germany--	K. H. Schneider
H. Bruvik----	Norway----------	J. Svoen
H. Erni------	Switzerland-----	G. Köppl
J. M. Frisson	Belgium---------	R. Blanquet,
		G. Dienne
C. R. Heising	U. S. A.	---------
R. Iglesias--	Spain---------	A. Ruiz De Lobera
L. S. Irwing-	Australia-------	J. R. Mortiss
A. Janssen---	Netherlands--	L. A. J. M. Wiercx
M. Kanzantzis	Greece	
R. Kearsley--	Sweden----------	B. Bentsberg
P. Kopecný---	Czechoslovakia	--------
*G. Mazza----	Italy	--------
K. Pönni-----	Finland	--------
J. F. Reid---	United Kingdom-	M. V. Bradbury
		E. A. Lane
N. V. Shillin	U. S. S. R.	---------
M. Tanabe----	Japan-----------	J. Senda,
		M. Yasuda,
		J. Fujimoto
E. Thuries---	France----------	J. Passaquin
J. Verdon ---	France	--------
------------	Canada----------	J. Mastrocola
------------	Poland----------	M. Pomianowski

* Also Secretary of Working Group 11, IEC 17A,
  "Mechanical Problems and Reliability"

Appendix K
Report of Circuit Breaker Reliability Survey
of Industrial and Commercial Installations

By
A. T. Norris
IEEE Industrial and Commercial
Power Systems Technical Conference
Chicago, IL, May 8-11, 1989

Reprinted from IEEE Conference Record
89CH27738-3, pp. 1-16

REPORT OF CIRCUIT BREAKER RELIABILITY SURVEY
OF INDUSTRIAL AND COMMERCIAL INSTALLATIONS

Andrew Norris (coordinating Author)[1]

CIRCUIT BREAKER RELIABILITY WORKING GROUP
POWER SYSTEMS RELIABILITY SUBCOMMITTEE
POWER SYSTEMS ENGINEERING COMMITTEE
INDUSTRIAL & COMMERCIAL POWER SYSTEMS DEPARTMENT
IEEE INDUSTRY APPLICATIONS SOCIETY

ABSTRACT

The Reliability Subcommittee of the IEEE Industry Applications Society initiated a survey of the reliability of circuit breakers in industrial and commercial installations in keeping with its commitment to update information on previous surveys. The survey was restricted to circuit breakers that are less than fifteen (15) years old, and excluded molded case breakers, in order to provide information on units of interest and to obtain information on new circuit breaker technologies.

A more detailed explanation on reasons for this survey is included in the appendix.

INTRODUCTION

The results of the survey conducted in 1985 on the reliability of circuit breakers in industrial and commercial installations are summarized in the attached tables. The data obtained includes information on estimated numbers of operations per year for both fault and non-fault situations. Information has also been collected on low voltage circuit breakers comparing static and electro-mechanical integral trip devices.

Each table is discussed to highlight results of the survey. It is the intent of this working group to present the results as updated information on industrial applications and the drawing of definite conclusions is left to the reader.

The reasons for conducting the survey were written down at the beginning and are included in the appendix. Some of these objectives were not achieved due to the small number of participants in the survey. It was not possible to determine the effect of preventive maintenance on failure rate. Insufficient data were submitted on vacuum and single-pressure SF-6 circuit breakers.

[1] The Author is with the University of Missouri - Columbia, Missouri 65211

Members of the Circuit Breaker Reliability Working are Andrew T. Norris (Chairman), James W. Aquilino, William F. Braun, Jr., Charles R. Heising, Don O. Koval, Lou D. Monaghan, Pat O'Donnell, A.D. Patton, and Peter W. Dwyer.

SURVEY RESPONSE

The survey questionnaire, along with the Reasons For Conducting a New Survey on Circuit Breaker Reliability, is included in the appendix.

Due to the low number of responses, 13 plant locations, no attempt was made to separate failures by industry types. While the number of respondents was less than hoped for, the questionnaires were all fully completed for the requested data, with only one (1) "unknown" entry listed which was for a failure duration.

The following list provides a summary of the survey response

No. of Plants	13
No. of Circuit Breakers	2137
Sample Size (unit years)	4097.17
Total no. of Failures	59

The small sample size of the data received limited the results that are being published to four equipment/voltage categories. A special note is made in the tables where the number of failures in a specific category is considered an inadequate sample size. Less than 8 failures has been considered as an inadequate sample size.

OVERALL SUMMARY OF RELIABILITY DATA

Table 1 summarizes the overall results by voltage class. The low number of failures (4) in the 601 volt to 15,000 volt circuit breaker class makes this failure rate data of questionable validity.

This survey shows an increase in the failure rate per unit-year, in the 0-600 volt class, of nearly 3 times the value shown in the 1973 survey. There is, however, a large reduction in the average and maximum failure durations of 30% and 99.5% respectively.

LOCATION

Table 2 shows the effect of outdoor vs. indoor location on the failure rate of 0 - 600 volt circuit breakers. The failure rate was 1.54 times higher for outdoor circuit breakers.

INTEGRAL TRIP

Table 3 compares the integral trip unit type on the failure rate for 0-600 volt circuit breakers. The failure rate of static type integral trip units is 36% of the electromechanical units.

FAILURE MODE

Table 4 shows the failure modes for circuit breakers reported in the survey. It is noted that there were only two instances of units that "failed to open on command", and no occurrences of "closes without command". In the 0-600 volt class all circuit breakers reported had an integral trip device. The circuit breakers with a static integral trip device were split between "failed to close on command" (44%), and "opens without command" (56%). Circuit breakers with electro-mechanical type of integral trip device had a very large portion (93%) of the failures reported to be "failed to close on command".

FAILURE INITIATING CAUSE

Table 5 shows the primary failure initiating cause reported for both 34.5-138kV and 345kV circuit breaker groups as "mechanical breakdown" as 56% and 65% respectively. The 0-600 volt circuit breaker group shows "malfunction of protective relay or tripping device" to be the major category at (93%) for units with electro-mechanical integral tripping. The 0-600 volt units with static type integral tripping reported a roughly even split between "transient overvoltage" and "malfunction of protective relay or tripping device".

FAILURE CONTRIBUTING CAUSE

Table 6 shows that "dust, salt spray, or other contaminant exposure" is the primary reported listing (at 93%) for failure contributing cause for 0-600 volt circuit breakers with electro-mechanical type integral trip. The 0-600 volt circuit breakers with static integral trip had "lack of preventive maintenance" reported for 56% of failures, with the remaining 44% shown as "persistent overload". Entries for other voltage classes are in much lower percentages, except for the "other" category in the 34.5-138kV and 345kV groups.

SUSPECTED FAILURE RESPONSIBILITY

In table 7 the data shows most 0-600 volt breakers with electro-mechanical type of integral trip as having "inadequate physical protection" (93%) as the suspected failure responsibility. The 0-600 volt breakers with static type integral trip reports 56% under "improper operation", and 44% under "inadequate

maintenance". The 34.5-138kV and 345kV voltage categories both show "defective component" as the main category.

FAILURE DISCOVERED DURING

Table 8 shows a very large percentage of failures in the 0-600 volt circuit breakers, a total of 96%, as being discovered "during normal operation". The 34.5-138kV class showed a significant percentage of failures (67%) as being discovered "during routing testing/maintenance", while the 345kV breakers were split between "during routing testing/maintenance" and "during normal operation" with 48% in each category.

FAILURES vs. MONTHS SINCE LAST MAINTENANCE

Table 9 shows that most failures occurred within 24 months of the last maintenance.

FAILURE REPAIR METHOD

Table 10 shows that a high percentage of circuit breakers in the 0-600 volt, 601-15,000 volt, and 34.5-138kV ratings were "repaired failed component in place or sent out for repair".

The 345kV group of circuit breakers shows the highest number (44%) as "replaced failed unit with spare". This large percentage is considered questionable since an inspection of the failed component entries showed in some cases that a failed component, such as an air compressor, was reported as "replaced failed unit with spare".

REPAIR URGENCY

It is of particular interest that, in Table 11, only 7% of the 59 failures reported for all voltage categories listed the repair urgency as requiring working on a round-the-clock bases. This may be due, at least in part, to the fact that two of the voltage classes (0-600 volt, and 601-15,000 volt) containing 45% of the total failures, and had maximum failure durations of 4 hours.

The 34.5-138kV and 345kV circuit breakers, with their longer failure durations, also show nearly all repair work as normal working hours.

POPULATION OF CIRCUIT BREAKERS vs. MAINTENANCE QUALITY AND NORMAL MAINTENANCE CYCLE

Table 12 shows the majority of respondents (53%) considered themselves as having a "fair" maintenance quality, while 39% considered their maintenance quality as "excellent". All of the respondents who listed their maintenance quality as excellent had a normal maintenance cycle of 0-24 months. The respondents with "fair" maintenance quality were split between categories with 37% (by unit-year) showing 0-24 month, 28% (by unit-year) showing more then 24 months and, interestingly enough, 35% with No preventive maintenance.

OVERALL CIRCUIT BREAKER OPERATIONS PER YEAR DATA

The listing of "overall circuit breaker operations data" has been entered in three different tables.

Table 13a shows the data entered in a non-weighted format. The fault, and non-fault, operations per year are based on non-weighted numbers. The non-weighted values were obtained by counting each population data line entry as one unit (regardless of how many circuit breakers or unit-years were reported in that line). The average number of operations for each entry line were summed and the result divided by the number of line entries.

Table 13b shows the data weighted by the number of circuit breakers. The fault and non-fault operations per year are based on the actual number of circuit breakers reported, regardless of time in service. The average number of operations for each entry line was multiplied times the number of circuit breakers reported for that line. The resulting values were summed and the total was then divided by the number of circuit breakers reported in that voltage category.

Table 13c shows the data weighted by the number of unit-years. The fault and non-fault operations per year are based on the number of circuit breakers reported times their number of years in service (unit-years). The unit-years for each circuit breaker times the average operations per year was summed and the result divided by the total number of unit-years reported in that voltage category.

With the exception of the 0-600 volt category, the average number of operations per year remained reasonably consistent over the three tables.

Table 1 -. OVERALL CIRCUIT BREAKER RELIABILITY DATA

	0-600 Volt Air Magnetic *	601-15,000 Volt Air Magnetic ***	34.5-138 kV Bulk Oil	345 kV Air Blast & SF-6 (2 presure)
Sample Size (number of units)	1695	315	64	51
Sample Size (unit-years)	2941.24	694.76	192.50	256.00
Total Fault Operations (for all unit-years)	225	343	103	434
Total Non-Fault Operations (for all unit-years)	24604	24914	4320	8200
Number of Failures	23	4 **	9	23
Failure Rate - Failures/Unit-Year	0.00782	0.00576 **	0.04675	0.08984
Failure Duration (Hours/Failure)		**		
Average	2.8	2.25	41.11	171.45
Minimum	0.5	1	1	1
Median	4	2	3	150
Maxumum	4	4	240	720

* Excludes Molded Case
** Small Sample Size - less than 8 failures (or data points)
*** Zero failures in 2.67 unit-years reported for Vacuum 601-15,000 volt (not included in this table)

NOTE: The "Total Fault Operations" and "Total Non-Fault Operations" were determined by taking the Unit-years (for each circuit breaker reported) times it's average number of operations (Fault or Non-Fault) per year, and adding the values for all circuit breakers in that catagory.

Table # 2 CIRCUIT BREAKERS, 0-600 VOLT *
OUTDOOR versus INDOOR LOCATION

	Outdoor	Indoor
Sample Size (unit-years)	873.57	2067.67
Number of Failures	9	14
Failure Rate - Failures/Unit-Year	0.0103	0.00677

* Excludes Molded Case

Table # 3 CIRCUIT BREAKERS, 0-600 VOLT *
EFFECT OF INTEGRAL TRIP TYPE

	Static	Electro-mechanical
Sample Size (unit-years)	1888.49	1052.75
Number of Failures	9	14
Failure Rate - Failures/Unit-Year	0.00477	0.0133

* Excludes Molded Case

TABLE # 4 - CIRCUIT BREAKERS
VOLTAGE VS. FAILURE MODE

	0-600 VOLT * Air Magnetic Static	0-600 VOLT * Electro-mech	601-15KV Air Magnetic	34.5KV-138KV Bulk Oil	345KV Air Blast & SF-6 (2 presure)
FAILED TO CLOSE ON COMMAND	4 44%	13 93%	-0-	3 33%	-0-
FAILED TO CLOSE AND LATCH	-0-	-0-	-0-	1 11%	-0-
FAILED TO OPEN ON COMMAND	-0-	-0-	2 50%	-0-	-0-
CLOSES WITHOUT COMMAND	-0-	-0-	-0-	-0-	-0-
OPENS WITHOUT COMMAND	5 56%	-0-	1 25%	-0-	7 30%
FAILED TO BREAK CURRENT WHEN OPENING	-0-	-0-	-0-	-0-	-0-
DAMAGED WHILE SUCCESSFULLY OPENING	-0-	-0-	1 25%	-0-	-0-
DAMAGED WHILE CLOSING	-0-	-0-	-0-	-0-	-0-
FAILED TO CARRY CURRENT	-0-	-0-	-0-	1 11%	-0-
FAULT TO GROUND, OR PHASE TO PHASE (NOT WHILE OPENING OR CLOSING)	-0-	-0-	-0-	-0-	-0-
FAULT ACROSS OPEN CONTACTS (NOT WHILE OPENING OR CLOSING)	-0-	-0-	-0-	-0-	-0-
LOSS OF VACUUM (FOR VACUUM BREAKERS)	-0-	-0-	-0-	-0-	-0-
OTHER FAILURE REQUIRING REMOVAL FROM SERVICE WITHIN 30 MINUTES	-0-	-0-	-0-	-0-	11 48%
OTHER FAILURE NOT REQUIRING REMOVAL FROM SERVICE	-0-	1 7%	-0-	3 33%	3 13%
UNKNOWN	-0-	-0-	-0-	1 11%	2 9%
TOTAL FAILURES	9 100%	14 100%	4 100%	9 100%	23 100%

* Excludes Molded Case

TABLE # 5 - CIRCUIT BREAKERS
VOLTAGE VS. FAILURE INITIATING CAUSE

| | 0-600 VOLT * | | 601-15KV ** | 34.5KV-138KV | 345KV |
	Air Magnetic Static	Electro-mech	Air Magnetic	Bulk Oil	Air Blast & SF-6 (2 presure)
TRANSIENT OVERVOLTAGE-SUCH AS LIGHTNING, SWITCHING SURGES, OR SYSTEM FAULTS	4 44%	-0-	1 25%	-0-	-0-
INSULATION BREAKDOWN	-0-	-0-	-0-	-0-	1 4%
MECHANICAL BURNOUT, FRICTION, OR SEIZING OF MOVING PARTS	-0-	-0-	-0-	-0-	1 4%
MECHANICAL BREAKDOWN - SUCH AS CRACKING, LOOSENING, ABRADING, OR DEFORMING OF STATIC OR STRUCTURAL PARTS	-0-	-0-	1 25%	5 56%	15 65%
PHYSICAL DAMAGE OR SHORTING FROM OUTSIDE SOURCE - SUCH AS VEHICULAR ACCIDENT	-0-	-0-	-0-	-0-	-0-
ELECTRICAL FAULT OR MALFUNCTION	-0-	-0-	1 25%	1 11%	3 13%
MALFUNCTION OF PROTECTIVE RELAY OR TRIPPING DEVICE	5 56%	13 93%	-0-	1 11%	-0-
OTHER AUXILIARY DEVICE MALFUNCTION	-0-	-0-	-0-	2 22%	-0-
LOW, OR NO, AUXILIARY VOLTAGE - FOR CIRCUITS SUCH AS AIR COMPRESSORS, AND SF-6 HEATERS	-0-	-0-	-0-	-0-	-0-
OTHER	-0-	1 7%	1 25%	-0-	3 13%
TOTAL FAILURES	9 100%	14 100%	4 100%	9 100%	23 100%

* Excludes Molded Case

347

TABLE # 6 - CIRCUIT BREAKERS
VOLTAGE VS. FAILURE CONTRIBUTING CAUSE

	0-600 VOLT * Air Magnetic Static	0-600 VOLT * Electro-mech	601-15KV ** Air Magnetic	34.5KV-138KV Bulk Oil	345KV Air Blast & SF-6 (2 presure)
OVERLOAD - PERSISTENT	4 44%	-0-	1 25%	-0-	-0-
EXTREME HEAT	-0-	-0-	-0-	-0-	-0-
EXTREME COLD	-0-	-0-	-0-	-0-	3 13%
SEVERE WEATHER - SUCH AS WIND, RAIN, SNOW, OR SLEET	-0-	-0-	-0-	-0-	-0-
ABNORMAL MOISTURE	-0-	-0-	-0-	-0-	-0-
AGRESSIVE CHEMICALS	-0-	-0-	-0-	-0-	-0-
DUST, SALT SPRAY, OR OTHER CONTAMINANT EXPOSURE	-0-	13 93%	-0-	-0-	-0-
NORMAL DETERIORATION FROM AGE	-0-	-0-	-0-	2 22%	1 4%
LUBRICANT LOSS, OR DEFICIENCY	-0-	-0-	1 25%	1 11%	-0-
IMPROPER OPERATING OR TEST PROCEDURE	-0-	-0-	-0-	-0-	1 4%
TRIPPING SOURCE DEFICIENT	-0-	-0-	-0-	-0-	-0-
LACK OF PREVENTIVE MAINTENANCE	5 56%	-0-	1 25%	1 11%	-0-
OTHER	-0-	1 7%	1 25%	5 56%	18 78%
TOTAL FAILURES	9 100%	14 100%	4 100%	9 100%	23 100%

* Excludes Molded Case

TABLE # 7 - CIRCUIT BREAKERS
VOLTAGE VS. SUSPECTED FAILURE RESPONSIBILITY

DEFECTIVE COMPONENT	0-600 VOLT * Air Magnetic Static	Electro-mech	601-15KV ** Air Magnetic	34.5KV-138KV Bulk Oil	345KV Air Blast & SF-6 (2 presure)
IMPROPER HANDLING/SHIPPING	-0-	-0-	1 25%	4 44%	13 57%
POOR INSTALLATION/TESTING	-0-	-0-	-0-	1 11%	1 4%
INADEQUATE MAINTENANCE	4 44%	-0-	-0-	1 11%	-0-
IMPROPER OPERATION	5 56%	-0-	1 25%	1 11%	-0-
IMPROPER APPLICATION	-0-	1 7%	-0-	-0-	-0-
INADEQUATE PHYSICAL PROTECTION	-0-	13 93%	-0-	-0-	-0-
OUTSIDE AGENCY (SUCH AS VEHICULAR ACCIDENT)	-0-	-0-	-0-	-0-	-0-
OTHER	-0-	-0-	2 50%	2 22%	4 17%
UNKNOWN	-0-	-0-	-0-	-0-	5 22%
TOTAL FAILURES	9 100%	14 100%	4 100%	9 100%	23 100%

* Excludes Molded Case

349

TABLE # 8 - CIRCUIT BREAKERS
VOLTAGE VS. "FAILURE DISCOVERED DURING"

	0-600 VOLT * Air Magnetic Static	0-600 VOLT * Electro-mech	601-15KV ** Air Magnetic	34.5KV-138KV Bulk Oil	345KV Air Blast & SF-6 (2 presure)
DURING ROUTINE TESTING/MAINTENANCE	-0-	1 7%	1 25%	6 67%	11 48%
DURING NORMAL OPERATION	9 100%	13 93%	3 75%	3 33%	11 48%
OTHER	-0-	-0-	-0-	-0-	1 4%
TOTAL FAILURES	9 100%	14 100%	4 100%	9 100%	23 100%

* Excludes Molded Case

** Small Sample Size - less than 8 failures (or data points)

TABLE # 9 - CIRCUIT BREAKERS
FAILURES VS. MONTHS SINCE LAST MAINTENANCE

	0-600 VOLT * Air Magnetic Static	0-600 VOLT * Electro-mech	601-15KV ** Air Magnetic	34.5KV-138KV Bulk Oil	345KV Air Blast & SF-6 (2 presure)
0 - 24 MONTHS ***	-0-	14 100%	2 50%	8 89%	17 74%
OVER 24 MONTHS	9 100%	-0-	2 50%	-0-	6 26%
NO PREVENTIVE MAINTENANCE	-0-	-0-	-0-	1 11%	-0-
TOTAL FAILURES	9 100%	14 100%	4 100%	9 100%	23 100%

* Excludes Molded Case

** Small Sample Size - less than 8 failures (or data points)

*** The survey requested data for 0-12 month and 12-24 month periods. Due to the uncertainty about which of these two periods should be used for entries of 12 months since maintenance, they were combined into a single entry of 0-24 months.

TABLE # 10 - CIRCUIT BREAKERS
VOLTAGE VS. FAILURE REPAIR METHOD

| | 0-600 VOLT * | | 601-15KV ** | 34.5KV-138KV | 345KV |
| | Air Magnetic | | Air Magnetic | Bulk Oil | Air Blast & |
	Static	Electro-mech			SF-6 (2 presure)
REPAIRED FAILED COMPONENT IN PLACE OR SENT OUT FOR REPAIR	8 89%	13 93%	3 75%	7 78%	7 30%
REPLACED FAILED UNIT WITH SPARE	1 11%	1 7%	1 25%	2 22%	*** 10 43%
OTHER	-0-	-0-	-0-	-0-	6 26%
TOTAL FAILURES	9 100%	14 100%	4 100%	9 100%	23 100%

* Excludes Molded Case

** Small Sample Size - less than 8 failures (or data points)

*** In some cases a failed component, not the complete breaker, was replaced with a spare.

TABLE # 11 - CIRCUIT BREAKERS
VOLTAGE VS. FAILURE REPAIR URGENCY

| | 0-600 VOLT * | | 601-15KV ** | 34.5KV-138KV | 345KV |
| | Air Magnetic | | Air Magnetic | Bulk Oil | Air Blast & |
	Static	Electro-mech			SF-6 (2 presure)
WORKING ROUND-THE-CLOCK	2 22%		1 25%	1 11%	-0-
NORMAL WORKING HOURS	7 78%	14 100%	3 75%	8 89%	23 100%
LOW PROIRITY	-0-	-0-	-0-	-0-	-0-
TOTAL FAILURES	9 100%	14 100%	4 100%	9 100%	23 100%

* Excludes Molded Case

** Small Sample Size - less than 8 failures (or data points)

TABLE # 12 - CIRCUIT BREAKERS
POPULATION OF CIRCUIT BREAKERS VERSUS
MAINTENANCE QUALITY & NORMAL MAINTENANCE CYCLE

| MAINTENAN QUALITY | MAINTENANCE, NORMAL CYCLE | | | TOTAL | |
| | 0 - 24 MONTHS * | MORE THEN 24 MONTHS | NO PREVENTIV MAINTENAN | | |
	POPULATION: UNIT-YEARS				
EXCELLENT	1198.25	383	-0-	1581.25	39%
FAIR	797.99	606.59	749.34	2153.92	53%
POOR	-0-	-0-	-0-	-0-	0%
NONE	-0-	-0-	362	362	9%

* The survey requested data for 0-12 month and 12-24 month periods. Due to the uncertainty about which of these two periods should be used for entries of 12 months since maintenance, they were combined into a single entry of 0-24 months.

Table 13a - OVERALL CIRCUIT BREAKER OPERATIONS DATA (Non-weighted)

| | 0-600 Volt * | 601-15,000 Volt * | 34.5-138 kV | 345 kV |
	Air Magnetic	Air Magnetic	Bulk Oil	Air Blast & SF-6 (2 presure)
Fault Operations/Year				
Average	0.175	0.3481	0.6945	1.1325
Minimum	0	0	0.05	0.2
Median	0.05	0.0769	0.75	0.2
Maximum	1	1	2	2
Non-Fault Operations/Year				
Average	19.2834	47.5357	24.125	30
Minimum	0	0.5	3	10
Median	1.667	5	15	30
Maximum	100	400	100	50

* Excludes Molded Case

To get the non-weighted values for Average Fault (and Non-Fault) Operations per year, each line entry was counted as one unit (regaurdless of how many circuit breakers were reported in that line). The average number of operations for each entry line were summed and the result divided by the number of line entries. Twenty (20) line entries would be counted as 20 units, even though each line might represent 5 circuit breakers.

Table 13b - OVERALL CIRCUIT BREAKER OPERATIONS DATA (weighted by number of breakers)

	0-600 Volt Air Magnetic*	601-15,000 Volt Air Magnetic	34.5-138 kV Bulk Oil	345 kV Air Blast & SF-6 (2 presure)
Fault Operations/Year				
Average	0.0174	0.5898	0.4877	1.2341
Minimum	0	0	0.05	0.2
Median	0	1	0.75	0.2
Maxumum	1	1	2	2
Non-Fault Operations/Year				
Average	5.0932	27.1841	20.0156	35.098
Minimum	0	0.5	3	10
Median	5	25	20	30
Maxumum	100	400	100	50

* Excludes Molded Case

To get the weighted values (weighted by number of circuit breakers) for Average Fault, and Non-Fault operations, the number of operations for each entry line is multiplied by the number of circuit breakers reported in that line. The product (number of circuit breakers times average operations) from each line was summed and the result divided by the total number of circuit breakers reported in that catagory.

Table 13c - OVERALL CIRCUIT BREAKER OPERATIONS DATA (weighted by number of unit-years)

	0-600 Volt Air Magnetic*	601-15,000 Volt Air Magnetic	34.5-138 kV Bulk Oil	345 kV Air Blast & SF-6 (2 presure)
Fault Operations/Year				
Average	0.0767	0.4936	0.5375	1.6948
Minimum	0	0	0.05	0.2
Median	0.02	0.5	0.5	0.2
Maxumum	1	1	2	2
Non-Fault Operations/Year				
Average	8.3652	35.86	22.439	32.0313
Minimum	0	0.5	3	10
Median	1.6667	5	20	30
Maxumum	100	400	100	50

* Excludes Molded Case

To get the weighted values (weighted by number of unit-years) for Average Fault, and Non-Fault operations, the number of operations for each survey line entry is multiplied by the number of unit-years (circuit breakers reported in that line times the number years in service). The product (number of unit-years times average operations) from each line was summed and the result divided by the total number of unit-years reported in that catagory.

APPENDIX

REASONS FOR CONDUCTING A NEW SURVEY ON CIRCUIT BREAKER RELIABILITY

by Circuit breaker Reliability Working Group
C. R. Heising, Coordinating Author
A. T. Norris, Chairman
J. W. Aquilino   L. G. Monaghan
R. N. Bell   P. O'Donnell
W. F. Braun   A. D. Patton

The main purpose of this reliability survey is to identify failure data and the effect of pertinent factors on important classes and types of circuit breakers, thus providing the designer and planner the valuable basic information needed to install a reliable and economic industrial or commercial power system.

Previous IEEE-IAS circuit breaker reliability surveys of industrial & commercial installations were published in 1962 and in 1973/74. The latter has been included in IEEE Standard No. 493-1980 - "Recommended Practice for the Design of Reliable Industrial & Commercial Power Systems." Pertinent information from the new survey will be included in future revisions of IEEE Standard No. 493.

Some of the important objectives in this new survey are: 1. Obtain failure mode data, 2. Obtain estimates of the number of operating cycles per year, 3. Obtain data on static trip devices for low voltage circuit breakers, 4. Obtain information on the effect of preventive maintenance on failure rate, 5. Obtain better information on suspected failure responsibility, failure initiating cause, and failure contributing cause, and 6. Obtain pertinent information on new circuit breaker technologies.

33% or more of the failures reported in the 1973/74 survey did not contain information on suspected failure responsibility, failure initiating cause, and failure contributing cause. It is hoped that this can be improved upon in the new survey. This is considered important information when trying to improve the reliability of circuit breakers used on industrial & commercial power systems. In the 1973/74 survey 23% of the failures were blamed on the manufacturer and 23% were blamed on inadequate maintenance and 36% were unknown. These were the three largest causes of failures. Inadequate maintenance is an area that an industrial or commercial user can do something about; and any pertinent information on this subject will be requested.

The 1973/74 survey did not collect information on the estimated number of operating cycles per year. This is important information when trying to estimate the probability of a circuit breaker successfully operating when commended to do so. This information will permit a reliability assessment versus duty application.

The 1973/74 survey did not collect low voltage circuit breaker data on whether or not a static trip device was used. This information is of interest to designers of power systems where there is much concern about failure rate of solid state versus electromechanical trip devices.

Approximately 30% of the circuit breakers in the 1973/74 survey were over ten years old. Circuit breakers more than 15 years old may not be typical of what is being used in the design of new power systems.

Various classes and types of circuit breakers in the 1973/74 survey had significantly different distributions of the various failure modes. Updated information on this subject is of interest to designers of power systems.

Reliability information on medium and high voltage circuit breakers using the newer technologies is of interest to

designers of power systems. This includes vacuum and SF6-puffer circuit breakers.

Switchgear bus is not included in this survey. A separate survey was published on this subject in 1979. Protective relays, fuses, and switches are not included in this survey. A survey in 1978 on these equipment categories asked for information that many industrial and commercial users did not have readily available; and the survey was unsuccessful. A limited amount of information is contained in the 1973/74 survey on disconnect switches, relays, and fuses.

CIRCUIT BREAKERS

COMPANY NAME AND PLANT: _____

INDUSTRY TYPE: _____

PERIOD REPORTED - FROM: MONTH_____ YEAR_____

TO: MONTH_____ YEAR_____

LOCATION: _____

TOTAL POPULATION

A	B	C	D	E	F	G	H	I	J	K	L	M
IDENTIFICATION NUMBER	CIRCUIT BREAKER TYPE (Insert Code)	NUMBER OF BREAKERS	USED PRIMARILY AS MOTOR STARTER (1-YES, N-NO)	LINE-TO-LINE VOLTAGE (KV)	LOCATION (I-INDOOR, O-OUTDOOR)	INTEGRAL TRIP DEVICE* (Y-YES, N-NO)	INTEGRAL TRIP IS (S-STATIC, EM-ELECTRO MECH)	EST. AVERAGE # OF NON-FAULT OPERATIONS/YEAR/BREAKER**	EST. AVERAGE # OF FAULT OPERATIONS/YEAR/BREAKER**	MAINTENANCE CYCLE (Insert Code)	MAINTENANCE QUALITY (Insert Code)	BRIEF DESCRIPTION OF MAINTENANCE

* IF TRIP INITIATION UNIT IS AN INTEGRAL PART OF THE BREAKER, INCLUDE ANY FAILURE OF THE TRIP UNIT AS A BREAKER FAILURE.

** CONSIDER EACH OPEN/CLOSE CYCLE AS ONE (1) OPERATION. INCLUDE OPERATIONS DURING MAINTENANCE.

CIRCUIT BREAKER

COMPANY NAME AND PLANT: _____

FAILED UNIT DATA - Fill in One Line for Each Failure

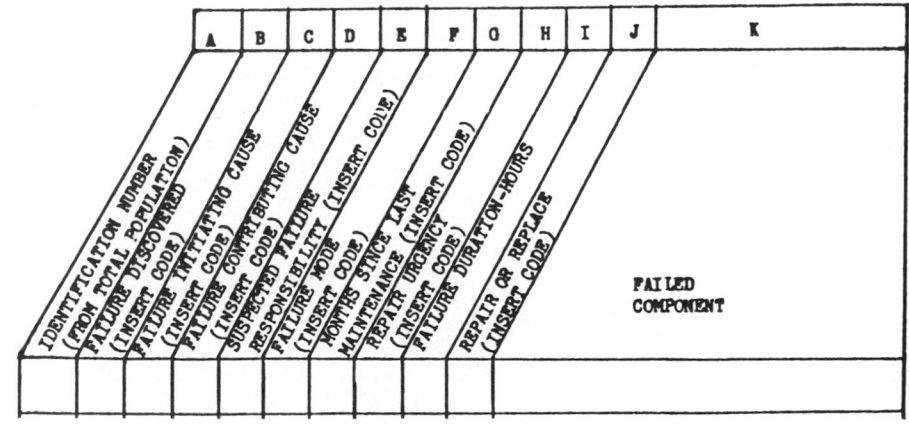

A	B	C	D	E	F	G	H	I	J	K
IDENTIFICATION NUMBER (FROM TOTAL POPULATION)	FAILURE DISCOVERED (INSERT CODE)	FAILURE INITIATING CAUSE (INSERT CODE)	FAILURE CONTRIBUTING CAUSE (INSERT CODE)	SUSPECTED FAILURE RESPONSIBILITY (INSERT CODE)	FAILURE MODE (INSERT CODE)	MONTHS SINCE LAST MAINTENANCE (INSERT CODE)	REPAIR URGENCY (INSERT CODE)	FAILURE DURATION-HOURS	REPAIR OR REPLACE (INSERT CODE)	FAILED COMPONENT

CIRCUIT BREAKER RELIABILITY SURVEY
SURVEY CODE

Total Population Form

Circuit breaker Type (B)
1. Air Magnetic
2. Vacuum
3. Bulk Oil
4. Air Blast
5. "Puffer" Type SF-6
6. All SF-6 other then "Puffer"
7. Other

Normal Maintenance Cycle (K)
1. 0-12 months
2. 12-24 months
3. Over 24 months
4. No preventive maintenance

Maintenance Quality (L)
Your estimate of Preventive
Maintenance Quality:
1. Excellent
2. Fair
3. Poor
4. None

Failed Unit Form

Failure Discovered (B)
1. During Routine Testing/
Maintenance
2. During Normal Operation
3. Other

Failure Initiating Cause (C)
1. Transient overvoltage - such as
lightning, switching surges, or
system faults.
2. Insulation Breakdown.
3. Mechanical burnout, friction, or
seizing of moving parts.

4. Mechanical breakdown - such as
cracking, loosening, abrading,
or deforming of static or
structural parts.
5. Physical damage or shorting from
outside source - such as
vehicular accident.
6. Electrical fault or malfunction.
7. Malfunction of protective relay
or tripping device.
8. Other auxiliary device
Malfunction.
9. Low, or no, auxiliary voltage -
for circuits such as air
compressors, and SF-6 heaters.
10. Other

Failure Contributing Cause (D)
1. Overload - persistent
2. Extreme hear (ambient temperature
if known _____ deg. C)
3. Extreme Cold (ambient temperature
if known _____ deg. C)
4. Severe weather - such as wind,
rain, snow, or sleet.
5. Abnormal moisture.
6. Aggressive chemicals.
7. Dust, salt spray, or other
contaminant exposure.
8. Normal deterioration from age.
9. Lubricant loss, or deficiency.

10. Improper operating or .test
procedure.
11. Tripping source deficient.
12. Lack or preventive maintenance.
13. Other

Suspected Failure Responsibility (E)
1. Defective component
2. Improper handling/shipping
3. Poor installation/testing
4. Inadequate maintenance
5. Improper operation
6. Improper application
7. Inadequate physical protection
8. Outside agency (such as vehicular
accident)
9. Other
10. Unknown

Failure Mode (F)
1. Failed to close on command
2. Failed to close and latch
3. Failed to open on command
4. Closes without command
5. Opens without command
6. Failed to break current when
opened
7. Damaged while successfully
opening
8. Damaged while closing
9. Failed to carry current
10. Fault to ground, or phase to
phase (not while opening or
closing)
11. Fault across open contacts (not
while opening or closing)
12. Loss of vacuum (for vacuum
breakers)
13. Other failure requiring removal
from service within 30 minutes
14. Other failure not requiring
immediate removal from service
15. Unknown

Months Since Last Maintenance (G)
1. 0 - 12 months
2. 12 - 24 months
3. Over 24 months
4. No preventive maintenance

Repair Urgency (H)
1. Working round-the-clock
2. Normal working hours
3. Low priority

Repair Or Replace (J)
1. Repaired failed component in
place or sent out for repair
2. Replaced failed unit with spare
3. Other

REFERENCES

[1] ANSI/IEEE Standard 493-1980, "IEEE
Recommended Practice For Design of
Reliable Industrial and Commercial Power
Systems".

Appendix L
Reliability Survey of 600 to 1800 kW Diesel
and Gas-Turbine Generating Units

By
Clayton Smith and Michael Donovan
Conference Record of the IEEE Industrial and
Commercial Power Systems Technical Conference
Chicago, IL, May 8–11, 1989

IEEE Transactions on Industry Applications
Jul./Aug. 1990, pp. 741–755

Reliability Survey of 600 to 1800 kW Diesel and Gas-Turbine Generating Units

CLAYTON A. SMITH, MICHAEL D. DONOVAN, ASSOCIATE MEMBER, IEEE, AND MICHAEL J. BARTOS

Abstract — In 1988 the U.S. Army Engineering and Housing Support Center (EHSC) sponsored a study of the reliability, availability, and maintainability (RAM) characteristics of diesel and gas-turbine power systems producing less than 2 MW. The study, conducted by ARINC Research Corporation, included collection and examination of source data for power systems at commercial and military facilities operating in continuous or standby service. A data base of system, subsystem, and component RAM data was established. These data will be useful in the design of primary and standby power systems for military or commercial facilities.

INTRODUCTION

THE U.S. Army Engineering and Housing Support Center (EHSC) sponsored a study [7] of the reliability, availability, and maintainability (RAM) characteristics of small diesel and gas-turbine power systems. The study, conducted by ARINC Research, produced a data base of system, subsystem, and component RAM data for industrial and military power systems in both continuous and standby service. An updated RAM data base was needed to support the analysis of power systems at command, control, communications, and intelligence (C^{3}I) installations worldwide. EHSC wanted higher confidence in the validity of the power-system reliability data used to analyze C^{3}I system reliability. Currently available RAM data were outdated and were not tailored to EHSC's specific requirements. Further, these data did not permit identifying component failure rates in alternative prime-mover designs.

The primary objective was to obtain data reflecting the reliability improvements resulting from advances in power-plant (prime-mover) technology since completion of the last comprehensive RAM study more than 15 years earlier. An additional objective was to provide data on the major components that failed in each system, together with data on the reliability of the prime mover. The information will be used in the evaluation of C^{3}I power-generation systems.

The prime movers of interest were diesel and gas-turbine generators ranging from 600 to 1800 kW. The diesel-generator configurations evaluated included both packaged

Paper ICPSD 89-02, approved by the Power Systems Engineering Committee of the IEEE Industry Applications Society for presentation at the 1989 Industrial and Commercial Power Systems Annual Technical Conference, Detroit, MI, April 29-May 3. Manuscript released for publications January 9, 1990.

C. A. Smith and M. D. Donovan are with ARINC Research Corporation, 2551 Riva Road, Annapolis, MD 21401.

M. J. Bartos is with the U.S. Army Engineering and Housing Support Center, Fort Belvoir, VA 20060.

IEEE Log Number 9035434.

systems and units with auxiliary support systems. Each of these types was categorized as standby and continuous duty. Because most gas-turbine systems in the size range of interest are configured as packaged units, the gas-turbine generators were categorized only by type of duty. Thus six categories were addressed:

- continuous-duty auxiliary diesels,
- standby auxiliary diesels,
- continuous-duty package diesels,
- standby package diesels,
- continuous-duty gas turbines,
- standby gas turbines.

METHODOLOGY

The data collection comprised five tasks: 1) review existing data bases and reports, 2) identify data sources, 3) collect field data, 4) reduce data and prepare data base, and 5) calculate RAM statistics. These tasks are described in the following subsections.

Review Existing Data Bases and Reports

The results of previous and ongoing efforts in the collection of RAM data were reviewed to determine their applicability to the selected diesel and gas-turbine categories. Data bases such as the Government Industry Data Exchange Program (GIDEP) [1] and the Institute of Nuclear Power Operations (INPO) Nuclear Plant Reliability Data System (NPRDS) [5] were investigated, but they were found to contain minimal detail on power plants in the size ranges addressed by the study. Several manufacturers provided the results of studies on reliability, starting reliability, and unit availability conducted in preparation for customer presentations or proposals. The RAM measures from these studies were not included in the data base, because the objectivity and accuracy of the data could not be validated.

Identify Candidate Data Sources

Three methods were used to identify as candidate data sources the industrial and military facilities that operated diesel and gas-turbine power systems in the specified categories. Equipment manufacturers and distributors were asked to provide lists of customers having power systems that met the category definitions. U.S. military and Government agencies were similarly requested to provide names of equipment operators and sources of maintenance data. In addition, industrial directories were used to identify facilities representing typi-

USER SURVEY: DIESEL AND GAS-TURBINE GENERATORS

User/Company: _____

Address: _____

Contact: _____

Telephone: (__) _____

Date: _____

Application: _____

Staffing (No. of personnel and titles): _____

_____

_____

Items to Address:

How many units do you have on-site: _____

What are their ratings? _____

Are units standby or in continuous use? _____

	Yes	No
Is there a central data bank for maintenance information?	_____	_____
Do you collect maintenance data?	_____	_____
Do you collect operating data?	_____	_____
Do you record attempted and successful starts?	_____	_____
Do you keep logs for scheduled maintenance?	_____	_____
Do you have records of failure events?	_____	_____
Have there been at least five failures to the unit?	_____	_____
Do you track administrative and logistic time?	_____	_____
Can these data be sent to us for this effort?	_____	_____
Can ARINC Research obtain permission to review these records?	_____	_____
Is there a maintenance program in use?	_____	_____
If yes, is it the manufacturer's program?	_____	_____
Are spares kept on site?	_____	_____

Remarks: (Include brief history and line diagram of plant)

User Code: _____

Fig. 1.   User survey form.

cal power-system users, such as computer centers, small utility sites, and cogeneration plants. The candidate data sources identified through these surveys were listed in a project data base for sorting and screening during the data collection task.

Collect Data

Potential data sources were screened by means of a structured telephone survey technique, using the questions shown in Fig. 1, to identify candidate power plants for data collection. The objective of this screening was to determine the applicability, availability, and quality of operational, maintenance, and failure data. Plants were selected from a wide variety of applications (e.g., electric utilities, cogenerators, hospitals, airfields, military installations, and computer and communication facilities) to represent a range of variables such as manufacturer, plant usage, age, environment, and maintenance practices. Where possible, plants with at least ten years of operation and maintenance history were selected.

Selected power-plant operators with formal data collection systems were requested to mail facility descriptions and historical records of their operation and maintenance logs. Follow-up technical questions to clarify data interpretation were directed, via telephone conversations, to senior facility personnel.

The problem most frequently encountered in obtaining data from participating facilities was the level of effort required by the plant staff to assemble and reproduce the necessary records. To ensure the acquisition of representative data, site visits were made to facilities that could not respond to the mailing requests. Technical personnel experienced in plant operation and maintenance conducted these visits. In addition to records collection, visits typically included structured interviews with senior operations and maintenance personnel to obtain additional insights into failure events and maintenance tasks.

Twenty-two plants participated in the study, providing data on 71 power systems. The data represented 708 unit-years of operating experience, and all plants provided data for periods of 3 years or longer.

Develop Data Base

The source data on maintenance and failure events were arranged in a consistent record format for computer entry and validation. Data reduction was performed by examining events

TABLE I
RAM Statistics

RAM Measures	Formula Based on Period Hours	Formula Based on Operating Hours
Failure Rate (FR) (Failures per year)	$\dfrac{\text{No. of Failures}}{\text{Period Hours}}$ X 8,760	$\dfrac{\text{No. of Failures}}{\text{Operating Hours}}$ X 8,760
Mean Time Between Failures (MTBF) (Hours)	$\dfrac{\text{Period Hours}}{\text{No. of Failures}}$	$\dfrac{\text{Operating Hours}}{\text{No. of Failures}}$
Mean Time To Repair (MTTR) (Hours)	$\dfrac{\text{Total Repair Time}}{\text{No. of Failures}}$	$\dfrac{\text{Total Repair Hours}}{\text{No. of Failures}}$
Mean Time Between Planned Outages (MTBPO) (Hours)	$\dfrac{\text{Period Hours}}{\text{No. of Planned Outages}}$	$\dfrac{\text{Operating Hours}}{\text{No. of Planned Outages}}$
Mean Time To Maintain (MTTM) (Hours)	$\dfrac{\text{Planned Outage Hours}}{\text{No. of Planned Outages}}$	$\dfrac{\text{Planned Outage Hours}}{\text{No. of Planned Outages}}$
Mean Time Between Outages (MTBO) (Hours)	$\dfrac{\text{Period Hours}}{\text{No. of Outages}}$	$\dfrac{\text{Operating Hours}}{\text{No. of Outages}}$
Mean Downtime (MDT) (Hours)	$\dfrac{\text{Repair Hours + Planned Outage Hours}}{\text{No. of Outages}}$	$\dfrac{\text{Repair Hours + Planned Outage Hours}}{\text{No. of Outages}}$
Mean Time Between Corrective Maintenance (MTBCM) (Hours)	$\dfrac{\text{Period Hours}}{\text{No. of CMs}}$	$\dfrac{\text{Operating Hours}}{\text{No. of CMs}}$
Mean Time To Perform Corrective Maintenance (MTTCM) (Hours)	$\dfrac{\text{Corrective Maintenance Hours}}{\text{No. of CMs}}$	$\dfrac{\text{Corrective Maintenance Hours}}{\text{No. of CMs}}$
Availability, Operational (AO)	$\dfrac{\text{Period Hours – Repair Time – Planned Outage Hours}}{\text{Period Hours}}$	$\dfrac{\text{Operating Hours}}{\text{Operating Hours + Repair Hours + Planned Outage Hours}}$
Availability, Inherent (AI)	$\dfrac{\text{Period Hours – Repair Hours}}{\text{Period Hours}}$	$\dfrac{\text{Operating Hours}}{\text{Operating Hours + Repair Hours}}$
Reliability for 24 hours (R24)	$e-24/\text{MTBF}$	$e-24/\text{MTBF}$
Reliability for 720 hours (R720)	$e-720/\text{MTBF}$	$e-720/\text{MTBF}$

in the operating and maintenance records to identify the subsystem and component, the type of outage, the impact of the failure, and the action required to complete the maintenance. This information was coded according to the equipment, failure-impact, outage-type, and action codes listed in Appendix I.

Summary descriptions of each maintenance event were also prepared to provide insight into failure modes. Operating data for each unit—such as period hours, operating hours, starts, and start failures—were extracted from operating logs.

The event records produced by the data-reduction process were entered into a microcomputer data base. The data base architecture, developed with a commercially available data base management system, included features for automated checking for data-entry errors or inconsistencies. Following data entry, samples of records were randomly selected for validation against the raw data, ensuring consistency in application of the event coding scheme during data reduction.

Calculate RAM Statistics

The maintenance-event and operational data and the formulas shown in Table I were used to calculate RAM statistics for each of the six categories of power systems. The terms used in the formulas are defined in Table II.

RAM statistics were also calculated for subsystems and components in each category on the basis of both period hours and unit operating hours. Subsystem and component measures included failure rate (FR), mean time between failures (MTBF), mean time between corrective maintenance (MTBCM), mean time to perform corrective maintenance (MTTCM), and operational availability (AO).

The RAM statistics are intended for use by EHSC for a variety of analyses, evaluations, and planning studies for C^3I facility support systems. To meet the requirements of these applications, the RAM statistics were calculated using both period hours (i.e., calendar time) and operating hours.

RESULTS

The data base developed contains more than 6000 maintenance events, representing 708 unit-years and nearly one million operating hours. Data from units within each of the six categories were combined, because units within the same category are of similar technology and utilization. The unit-level RAM statistics for the six major categories from this data base are compiled in Tables III and IV. Data for subsystems and components within these categories are presented in Appendix II.

TABLE II
DEFINITIONS OF TERMS

Concurrent Maintenance Event (CC)	Maintenance action taken while unit is already in an outage
Corrective Maintenance Event (CM)	An event in which some equipment had to be repaired (outage-causing or not)
Corrective Maintenance Time	Time, in hours, required to complete a CM
Failure	An unexpected event that results in the interruption of electrical power at the generator output terminals
Forced-Outage Event (FO)	Failure
Noncurtailing Event (NC)	Maintenance action taken while the unit is available to produce power
Operating Hours	Number of hours the unit is producing power
Outage Event	Any interruption of electrical power at the generator output terminals
Period Hours (PH)	Number of calendar hours in a year (8,760)
Planned Outage Event (PO)	Outage taken for any scheduled reason (e.g., inspections, overhauls, cleaning)
Planned Outage Time	Time, in hours, taken to complete any planned outage event
Repair Time	Time, in hours, required to repair any failure
Unit-Years	Calendar hours in a year (8,760) multiplied by the number of units

TABLE III
COMPOSITE RAM STATISTICS BASED ON PERIOD HOURS

RAM Measures	Diesel Auxiliary		Diesel Package		Gas Turbine	
	Continuous	Standby	Continuous	Standby	Continuous	Standby
Number of Units	7	5	9	15	15	20
Period Hours	674,520	1,357,800	814,776	1,068,594	333,888	1,951,224
Number of Events	1,702	1,408	1,535	498	509	319
Unit Failures	302	198	408	118	174	70
Unit Outages (Planned and Forced)	1,311	615	959	365	385	278
Number of Corrective Maintenance Events	409	630	812	243	253	102
Failure Rate (Failures per Unit-Year)	3.9	1.2	4.3	0.9	4.5	0.3
MTBF (Hours)	2,233.5	6,857.4	1,997.0	9,055.8	1,918.8	27,874.6
MTTR (Hours)	2.9	2.8	6.4	3.9	7.2	111.6
MTBPO (Hours)	668.5	3,256.1	1,478.7	4,326.2	1,582.4	9,380.8
MTTM (Hours)	1.3	3.8	12.5	7.8	21.1	10.6
MTBO (Hours)	514.5	2,207.8	849.6	2,927.7	867.2	7,018.8
MDT (Hours)	1.7	3.5	9.9	6.5	14.8	36.1
MTBCM (Hours)	1,699.1	2,155.2	923.3	4,897.5	1,319.7	19,129.6
MTTCM (Hours)	2.8	2.9	4.3	2.9	5.7	77.4
AI	0.9986	0.9995	0.9967	0.9995	0.9962	0.9959
AO	0.9965	0.9984	0.9882	0.9977	0.9828	0.9948
R24	0.9893	0.9965	0.9880	0.9973	0.9875	0.9991
R720	0.7244	0.9003	0.6973	0.9235	0.6871	0.9745

Observations

The objective of the study was to compile a data base for use in the evaluation of power systems in C^3I facilities; thus no detailed analysis of the data was performed. However, some observations can be made from examination of the calculation results.

Table III indicates that, on the basis of period hours, units in the continuous-duty categories have similar failure rates. The period-based failure rates for the standby categories are much lower, because the low utilization of units in this category provides fewer opportunities for failures to occur.

The gas turbines exhibit the lowest failure rates of units in standby service. However, this is negated by long repair times. The raw data show that the large repair-time value is attributable to a relatively small number of long-duration events, including a main bearing failure (200 h), a reduction-gear failure (350 h), seven broker starter shafts (150 h each), and two events in which the turbine had to be sent back to the manufacturer (3000 h each). The starter-shaft problem was an initial design problem and has not occurred since the implementation of a design change to the part.

For the continuous-duty diesels with auxiliary systems, the failure rate based on operating hours, shown in Table IV, is significantly higher than that of the other categories in continuous service. This difference is attributable to the relatively low utilization of these diesels at the plants reporting in this category. These units were classified as continuous because they were scheduled for operation on a regular basis. However, most of them were operated in a cycling mode and operated only for several hours each day. The high failure rate results

TABLE IV
COMPOSITE RAM STATISTICS BASED ON OPERATING HOURS

RAM Measures	Diesel Auxiliary		Diesel Package		Gas Turbine	
	Continuous	Standby	Continuous	Standby	Continuous	Standby
Number of Units	7	5	9	15	15	20
Operating Hours	80,174	323,242	300,698	64,364	204,037	13,591
Number of Events	1,702	1,408	1,535	498	509	319
Unit Failures	302	198	408	118	174	70
Unit Outages (Planned and Forced)	1,311	615	959	365	385	278
Number of Corrective Maintenance Events	409	630	872	243	253	102
Failure Rate (Failures per Unit-Year)	32.9	5.3	11.8	16.0	7.4	45.1
MTBF (Hours)	264.4	1,632.5	737.0	545.4	1,172.6	194.1
MTTR (Hours)	2.9	2.8	6.4	3.9	7.2	111.6
MTBPO (Hours)	79.4	775.1	545.7	260.5	967.0	65.3
MTTM (Hours)	1.3	3.8	12.5	7.8	21.1	10.6
MTBO (Hours)	61.1	525.5	313.5	176.3	529.9	48.8
MDT (Hours)	1.7	3.5	9.9	6.5	14.8	36.1
MTBCM (Hours)	196.0	513.0	344.8	264.8	806.4	133.2
MTTCM (Hours)	2.8	2.9	4.3	2.9	5.7	77.4
AI	0.9889	N/A	0.9912	N/A	0.9938	N/A
AO	0.9713	N/A	0.9682	N/A	0.9720	N/A
R24	0.9132	0.9854	0.9680	0.9569	0.9797	0.8837
R720	0.0657	0.6434	0.3765	0.2671	0.5412	0.0245

from dividing the large number of failures induced in this type of operation by the relatively small number of operating hours.

Similarly, for the gas turbines in standby service, the high failure rate based on operating hours can be attributed to the relatively low utilization of these units. Most of the units in this category are used as emergency power supplies in computer or communications facilities. They are typically tested on a weekly or monthly basis and run for less than 1 h, with failures most likely to occur during the start sequence. On the basis of this limited operating time, the failure rate is high.

The subsystem and component data presented in Appendix II provide information on the causes of unit failures and unavailability. For example, problems with the standby gas turbines reside mostly in the starting system, particularly the battery. The fuel system, the generator, and the controls add to the overall failure rate. It is also of interest that much of the unavailability is due to inspection and cleaning actions, even though these actions do not contribute to the overall failure rate. For the continuous-duty auxiliary diesels, the failure rate is due largely to the engine itself, specifically the cylinder heads and the crankcase. Tracking these same components through all of the diesel categories shows them to have consistently the highest failure rate.

SUMMARY

Information collected through this study is useful in the design assessment of primary and standby power systems for military or commercial facilities. The unit-level RAM statistics for the six categories provide a baseline for comparison of RAM measures for a specific plant against a representative population similar in configuration and type of service. The subsystem and component data, in conjunction with appropriate modeling tools, provide a means for forecasting the availability performance of specific plant designs. Since the data base includes all component maintenance events rather

TABLE V
ACTION, FAILURE-IMPACT, AND OUTAGE-TYPE CODES

Action Codes

CL	—	Cleaned
FL	—	Fixed Leak
IN	—	Inspection
MD	—	Modification
NA	—	No Action Taken
OV	—	Overhaul
PM	—	Preventive Maintenance
RA	—	Repaired
RC	—	Recalibrated
TS	—	Tested

Failure Impact

0	—	No Failure
1	—	Failure Affected Only the Component
2	—	Failure Affected Component and Subsystem
3	—	Failure Affected Component, Subsystem, and Unit

Outage Type

CC	—	Concurrent Maintenance
FO	—	Forced Outage
FS	—	Failure to Start
NC	—	Noncurtailing Maintenance
PO	—	Planned Outage

than just outage failures, it provides information that will be useful in maintenance and logistic planning for power systems.

APPENDIX I

CODES

Failure-impact, outage-type, and the action codes are listed in Table V.

APPENDIX II

SUBSYSTEM AND COMPONENT DATA

Tables VI–XI reflect the RAM statistics based on equipment failure maintenance events for the units within the six categories. Components or subsystems that do not appear in a table did not experience any failure or maintenance events.

TABLE VI
Subsystem and Component RAM Measures for Continuous-Duty Auxiliary
Diesels

Equipment	Period Hours					Operating Hours			
	Failures per Year	MTBF (Hours)	MTBCM (Hours)	MTTCM (Hours)	Operational Availability	Failures per Year	MTBF (Hours)	MTBCM (Hours)	MTTCM (Hours)
CONTROL & INSTRUMENTATION (DS-CTI)	0.08	112420.0	112420.0	1.7	1.0000	0.66	13362.3	13362.3	1.7
CIRCUIT BREAKERS (DS-CTI01)	0.05	168630.0	168630.0	1.5	1.0000	0.44	20043.5	20043.5	1.5
ELECTRICAL MODULE (DS-CTI02)	0.01	674520.0	674520.0	1.0	1.0000	0.11	80174.0	80174.0	1.0
SWITCHES (DS-CTI04)	0.01	674520.0	674520.0	3.0	1.0000	0.11	80174.0	80174.0	3.0
COOLING WATER SYSTEM (DS-CWT)	0.13	67452.0	44968.0	1.7	1.0000	1.09	8017.4	5344.9	1.7
COOLING WATER PUMP (DS-CWT02)	0.00	0.0	674520.0	0.5	1.0000	0.00	0.0	80174.0	0.5
ENGINE COOLING (DS-CWT03)	0.01	674520.0	674520.0	4.0	1.0000	0.11	80174.0	80174.0	4.0
THERMOSTAT (DS-CWT05)	0.00	0.0	674520.0	1.0	1.0000	0.00	0.0	80174.0	1.0
VALVES (DS-CWT07)	0.01	674520.0	337260.0	0.8	1.0000	0.11	80174.0	40087.0	0.8
WATER LINE (DS-CWT09)	0.06	134904.0	96360.0	0.9	1.0000	0.55	16034.8	11453.4	0.9
HEAT EXCHANGER (DS-CWT10)	0.03	337260.0	337260.0	5.0	1.0000	0.22	40087.0	40087.0	5.0
WATER HEADER (DS-CWT12)	0.01	674520.0	674520.0	2.0	1.0000	0.11	80174.0	80174.0	2.0
WATER MANIFOLD (DS-CWT13)	0.00	0.0	0.0	0.0	1.0000	0.00	0.0	0.0	0.0
DIESEL ENGINE (DS-ENG)	2.25	3899.0	3122.8	3.6	0.9984	18.90	463.4	371.2	3.6
BEARINGS (DS-ENG01)	0.08	112420.0	74946.7	2.9	0.9999	0.66	13362.3	8908.2	2.9
CYLINDER (DS-ENG02)	0.32	26980.8	20440.0	2.0	0.9999	2.73	3207.0	2429.5	2.0
CYLINDER HEADS (DS-ENG03)	0.99	8875.3	8225.9	4.0	0.9995	8.30	1054.9	977.7	4.0
DRIVE SHAFT (DS-ENG04)	0.00	0.0	0.0	0.0	1.0000	0.00	0.0	0.0	0.0
PISTONS (DS-ENG06)	0.26	33726.0	28105.0	4.7	0.9998	2.19	4008.7	3340.6	4.7
TURBO CHARGER (DS-ENG07)	0.01	674520.0	337260.0	4.0	1.0000	0.11	80174.0	40087.0	4.0
VALVES (DS-ENG08)	0.03	337260.0	224840.0	2.0	1.0000	0.22	40087.0	26724.7	2.0
RINGS (DS-ENG09)	0.31	28105.0	18230.3	3.5	0.9996	2.62	3340.6	2166.9	3.5
TIMING (DS-ENG10)	0.05	168630.0	134904.0	1.0	0.9998	0.44	20043.5	16034.8	1.0
INTAKE MANIFOLD (DS-ENG11)	0.08	112420.0	112420.0	3.2	1.0000	0.66	13362.3	13362.3	3.2
CRANKCASE (DS-ENG12)	0.01	674520.0	224840.0	15.0	0.9999	0.11	80174.0	26724.7	15.0
RODS (DS-ENG14)	0.01	674520.0	224840.0	3.7	1.0000	0.11	80174.0	26724.7	3.7
CAM (DS-ENG15)	0.08	112420.0	96360.0	2.0	0.9999	0.66	13362.3	11453.4	2.0
CHAIN DRIVE (DS-ENG17)	0.01	674520.0	337260.0	1.0	1.0000	0.11	80174.0	40087.0	1.0
TAPPET (DS-ENG18)	0.00	0.0	0.0	0.0	1.0000	0.00	0.0	0.0	0.0
EXHAUST SYSTEM (DS-EXH)	0.04	224840.0	96360.0	2.1	0.9997	0.33	26724.7	11453.4	2.1
EXHAUST SYSTEM (DS-EXH)	0.00	0.0	0.0	0.0	1.0000	0.00	0.0	0.0	0.0
EXPANSION JOINTS (DS-EXH03)	0.00	0.0	337260.0	4.0	1.0000	0.00	0.0	40087.0	4.0
PORTS (DS-EXH05)	0.01	674520.0	674520.0	0.5	0.9997	0.11	80174.0	80174.0	0.5
EXHAUST MANIFOLD (DS-EXH06)	0.03	337260.0	337260.0	0.5	1.0000	0.22	40087.0	40087.0	0.5
EXHAUST VALVE (DS-EXH07)	0.00	0.0	674520.0	1.0	1.0000	0.00	0.0	80174.0	1.0
MUFFLER (DS-EXH10)	0.00	0.0	674520.0	4.0	1.0000	0.00	0.0	80174.0	4.0
FUEL SYSTEM (DS-FLS)	0.91	9636.0	8225.9	2.1	0.9989	7.65	1145.3	977.7	2.1
DEAERATOR TANK (DS-FLS02)	0.00	0.0	674520.0	1.0	1.0000	0.00	0.0	80174.0	1.0
FUEL FILTER (DS-FLS03)	0.00	0.0	337260.0	1.0	1.0000	0.00	0.0	40087.0	1.0
GOVERNOR (DS-FLS04)	0.12	74946.7	67452.0	3.2	0.9997	0.98	8908.2	8017.4	3.2
PUMPS (DS-FLS06)	0.35	24982.2	21758.7	1.8	0.9999	2.95	2969.4	2586.3	1.8
VALVES (DS-FLS07)	0.01	674520.0	674520.0	1.5	1.0000	0.11	80174.0	80174.0	1.5
INJECTOR (DS-FLS08)	0.21	42157.5	39677.6	2.5	0.9994	1.75	5010.9	4716.1	2.5
FUEL LINE (DS-FLS09)	0.22	39677.6	37473.3	1.8	1.0000	1.86	4716.1	4454.1	1.8
FUEL OIL REGULATOR (DS-FLS10)	0.00	0.0	337260.0	2.0	1.0000	0.00	0.0	40087.0	2.0
GENERATOR (DS-GNR)	0.09	96360.0	74946.7	2.2	0.9999	0.76	11453.4	8908.2	2.2
GENERATOR (DS-GNR)	0.00	0.0	0.0	0.0	1.0000	0.00	0.0	0.0	0.0
BEARINGS (DS-GNR01)	0.00	0.0	0.0	0.0	1.0000	0.00	0.0	0.0	0.0
FIELD (DS-GNR05)	0.05	168630.0	112420.0	1.8	1.0000	0.44	20043.5	13362.3	1.8
FLYWHEEL (DS-GNR10)	0.04	224840.0	224840.0	3.0	1.0000	0.33	26724.7	26724.7	3.0
INSULATION (DS-GNR11)	0.00	0.0	0.0	0.0	1.0000	0.00	0.0	0.0	0.0
COLLECTOR RINGS (DS-GNR12)	0.00	0.0	0.0	0.0	1.0000	0.00	0.0	0.0	0.0
LUBE OIL/HYDRAULIC SYSTEM (DS-LBO)	0.25	35501.1	19272.0	2.5	0.9998	2.08	4219.7	2290.7	2.5
LUBE OIL/HYDRAULIC SYSTEM (DS-LBO)	0.00	0.0	0.0	0.0	1.0000	0.00	0.0	0.0	0.0
COOLER (DS-LBO02)	0.16	56210.0	56210.0	3.8	0.9999	1.31	6681.2	6681.2	3.8
FILTER (DS-LBO04)	0.00	0.0	74946.7	0.8	1.0000	0.00	0.0	8908.2	0.8
PIPING (DS-LBO06)	0.01	674520.0	337260.0	2.0	1.0000	0.11	80174.0	40087.0	2.0
STRAINER (DS-LBO10)	0.01	674520.0	134904.0	1.4	1.0000	0.11	80174.0	16034.8	1.4
LUBRICATOR (DS-LBO12)	0.06	134904.0	96360.0	3.3	1.0000	0.55	6034.8	11453.4	3.3
STARTING SYSTEM (DS-STS)	0.18	48180.0	17295.4	1.6	0.9999	1.53	5726.7	2055.7	1.6
AIR FILTER (DS-STS04)	0.00	0.0	26980.8	1.4	1.0000	0.00	0.0	3207.0	1.4
AIR CYLINDER (DS-STS05)	0.00	0.0	0.0	0.0	1.0000	0.00	0.0	0.0	0.0
STARTING AIR ELBOW (DS-STS06)	0.13	67452.0	67452.0	2.2	1.0000	1.09	8017.4	8017.4	2.2
AIR LINE (DS-STS07)	0.03	337260.0	337260.0	1.0	1.0000	0.22	40087.0	40087.0	1.0
VALVES (DS-STS08)	0.03	337260.0	337260.0	2.5	1.0000	0.22	40087.0	40087.0	2.5
GOVERNOR BOOSTER (DS-STS13)	0.00	0.0	0.0	0.0	1.0000	0.00	0.0	0.0	0.0

TABLE VII

Subsystem and Component RAM Measures for Stability Auxiliary Diesels

Equipment	Period Hours					Operating Hours			
	Failures per Year	MTBF (Hours)	MTBCM (Hours)	MTTCM (Hours)	Operational Availability	Failures per Year	MTBF (Hours)	MTBCM (Hours)	MTTCM (Hours)
CONTROL & INSTRUMENTATION (DS-CTI)	0.01	678900.0	452600.0	1.7	1.0000	0.05	161621.0	107747.3	1.7
CIRCUIT BREAKERS (DS-CTI01)	0.01	1357800.0	1357800.0	2.0	1.0000	0.03	323242.0	323242.0	2.0
GAUGES (DS-CTI03)	0.01	1357800.0	678900.0	1.5	1.0000	0.03	323242.0	161621.0	1.5
SWITCHES (DS-CTI04)	0.00	0.0	0.0	0.0	1.0000	0.00	0.0	0.0	0.0
COOLING WATER SYSTEM (DS-CWT)	0.05	193971.4	135780.0	1.8	1.0000	0.19	46177.4	32324.2	1.8
AIR COOLER (DS-CWT01)	0.01	1357800.0	1357800.0	1.0	1.0000	0.03	323242.0	323242.0	1.0
COOLING WATER PUMP (DS-CWT02)	0.01	678900.0	271560.0	1.8	1.0000	0.05	161621.0	64648.4	1.8
COOLING TOWERS (DS-CWT08)	0.01	1357800.0	1357800.0	2.0	1.0000	0.03	323242.0	323242.0	2.0
HEAT EXCHANGER (DS-CWT10)	0.02	452600.0	452600.0	1.8	1.0000	0.08	107747.3	107747.3	1.8
DIESEL ENGINE (DS-ENG)	0.64	13715.2	4437.3	3.8	0.9988	2.68	3265.1	1056.3	3.8
DIESEL ENGINE (DS-ENG)	0.01	1357800.0	1357800.0	0.0	0.9998	0.03	323242.0	323242.0	0.0
BEARINGS (DS-ENG01)	0.10	84862.5	38794.3	2.4	1.0000	0.43	20202.6	9235.5	2.4
CYLINDER (DS-ENG02)	0.05	193971.4	150866.7	2.8	1.0000	0.19	46177.4	35915.8	2.8
CYLINDER HEADS (DS-ENG03)	0.12	71463.2	52223.1	3.4	0.9999	0.51	17012.7	12432.4	3.4
PISTONS (DS-ENG06)	0.19	45260.0	9236.7	4.4	0.9992	0.81	10774.7	2198.9	4.4
VALVES (DS-ENG08)	0.01	1357800.0	271560.0	2.5	1.0000	0.03	323242.0	64648.4	2.5
RINGS (DS-ENG09)	0.15	59034.8	16972.5	3.6	0.9999	0.62	14054.0	4040.5	3.6
INTAKE MANIFOLD (DS-ENG11)	0.01	1357800.0	1357800.0	2.0	1.0000	0.03	323242.0	323242.0	2.0
CRANKCASE (DS-ENG12)	0.00	0.0	1357800.0	4.0	1.0000	0.00	0.0	323242.0	4.0
CAM (DS-ENG15)	0.01	1357800.0	1357800.0	8.0	1.0000	0.03	323242.0	323242.0	8.0
EXHAUST SYSTEM (DS-EXH)	0.03	339450.0	79870.6	1.9	0.9999	0.11	80810.5	19014.2	1.9
EXHAUST SYSTEM (DS-EXH)	0.00	0.0	0.0	0.0	1.0000	0.00	0.0	0.0	0.0
EXHAUST MANIFOLD (DS-EXH06)	0.00	0.0	1357800.0	2.0	1.0000	0.00	0.0	323242.0	2.0
EXHAUST VALVE (DS-EXH07)	0.02	52600.0	90520.0	1.9	1.0000	0.08	107747.3	21549.5	1.9
HEADER (DS-EXH09)	0.00	0.0	0.0	0.0	1.0000	0.00	0.0	0.0	0.0
MUFFLER (DS-EXH10)	0.01	1357800.0	1357800.0	1.5	1.0000	0.03	323242.0	323242.0	1.5
FUEL SYSTEM (DS-FLS)	0.41	21552.4	7339.5	2.7	0.9998	1.71	5130.8	1747.3	2.7
FUEL SYSTEM (DS-FLS)	0.00	0.0	0.0	0.0	1.0000	0.00	0.0	0.0	0.0
FUEL FILTER (DS-FLS03)	0.02	452600.0	21552.4	1.1	1.0000	0.08	107747.3	5130.8	1.1
GOVERNOR (DS-FLS04)	0.05	193971.4	169725.0	3.0	1.0000	0.19	6177.4	40405.2	3.0
PUMPS (DS-FLS06)	0.06	135780.0	113150.0	2.2	1.0000	0.27	32324.2	26936.8	2.2
INJECTOR (DS-FLS08)	0.25	35731.6	13997.9	3.7	0.9999	1.03	8506.4	3332.4	3.7
FUEL LINE (DS-FLS09)	0.03	271560.0	271560.0	2.6	1.0000	0.14	64648.4	64648.4	2.6
FUEL OIL REGULATOR (DS-FLS10)	0.00	0.0	0.0	0.0	1.0000	0.00	0.0	0.0	0.0
GENERATOR (DS-GNR)	0.06	135780.0	123436.4	2.3	1.0000	0.27	32324.2	29385.6	2.3
GENERATOR (DS-GNR)	0.00	0.0	0.0	0.0	1.0000	0.00	0.0	0.0	0.0
BEARINGS (DS-GNR01)	0.02	452600.0	339450.0	2.2	1.0000	0.08	107747.3	80810.5	2.2
FIELD (DS-GNR05)	0.04	226300.0	226300.0	2.2	1.0000	0.16	53873.7	53873.7	2.2
COLLECTOR RINGS (DS-GNR12)	0.01	1357800.0	1357800.0	3.0	1.0000	0.03	323242.0	323242.0	3.0
LUBE OIL/HYDRAULIC SYSTEM (DS-LBO)	0.06	135780.0	64657.1	1.2	0.9999	0.27	32324.2	15392.5	1.2
LUBE OIL/HYDRAULIC SYSTEM (DS-LBO)	0.00	0.0	0.0	0.0	1.0000	0.00	0.0	0.0	0.0
COOLER (DS-LBO02)	0.00	0.0	0.0	0.0	1.0000	0.00	0.0	0.0	0.0
FILTER (DS-LBO04)	0.01	1357800.0	113150.0	0.9	1.0000	0.03	323242.0	26936.8	0.9
PUMP (DS-LBO05)	0.05	193971.4	193971.4	1.6	1.0000	0.19	6177.4	46177.4	1.6
TANK (DS-LBO08)	0.01	1357800.0	1357800.0	2.0	1.0000	0.03	323242.0	323242.0	2.0
STRAINER (DS-LBO10)	0.00	0.0	0.0	0.0	1.0000	0.00	0.0	0.0	0.0
OIL SWITCH (DS-LBO14)	0.01	1357800.0	1357800.0	1.0	1.0000	0.03	323242.0	323242.0	1.0
STARTING SYSTEM (DS-STS)	0.02	452600.0	17407.7	1.0	1.0000	0.08	107747.3	4144.1	1.0
AIR FILTER (DS-STS04)	0.00	0.0	18104.0	1.0	1.0000	0.00	0.0	4309.9	1.0
VALVES (DS-STS08)	0.02	452600.0	452600.0	1.7	1.0000	0.08	107747.3	107747.3	1.7

TABLE VIII
Subsystem and Component RAM Measures for Continuous-Duty Package
Diesels

Equipment	Period Hours					Operating Hours			
	Failures per Year	MTBF (Hours)	MTBCM (Hours)	MTTCM (Hours)	Operational Availability	Failures per Year	MTBF (Hours)	MTBCM (Hours)	MTTCM (Hours)
BALANCE OF PLANT (DS–BOP)	0.02	407388.0	271592.0	0.3	1.0000	0.06	150349.0	100232.7	0.3
BALANCE OF PLANT (DS–BOP)	0.00	0.0	0.0	0.0	1.0000	0.00	0.0	0.0	0.0
COMBUSTION GAS MONITORING (DS–BOP01)	0.01	814776.0	814776.0	0.0	1.0000	0.03	300698.0	300698.0	0.0
ENCLOSURES (DS–BOP02)	0.00	0.0	814776.0	1.0	1.0000	0.00	0.0	300698.0	1.0
FIRE SUPPRESSION/DETECTION (DS–BOP03)	0.01	814776.0	814776.0	0.0	1.0000	0.03	300698.0	300698.0	0.0
CONTROL & INSTRUMENTATION (DS–CTI)	0.12	74070.5	28095.7	3.0	0.9999	0.32	27336.2	10368.9	3.0
CONTROL & INSTRUMENTATION (DS–CTI)	0.00	0.0	407388.0	1.5	1.0000	0.00	0.0	150349.0	1.5
CIRCUIT BREAKERS (DS–CTI01)	0.00	0.0	814776.0	1.0	1.0000	0.00	0.0	300698.0	1.0
ELECTRICAL MODULE (DS–CTI02)	0.04	203694.0	162555.2	4.5	1.0000	0.12	75174.5	60139.6	4.5
GAUGES (DS–CTI03)	0.04	203694.0	47928.0	1.2	1.0000	0.12	75174.5	17688.1	1.2
SWITCHES (DS–CTI04)	0.01	814776.0	407388.0	5.7	1.0000	0.03	300698.0	150349.0	5.7
WIRING (DS–CTI05)	0.02	407388.0	407388.0	14.0	1.0000	0.06	150349.0	150349.0	14.0
COOLING WATER SYSTEM (DS–CWT)	0.43	20369.4	10720.7	1.6	0.9998	1.17	7517.4	3956.6	1.6
COOLING WATER SYSTEM (DS–CWT)	0.01	814776.0	814776.0	5.6	1.0000	0.03	300698.0	300698.0	5.6
COOLING WATER PUMP (DS–CWT02)	0.13	67898.0	37035.3	2.2	0.9999	0.35	25058.2	13668.1	2.2
ENGINE COOLING (DS–CWT03)	0.22	40738.8	23964.0	1.2	0.9999	0.58	15034.9	8844.1	1.2
THERMOSTAT (DS–CWT05)	0.01	814776.0	162955.2	1.3	1.0000	0.03	300698.0	60139.6	1.3
TURBO CHARGER COOLING (DS–CWT06)	0.00	0.0	814776.0	2.0	1.0000	0.00	0.0	300698.0	2.0
VALVES (DS–CWT07)	0.01	814776.0	407388.0	1.5	1.0000	0.03	300698.0	150349.0	1.5
COOLING TOWERS (DS–CWT08)	0.01	814776.0	407388.0	1.0	1.0000	0.03	300698.0	150349.0	1.0
WATER LINE (DS–CWT09)	0.02	407388.0	135796.0	2.1	1.0000	0.06	150349.0	50116.3	2.1
HEAT EXCHANGER (DS–CWT10)	0.01	814776.0	814776.0	1.0	1.0000	0.03	300698.0	300698.0	1.0
WATER HEADER (DS–CWT12)	0.00	0.0	814776.0	2.0	1.0000	0.00	0.0	300698.0	2.0
WATER MANIFOLD (DS–CWT13)	0.01	814776.0	814776.0	1.0	1.0000	0.03	300698.0	300698.0	1.0
DIESEL ENGINE (DS–ENG)	1.91	4577.4	3409.1	8.5	0.9902	5.19	1689.3	1258.2	8.5
DIESEL ENGINE (DS–ENG)	0.04	203694.0	135796.0	9.7	0.9950	0.12	75174.5	50116.3	9.7
BEARINGS (DS–ENG01)	0.09	101847.0	81477.6	8.9	0.9999	0.23	37587.2	30069.8	8.9
CYLINDER (DS–ENG02)	0.30	29099.1	19399.4	4.3	0.9999	0.82	10739.2	7159.5	4.3
CYLINDER HEADS (DS–ENG03)	0.77	11316.3	10445.8	10.7	0.9968	2.10	4176.4	3855.1	10.7
DRIVE SHAFT (DS–ENG04)	0.02	407388.0	271592.0	30.0	0.9999	0.06	150349.0	100232.7	30.0
PISTONS (DS–ENG06)	0.22	40738.8	32591.0	2.9	0.9999	0.58	15034.9	12027.9	2.9
TURBO CHARGER (DS–ENG07)	0.14	62675.1	35425.0	3.6	0.9995	0.38	23130.6	13073.8	3.6
VALVES (DS–ENG08)	0.02	407388.0	203694.0	3.8	1.0000	0.06	150349.0	75174.5	3.8
RINGS (DS–ENG09)	0.04	203694.0	81477.6	8.3	1.0000	0.12	75174.5	30069.8	8.3
TIMING (DS–ENG10)	0.00	0.0	0.0	0.0	1.0000	0.00	0.0	0.0	0.0
INTAKE MANIFOLD (DS–ENG11)	0.05	162955.2	135796.0	9.9	0.9999	0.15	60139.6	50116.3	9.9
CRANKCASE (DS–ENG12)	0.10	90530.7	90530.7	5.6	0.9998	0.26	33410.9	33410.9	15.6
RODS (DS–ENG14)	0.02	407388.0	407388.0	21.5	0.9999	0.06	50349.0	150349.0	21.5
CAM (DS–ENG15)	0.08	116396.6	58198.3	14.1	0.9998	0.20	42956.9	21478.4	14.1
CHAIN DRIVE (DS–ENG17)	0.01	814776.0	407388.0	21.0	1.0000	0.03	300698.0	150349.0	21.0
TAPPET (DS–ENG18)	0.01	814776.0	162955.2	9.7	1.0000	0.03	300698.0	60139.6	9.7
EXHAUST SYSTEM (DS–EXH)	0.12	74070.5	40738.8	5.2	0.9997	0.32	27336.2	15034.9	5.2
EXHAUST SYSTEM (DS–EXH)	0.01	814776.0	407388.0	5.0	0.9998	0.03	300698.0	150349.0	5.0
EXHAUST DUCTING (DS–EXH01)	0.00	0.0	814776.0	2.0	1.0000	0.00	0.0	300698.0	2.0
EXPANSION JOINTS (DS–EXH03)	0.01	814776.0	203694.0	2.9	1.0000	0.03	300698.0	75174.5	2.9
EXHAUST MANIFOLD (DS–EXH06)	0.06	135796.0	116396.6	9.2	0.9999	0.17	50116.3	42956.9	9.2
EXHAUST VALVE (DS–EXH07)	0.03	271592.0	135796.0	2.8	1.0000	0.09	100232.7	50116.3	2.8
MUFFLER (DS–EXH10)	0.00	0.0	0.0	0.0	1.0000	0.00	0.0	0.0	0.0

TABLE VIII (Continued)

Equipment	Period Hours					Operating Hours			
	Failures per Year	MTBF (Hours)	MTBCM (Hours)	MTTCM (Hours)	Operational Availability	Failures per Year	MTBF (Hours)	MTBCM (Hours)	MTTCM (Hours)
FUEL SYSTEM (DS–FLS)	1.19	7340.3	3366.8	3.5	0.9992	3.23	2709.0	1242.6	3.5
FUEL SYSTEM (DS–FLS)	0.01	814776.0	814776.0	1.0	1.0000	0.03	300698.0	300698.0	1.0
DAY TANKS (DS–FLS01)	0.01	814776.0	203694.0	1.5	1.0000	0.03	300698.0	75174.5	1.5
FUEL FILTER (DS–FLS03)	0.02	407388.0	12730.9	1.2	1.0000	0.06	150349.0	4698.4	1.2
GOVERNOR (DS–FLS04)	0.27	32591.0	23279.3	5.5	0.9997	0.73	12027.9	8591.4	5.5
PUMPS (DS–FLS06)	0.24	37035.3	23279.3	3.3	0.9999	0.64	13668.1	8591.4	3.3
VALVES (DS–FLS07)	0.06	135796.0	101847.0	2.1	1.0000	0.17	50116.3	37587.2	2.1
INJECTOR (DS–FLS08)	0.40	22021.0	13357.0	6.0	0.9998	1.08	8127.0	4929.5	6.0
FUEL LINE (DS–FLS09)	0.18	47928.0	23964.0	2.1	0.9999	0.50	17688.1	8844.1	2.1
GEARBOX (DS–GBX)	0.01	814776.0	814776.0	12.0	1.0000	0.03	300698.0	300698.0	12.0
GEARBOX (DS–GBX)	0.01	814776.0	814776.0	12.0	1.0000	0.03	300698.0	300698.0	12.0
GENERATOR (DS–GNR)	0.09	101847.0	74070.5	7.7	0.9999	0.23	37587.2	27336.2	7.7
GENERATOR (DS–GNR)	0.04	203694.0	203694.0	18.9	0.9999	0.12	75174.5	75174.5	18.9
COOLING FANS (DS–GNR04)	0.00	0.0	0.0	10.0	1.0000	0.00	0.0	0.0	0.0
FIELD (DS–GNR05)	0.03	271592.0	203694.0	1.0	1.0000	0.09	100232.7	75174.5	1.0
FLYWHEEL (DS–GNR10)	0.01	814776.0	271592.0	1.8	1.0000	0.03	300698.0	100232.7	1.8
LUBE OIL/HYDRAULIC SYSTEM (DS–LBO)	0.30	29099.1	5580.7	2.0	0.9997	0.82	10739.2	2059.6	2.0
LUBE OIL/HYDRAULIC SYSTEM (DS–LBO)	0.00	0.0	814776.0	4.0	1.0000	0.00	0.0	300698.0	4.0
HEATER (DS–LBO01)	0.00	0.0	0.0	0.0	1.0000	0.00	0.0	0.0	0.0
COOLER (DS–LBO02)	0.01	814776.0	814776.0	74.2	0.9999	0.03	300698.0	300698.0	74.2
COOLER FAN (DS–LBO03)	0.01	814776.0	814776.0	2.0	1.0000	0.03	300698.0	300698.0	2.0
FILTER (DS–LBO04)	0.02	407388.0	8761.0	1.2	1.0000	0.06	150349.0	3233.3	1.2
PUMP (DS–LBO05)	0.12	74070.5	42882.9	1.7	0.9999	0.32	27336.2	15826.2	1.7
PIPING (DS–LBO06)	0.10	90530.7	40738.8	2.4	1.0000	0.26	33410.9	15034.9	2.4
TANK (DS–LBO08)	0.00	0.0	0.0	0.0	1.0000	0.00	0.0	0.0	0.0
VALVES (DS–LBO09)	0.02	407388.0	135796.0	1.5	1.0000	0.06	150349.0	50116.3	1.5
STRAINER (DS–LBO10)	0.02	407388.0	203694.0	1.2	1.0000	0.06	150349.0	75174.5	1.2
OIL SWITCH (DS–LBO14)	0.00	0.0	814776.0	1.0	1.0000	0.00	0.0	300698.0	1.0
STARTING SYSTEM (DS–STS)	0.19	45265.3	7686.6	1.6	0.9999	0.52	16705.4	2836.8	1.6
STARTING SYSTEM (DS–STS)	0.00	0.0	814776.0	2.5	1.0000	0.00	0.0	300698.0	2.5
STARTING AIR COMPRESSOR (DS–STS02)	0.03	271592.0	271592.0	4.2	1.0000	0.09	100232.7	100232.7	4.2
AIR FILTER (DS–STS04)	0.00	0.0	10720.7	1.3	1.0000	0.00	0.0	3956.6	1.3
STARTING AIR ELBOW (DS–STS06)	0.01	814776.0	814776.0	1.0	1.0000	0.03	300698.0	300698.0	1.0
AIR LINE (DS–STS07)	0.01	814776.0	814776.0	1.0	1.0000	0.03	300698.0	300698.0	1.0
VALVES (DS–STS08)	0.01	814776.0	407388.0	1.2	1.0000	0.03	300698.0	150349.0	1.2
AIR STARTS (DS–STS10)	0.12	74070.5	42882.9	2.5	1.0000	0.32	27336.2	15826.2	2.5
AIR INTAKE (DS–STS11)	0.00	0.0	0.0	0.0	1.0000	0.00	0.0	0.0	0.0
AIR DISTRIBUTOR (DS–STS12)	0.01	814776.0	814776.0	1.5	1.0000	0.03	300698.0	300698.0	1.5
BATTERY (DS–STS15)	0.00	0.0	407388.0	1.2	1.0000	0.00	0.0	150349.0	1.2

TABLE IX
Subsystem and Component RAM Measures for Standby Package Diesels

Equipment	Period Hours					Operating Hours			
	Failures per Year	MTBF (Hours)	MTBCM (Hours)	MTTCM (Hours)	Operational Availability	Failures per Year	MTBF (Hours)	MTBCM (Hours)	MTTCM (Hours)
CONTROL & INSTRUMENTATION (DS–CTI)	0.05	178099.0	152656.3	1.2	1.0000	0.82	10727.3	9194.9	1.2
CIRCUIT BREAKERS (DS–CTI01)	0.00	0.0	0.0	0.0	1.0000	0.00	0.0	0.0	0.0
GAUGES (DS–CTI03)	0.04	213718.8	178099.0	1.2	1.0000	0.68	12872.8	10727.3	1.2
THERMOCOUPLES (DS–CTI06)	0.01	1068594.0	1068594.0	1.0	1.0000	0.14	64364.0	64364.0	1.0
COOLING WATER SYSTEM (DS–CWT)	0.07	133574.2	56241.8	1.9	1.0000	1.09	8045.5	3387.6	1.9
COOLING WATER PUMP (DS–CWT02)	0.04	213718.8	89049.5	1.8	1.0000	0.68	12872.8	5363.7	1.8
ENGINE COOLING (DS–CWT03)	0.00	0.0	0.0	0.0	1.0000	0.00	0.0	0.0	0.0
VALVES (DS–CWT07)	0.00	0.0	1068594.0	1.0	1.0000	0.00	0.0	64364.0	1.0
COOLING TOWERS (DS–CWT08)	0.01	1068594.0	1068594.0	2.0	1.0000	0.14	64364.0	64364.0	2.0
WATER LINE (DS–CWT09)	0.01	1068594.0	534297.0	1.0	1.0000	0.14	64364.0	32182.0	1.0
HEAT EXCHANGER (DS–CWT10)	0.01	1068594.0	534297.0	5.0	1.0000	0.14	64364.0	32182.0	5.0
WATER HEADER (DS–CWT12)	0.00	0.0	1068594.0	1.0	1.0000	0.00	0.0	64364.0	1.0
DIESEL ENGINE (DS–ENG)	0.26	33393.6	18424.0	4.1	0.9995	4.36	2011.4	1109.7	4.1
DIESEL ENGINE (DS–ENG)	0.00	0.0	0.0	0.0	0.9997	0.00	0.0	0.0	0.0
BEARINGS (DS–ENG01)	0.01	1068594.0	213718.8	2.0	1.0000	0.14	64364.0	12872.8	2.0
CYLINDER (DS–ENG02)	0.05	178099.0	152656.3	2.3	1.0000	0.82	10727.3	9194.9	2.3
CYLINDER HEADS (DS–ENG03)	0.08	106859.4	62858.5	5.6	0.9999	1.36	6436.4	3786.1	5.6
PISTONS (DS–ENG06)	0.02	356198.0	178099.0	4.0	1.0000	0.41	21454.7	10727.3	4.0
TURBO CHARGER (DS–ENG07)	0.01	1068594.0	534297.0	6.0	1.0000	0.14	64364.0	32182.0	6.0
VALVES (DS–ENG08)	0.01	1068594.0	534297.0	3.0	1.0000	0.14	64364.0	32182.0	3.0
RINGS (DS–ENG09)	0.07	133574.2	97144.9	5.4	1.0000	1.09	8045.5	5851.3	5.4
TIMING (DS–ENG10)	0.00	0.0	1068594.0	1.0	1.0000	0.00	0.0	64364.0	1.0
INTAKE MANIFOLD (DS–ENG11)	0.00	0.0	534297.0	1.0	1.0000	0.00	0.0	32182.0	1.0
CRANKCASE (DS–ENG12)	0.01	1068594.0	356198.0	1.3	1.0000	0.14	64364.0	21454.7	1.3
RODS (DS–ENG14)	0.00	0.0	1068594.0	2.0	1.0000	0.00	0.0	64364.0	2.0
CAM (DS–ENG15)	0.00	0.0	0.0	0.0	1.0000	0.00	0.0	0.0	0.0
CHAIN DRIVE (DS–ENG17)	0.00	0.0	0.0	0.0	1.0000	0.00	0.0	0.0	0.0
TAPPET (DS–ENG18)	0.00	0.0	0.0	0.0	1.0000	0.00	0.0	0.0	0.0
ENGINE SWITCH GEAR (DS–ENG19)	0.01	1068594.0	1068594.0	5.0	1.0000	0.14	64364.0	64364.0	5.0
EXHAUST SYSTEM (DS–EXH)	0.02	356198.0	213718.8	1.8	1.0000	0.41	21454.7	12872.8	1.8
EXHAUST SYSTEM (DS–EXH)	0.02	534297.0	356198.0	1.7	1.0000	0.27	32182.0	21454.7	1.7
EXPANSION JOINTS (DS–EXH03)	0.01	1068594.0	1068594.0	3.0	1.0000	0.14	64364.0	64364.0	3.0
PORTS (DS–EXH05)	0.00	0.0	0.0	0.0	1.0000	0.00	0.0	0.0	0.0
EXHAUST MANIFOLD (DS–EXH06)	0.00	0.0	1068594.0	1.0	1.0000	0.00	0.0	64364.0	1.0
FUEL SYSTEM (DS–FLS)	0.24	36848.1	14841.6	2.3	0.9998	3.95	2219.4	893.9	2.3
FUEL SYSTEM (DS–FLS)	0.01	1068594.0	534297.0	1.0	1.0000	0.14	64364.0	32182.0	1.0
DAY TANKS (DS–FLS01)	0.01	1068594.0	1068594.0	1.0	1.0000	0.14	64364.0	64364.0	1.0
FUEL FILTER (DS–FLS03)	0.00	0.0	56241.8	1.0	1.0000	0.00	0.0	3387.6	1.0
GOVERNOR (DS–FLS04)	0.00	0.0	0.0	0.0	1.0000	0.00	0.0	0.0	0.0
PUMPS (DS–FLS06)	0.04	213718.8	178099.0	1.8	0.9999	0.68	12872.8	10727.3	1.8
VALVES (DS–FLS07)	0.03	267148.5	106859.4	2.1	1.0000	0.54	16091.0	6436.4	2.1
INJECTOR (DS–FLS08)	0.13	66787.1	38164.1	3.6	0.9999	2.18	4022.8	2298.7	3.6
FUEL LINE (DS–FLS09)	0.00	0.0	1068594.0	2.0	1.0000	0.00	0.0	64364.0	2.0
FUEL OIL REGULATOR (DS–FLS10)	0.00	0.0	0.0	0.0	1.0000	0.00	0.0	0.0	0.0
GAS JUMPER (DS–FLS11)	0.02	534297.0	213718.8	2.0	1.0000	0.27	32182.0	12872.8	2.0
GENERATOR (DS–GNR)	0.03	267148.5	213718.8	2.8	0.9987	0.54	16091.0	12872.8	2.8
GENERATOR (DS–GNR)	0.01	1068594.0	1068594.0	2.0	0.9987	0.14	64364.0	64364.0	2.0
FIELD (DS–GNR05)	0.02	356198.0	267148.5	3.0	1.0000	0.41	21454.7	16091.0	3.0
FLYWHEEL (DS–GNR10)	0.00	0.0	0.0	0.0	1.0000	0.00	0.0	0.0	0.0
LUBE OIL/HYDRAULIC SYSTEM (DS–LBO)	0.16	56241.8	20162.2	3.4	0.9998	2.59	3387.6	1214.4	3.4
LUBE OIL/HYDRAULIC SYSTEM (DS–LBO)	0.00	0.0	0.0	0.0	1.0000	0.00	0.0	0.0	0.0
HEATER (DS–LBO01)	0.05	178099.0	76328.1	1.9	1.0000	0.82	10727.3	4597.4	1.9
COOLER (DS–LBO02)	0.00	0.0	1068594.0	1.0	1.0000	0.00	0.0	64364.0	1.0
COOLER FAN (DS–LBO03)	0.01	1068594.0	1068594.0	15.0	1.0000	0.14	64364.0	64364.0	15.0
FILTER (DS–LBO04)	0.02	356198.0	56241.8	5.4	0.9999	0.41	21454.7	3387.6	5.4
PUMP (DS–LBO05)	0.02	356198.0	133574.2	1.9	1.0000	0.41	21454.7	8045.5	1.9
PIPING (DS–LBO06)	0.01	1068594.0	1068594.0	8.0	1.0000	0.14	64364.0	64364.0	8.0
TANK (DS–LBO08)	0.01	1068594.0	1068594.0	2.0	1.0000	0.14	64364.0	64364.0	2.0
VALVES (DS–LBO09)	0.01	1068594.0	1068594.0	2.0	1.0000	0.14	64364.0	64364.0	2.0
STRAINER (DS–LBO10)	0.02	534297.0	356198.0	1.7	1.0000	0.27	32182.0	21454.7	1.7
LUBRICATOR (DS–LBO12)	0.01	1068594.0	267148.5	1.2	1.0000	0.14	64364.0	16091.0	1.2
STARTING SYSTEM (DS–STS)	0.14	62858.5	44524.8	2.6	0.9999	2.31	3786.1	2681.8	2.6
STARTING SYSTEM (DS–STS)	0.02	534297.0	534297.0	2.5	1.0000	0.27	32182.0	32182.0	2.5
STARTING AIR COMPRESSOR (DS–STS02)	0.02	356198.0	356198.0	6.7	1.0000	0.41	21454.7	21454.7	6.7
AIR FILTER (DS–STS04)	0.00	0.0	1068594.0	1.0	1.0000	0.00	0.0	64364.0	1.0
VALVES (DS–STS08)	0.01	1068594.0	534297.0	3.0	1.0000	0.14	64364.0	32182.0	3.0
AIR STARTS (DS–STS10)	0.07	133574.2	89049.5	2.0	1.0000	1.09	8045.5	5363.7	2.0
AIR INTAKE (DS–STS11)	0.00	0.0	1068594.0	1.0	1.0000	0.00	0.0	64364.0	1.0
BATTERY (DS–STS15)	0.02	356198.0	356198.0	2.0	1.0000	0.41	21454.7	21454.7	2.0

TABLE X
Subsystem and Component RAM Measures for Continuous-Duty Gas
Turbines

Equipment	Period Hours					Operating Hours			
	Failures per Year	MTBF (Hours)	MTBCM (Hours)	MTTCM (Hours)	Operational Availability	Failures per Year	MTBF (Hours)	MTBCM (Hours)	MTTCM (Hours)
AIR INTAKE SYSTEM (GT–AIS)	0.00	0.0	47698.3	8.6	1.0000	0.00	0.0	29148.1	8.6
AIR INLET FILTER (GT–AIS01)	0.00	0.0	55648.0	2.0	1.0000	0.00	0.0	34006.2	2.0
DUCTING (GT–AIS03)	0.00	0.0	333888.0	48.0	1.0000	0.00	0.0	204037.0	48.0
BALANCE OF PLANT (GT–BOP)	0.05	166944.0	83472.0	2.3	0.9907	0.09	102018.5	51009.2	2.3
FIRE SUPPRESSION/DETECTION (GT–BOP03)	0.05	166944.0	83472.0	2.3	1.0000	0.09	102018.5	51009.2	2.3
TESTING (GT–BOP04)	0.00	0.0	0.0	0.0	1.0000	0.00	0.0	0.0	0.0
CLEANING (GT–BOP05)	0.00	0.0	0.0	0.0	0.9990	0.00	0.0	0.0	0.0
INSPECTION (GT–BOP06)	0.00	0.0	0.0	0.0	0.9918	0.00	0.0	0.0	0.0
COMBUSTION SYSTEM (GT–CMB)	0.21	41736.0	23849.1	1.5	0.9999	0.34	25504.6	14574.1	1.5
COMBUSTION SYSTEM (GT–CMB)	0.08	111296.0	111296.0	1.3	1.0000	0.13	68012.3	68012.3	1.3
FUEL NOZZLES (GT–CMB02)	0.13	66777.6	30353.5	1.6	1.0000	0.21	40807.4	18548.8	1.6
COMPRESSOR (GT–CMP)	0.10	83472.0	47698.3	1.1	1.0000	0.17	51009.2	29148.1	1.1
FLEXLINE (GT–CMP05)	0.00	333888.0	333888.0	1.0	1.0000	0.00	204037.0	204037.0	1.0
BLEEDVALVE (GT–CMP06)	0.10	83472.0	55648.0	1.2	1.0000	0.17	51009.2	34006.2	1.2
CONTROL & INSTRUMENTATION (GT–CTI)	0.63	13912.0	9274.7	1.2	0.9999	1.03	8501.5	5667.7	1.2
CONTROL & INSTRUMENTATION (GT–CTI)	0.03	333888.0	333888.0	1.0	1.0000	0.04	204037.0	204037.0	1.0
CIRCUIT BREAKERS (GT–CTI01)	0.05	166944.0	166944.0	1.0	1.0000	0.09	102018.5	102018.5	1.0
ELECTRICAL MODULE (GT–CTI02)	0.31	27824.0	23849.1	1.5	0.9999	0.52	17003.1	14574.1	1.5
GAUGES (GT–CTI03)	0.05	166944.0	37098.7	0.8	1.0000	0.09	102018.5	22670.8	0.8
SWITCHES (GT–CTI04)	0.16	55648.0	41736.0	1.1	1.0000	0.26	34006.2	25504.6	1.1
THERMOCOUPLE (GT–CTI07)	0.03	333888.0	166944.0	2.0	1.0000	0.04	204037.0	102018.5	2.0
EXHAUST SYSTEM (GT–EXH)	0.00	0.0	333888.0	1.0	1.0000	0.00	0.0	204037.0	1.0
EXHAUST FAN (GT–EXH03)	0.00	0.0	333888.0	1.0	1.0000	0.00	0.0	204037.0	1.0
FUEL SYSTEM (GT–FLS)	1.89	4637.3	3442.1	3.0	0.9992	3.09	2833.8	2103.5	3.0
FUEL SYSTEM (GT–FLS)	0.08	111296.0	111296.0	1.5	1.0000	0.13	68012.3	68012.3	1.5
AIR MANIFOLD (GT–FLS01)	0.00	0.0	333888.0	2.0	1.0000	0.00	0.0	204037.0	2.0
BOOST PUMP (GT–FLS02)	0.13	66777.6	66777.6	2.2	1.0000	0.21	40807.4	40807.4	2.2
FILTERS (GT–FLS04)	0.10	83472.0	55648.0	1.8	1.0000	0.17	51009.2	34006.2	1.8
GAS MANIFOLD (GT–FLS06)	0.00	0.0	0.0	0.0	1.0000	0.00	0.0	0.0	0.0
GOVERNOR (GT–FLS07)	0.60	14516.9	10770.6	5.9	0.9995	0.99	8871.2	6581.8	5.9
MAIN FUEL PUMP (GT–FLS08)	0.10	83472.0	66777.6	1.6	1.0000	0.17	51009.2	40807.4	1.6
ORIFICE (GT–FLS10)	0.03	333888.0	166944.0	2.0	1.0000	0.04	204037.0	102018.5	2.0
PRESSURE GAUGE (GT–FLS12)	0.03	333888.0	333888.0	1.0	1.0000	0.04	204037.0	204037.0	1.0
STRAINER (GT–FLS13)	0.05	166944.0	83472.0	1.2	1.0000	0.09	102018.5	51009.2	1.2
VALVES (GT–FLS14)	0.39	22259.2	16694.4	1.5	0.9999	0.64	13602.5	10201.9	1.5
PIPING (GT–FLS15)	0.18	47698.3	33888.8	2.6	0.9999	0.30	29148.1	20403.7	2.6
SEALS (GT–FLS16)	0.16	55648.0	47698.3	1.1	1.0000	0.26	34006.2	29148.1	1.1
FLOW METER (GT–FLS17)	0.03	333888.0	166944.0	1.0	1.0000	0.04	204037.0	102018.5	1.0
GEARBOX (GT–GBX)	0.03	333888.0	166944.0	1.5	1.0000	0.04	204037.0	102018.5	1.5
GEARBOX (GT–GBX)	0.00	0.0	333888.0	2.0	1.0000	0.00	0.0	204037.0	2.0
SEALS (GT–GBX04)	0.03	333888.0	333888.0	1.0	1.0000	0.04	204037.0	204037.0	1.0
GENERATOR (GT–GNR)	0.13	66777.6	41736.0	4.1	0.9999	0.21	40807.4	25504.6	4.1
GENERATOR (GT–GNR)	0.00	0.0	333888.0	8.0	1.0000	0.00	0.0	204037.0	8.0
BEARINGS (GT–GNR01)	0.00	0.0	0.0	0.0	1.0000	0.00	0.0	0.0	0.0
FIELD (GT–GNR05)	0.00	0.0	0.0	0.0	1.0000	0.00	0.0	0.0	0.0
STATOR (GT–GNR09)	0.00	0.0	0.0	0.0	1.0000	0.00	0.0	0.0	0.0
TURBINE COUPLING (GT–GNR10)	0.05	166944.0	83472.0	2.8	1.0000	0.09	102018.5	51009.2	2.8
VOLTAGE REGULATOR (GT–GNR11)	0.08	111296.0	111296.0	4.7	1.0000	0.13	68012.3	68012.3	4.7
LUBE OIL/HYDRAULIC SYSTEM (GT–LBO)	0.71	12366.2	8347.2	1.8	0.9998	1.16	7556.9	5100.9	1.8
AIR-TO-OIL COOLER (GT–LBO01)	0.13	66777.6	66777.6	2.3	0.9999	0.21	40807.4	40807.4	2.3
HYDRAULIC PUMP (GT–LBO02)	0.05	166944.0	166944.0	2.0	1.0000	0.09	102018.5	102018.5	2.0
LUBE OIL FILTER (GT–LBO03)	0.08	111296.0	30353.5	1.9	1.0000	0.13	68012.3	18548.8	1.9
OIL COOLER FAN (GT–LBO05)	0.08	111296.0	83472.0	2.0	1.0000	0.13	68012.3	51009.2	2.0
OIL MANIFOLDS (GT–LBO06)	0.03	333888.0	333888.0	1.0	1.0000	0.04	204037.0	204037.0	1.0
OIL TANK (GT–LBO07)	0.05	166944.0	166944.0	1.0	1.0000	0.09	102018.5	102018.5	1.0
PRE LUBE OIL PUMP (GT–LBO09)	0.10	83472.0	83472.0	2.6	1.0000	0.17	51009.2	51009.2	2.6
PIPING (GT–LBO12)	0.10	83472.0	55648.0	1.1	1.0000	0.17	51009.2	34006.2	1.1
SEALS (GT–LBO13)	0.03	333888.0	111296.0	1.0	1.0000	0.04	204037.0	68012.3	1.0
PRECIPITATOR (GT–LBO14)	0.05	166944.0	166944.0	2.5	1.0000	0.09	102018.5	102018.5	2.5
REDUCTION GEARBOX (GT–RGB)	0.03	333888.0	333888.0	2.0	1.0000	0.04	204037.0	204037.0	2.0
REDUCTION GEARBOX (GT–RGB)	0.03	333888.0	333888.0	2.0	1.0000	0.04	204037.0	204037.0	2.0

TABLE X (Continued)

Equipment	Period Hours					Operating Hours			
	Failures per Year	MTBF (Hours)	MTBCM (Hours)	MTTCM (Hours)	Operational Availability	Failures per Year	MTBF (Hours)	MTBCM (Hours)	MTTCM (Hours)
STARTING SYSTEM (GT-STS)	0.71	12366.2	9820.2	19.5	0.9980	1.16	7556.9	6001.1	19.5
STARTING SYSTEM (GT-STS)	0.08	111296.0	111296.0	0.7	1.0000	0.13	68012.3	68012.3	0.7
AIR PUMP (GT-STS01)	0.03	333888.0	111296.0	2.3	1.0000	0.04	204037.0	68012.3	2.3
FILTER (GT-STS02)	0.03	333888.0	333888.0	1.0	1.0000	0.04	204037.0	204037.0	1.0
REGULATOR (GT-STS03)	0.00	0.0	0.0	0.0	1.0000	0.00	0.0	0.0	0.0
BATTERY (GT-STS06)	0.00	0.0	0.0	0.0	1.0000	0.00	0.0	0.0	0.0
STARTING SHAFT (GT-STS07)	0.03	333888.0	333888.0	2.0	1.0000	0.04	204037.0	204037.0	2.0
STARTER MOTOR (GT-STS08)	0.13	66777.6	47698.3	83.6	0.9982	0.21	40807.4	29148.1	83.6
GARLOC SEAL (GT-STS11)	0.42	20868.0	17573.1	3.5	0.9998	0.69	12752.3	10738.8	3.5
TURBINE (GT-TRB)	0.08	111296.0	166944.0	121.0	0.9954	0.13	68012.3	102018.5	121.0
TURBINE (GT-TRB)	0.05	166944.0	333888.0	240.0	0.9954	0.09	102018.5	204037.0	240.0
CASING (GT-TRB02)	0.00	0.0	0.0	0.0	1.0000	0.00	0.0	0.0	0.0
BEARING (GT-TRB05)	0.03	333888.0	333888.0	2.0	1.0000	0.04	204037.0	204037.0	2.0

TABLE XI

SUBSYSTEM AND COMPONENT RAM MEASURES FOR STANDBY GAS TURBINES

Equipment	Period Hours					Operating Hours			
	Failures per Year	MTBF (Hours)	MTBCM (Hours)	MTTCM (Hours)	Operational Availability	Failures per Year	MTBF (Hours)	MTBCM (Hours)	MTTCM (Hours)
AIR INTAKE SYSTEM (GT-AIS)	0.01	975612.0	975612.0	1.0	1.0000	1.29	6795.5	6795.5	1.0
DUMPERS (GT-AIS04)	0.01	975612.0	975612.0	1.0	1.0000	1.29	6795.5	6795.5	1.0
BALANCE OF PLANT (GT-BOP)	0.00	0.0	0.0	0.0	0.9989	0.00	0.0	0.0	0.0
TESTING (GT-BOP04)	0.00	0.0	0.0	0.0	1.0000	0.00	0.0	0.0	0.0
CLEANING (GT-BOP05)	0.00	0.0	0.0	0.0	1.0000	0.00	0.0	0.0	0.0
INSPECTION (GT-BOP06)	0.00	0.0	0.0	0.0	0.9989	0.00	0.0	0.0	0.0
COMBUSTION SYSTEM (GT-CMB)	0.00	1951224.0	1951224.0	4.0	1.0000	0.64	13591.0	13591.0	4.0
FUEL NOZZLES (GT-CMB02)	0.00	1951224.0	1951224.0	4.0	1.0000	0.64	13591.0	13591.0	4.0
CONTROL & INSTRUMENTATION (GT-CTI)	0.04	216802.7	150094.2	8.3	0.9999	5.80	1510.1	1045.5	8.3
CONTROL & INSTRUMENTATION (GT-CTI)	0.00	1951224.0	1951224.0	1.0	1.0000	0.64	13591.0	13591.0	1.0
CIRCUIT BREAKERS (GT-CTI01)	0.00	0.0	1951224.0	0.5	1.0000	0.00	0.0	13591.0	0.5
ELECTRICAL MODULE (GT-CTI02)	0.02	487806.0	325204.0	7.2	1.0000	2.58	3397.8	2265.2	7.2
GAUGES (GT-CTI03)	0.00	1951224.0	975612.0	1.0	1.0000	0.64	13591.0	6795.5	1.0
SWITCHES (GT-CTI04)	0.01	975612.0	975612.0	29.0	1.0000	1.29	6795.5	6795.5	29.0
WIRING (GT-CTI05)	0.00	1951224.0	1951224.0	4.0	1.0000	0.64	13591.0	13591.0	4.0
EXHAUST SYSTEM (GT-EXH)	0.00	1951224.0	975612.0	5.5	1.0000	0.64	13591.0	6795.5	5.5
EXHAUST DUCTING (GT-EXH01)	0.00	0.0	1951224.0	1.0	1.0000	0.00	0.0	13591.0	1.0
EXHAUST FAN (GT-EXH03)	0.00	1951224.0	1951224.0	10.0	1.0000	0.64	13591.0	13591.0	10.0
FUEL SYSTEM (GT-FLS)	0.04	243903.0	130081.6	5.0	1.0000	5.16	1698.9	906.1	5.0
BOOST PUMP (GT-FLS02)	0.01	650408.0	650408.0	2.0	1.0000	1.93	4530.3	4530.3	2.0
FILTERS (GT-FLS04)	0.00	1951224.0	278746.3	1.1	1.0000	0.64	13591.0	1941.6	1.1
GOVERNOR (GT-FLS07)	0.00	1951224.0	1951224.0	2.0	1.0000	0.64	13591.0	13591.0	2.0
MAIN FUEL PUMP (GT-FLS08)	0.00	1951224.0	1951224.0	4.0	1.0000	0.64	13591.0	13591.0	4.0
STRAINER (GT-FLS13)	0.00	0.0	0.0	0.0	1.0000	0.00	0.0	0.0	0.0
VALVES (GT-FLS14)	0.01	975612.0	975612.0	27.5	1.0000	1.29	6795.5	6795.5	27.5
PIPING (GT-FLS15)	0.00	0.0	1951224.0	1.0	1.0000	0.00	0.0	13591.0	1.0
GENERATOR (GT-GNR)	0.04	216802.7	216802.7	33.3	0.9998	5.80	1510.1	1510.1	33.3
GENERATOR (GT-GNR)	0.00	1951224.0	1951224.0	72.0	1.0000	0.64	13591.0	13591.0	72.0
TURBINE COUPLING (GT-GNR10)	0.03	278746.3	278746.3	32.2	0.9999	4.51	1941.6	1941.6	32.2
VOLTAGE REGULATOR (GT-GNR11)	0.00	1951224.0	1951224.0	2.0	1.0000	0.64	13591.0	13591.0	2.0
LUBE OIL/HYDRAULIC SYSTEM (GT-LBO)	0.02	390244.8	177384.0	1.6	1.0000	3.22	2718.2	1235.5	1.6
LUBE OIL FILTER (GT-LBO03)	0.00	0.0	650408.0	2.0	1.0000	0.00	0.0	4530.3	2.0
VALVES (GT-LBO11)	0.00	0.0	1951224.0	1.0	1.0000	0.00	0.0	13591.0	1.0
PIPING (GT-LBO12)	0.02	487806.0	487806.0	1.8	1.0000	2.58	3397.8	3397.8	1.8
SEALS (GT-LBO13)	0.00	1951224.0	650408.0	1.3	1.0000	0.64	13591.0	4530.3	1.3
REDUCTION GEARBOX (GT-RGB)	0.00	1951224.0	1951224.0	360.0	0.9998	0.64	13591.0	13591.0	360.0
REDUCTION GEARBOX (GT-RGB)	0.00	1951224.0	1951224.0	360.0	0.9998	0.64	13591.0	13591.0	360.0
STARTING SYSTEM (GT-STS)	0.13	67283.6	45377.3	28.6	0.9994	18.69	468.7	316.1	28.6
STARTING SYSTEM (GT-STS)	0.00	1951224.0	1951224.0	2.0	1.0000	0.64	13591.0	13591.0	2.0
BATTERY (GT-STS06)	0.08	108401.3	60975.8	3.3	1.0000	11.60	755.1	424.7	3.3
STARTING SHAFT (GT-STS07)	0.02	390244.8	390244.8	93.5	0.9998	3.22	2718.2	2718.2	93.5
STARTER MOTOR (GT-STS08)	0.02	390244.8	390244.8	130.8	0.9997	3.22	2718.2	2718.2	130.8
TURBINE (GT-TRB)	0.02	390244.8	390244.8	1158.4	0.9970	3.22	2718.2	2718.2	1158.4
TURBINE (GT-TRB)	0.02	487806.0	487806.0	1398.0	0.9971	2.58	3397.8	3397.8	1398.0
BEARING (GT-TRB05)	0.00	1951224.0	1951224.0	200.0	0.9999	0.64	13591.0	13591.0	200.0

Discussion

R. H. Gauger (Holmes & Narver): This is an excellent survey and is the most comprehensive one available for the 600–1800-kW size range of diesel and gas-turbine-generating units. The results are not what I would have expected, and users of these data should be altered to differing results from surveys made by others. I have made a number of surveys of the reliability of diesel and gas-turbine-generating units of various sizes and will be making a comparison of the results with this new survey.

L. D. Monaghan (Hartford Steam Boiler Inspection and Insurance Company): My comments are directed at the corrective maintenance category. The corrective maintenance code should indicate why the corrective maintenance was necessary. The cause should address such things as lack of preventive maintenance or a manufacturer's defect. Knowing the reason for the maintenance would help the user of these data to differentiate between a manufacturing problem and an operational problem. Another suggestion is for the maintenance category to be subdivided into routine, preventive, and lack of maintenance.

Richard H. McFadden, Peter L. Appignani, and Gary DeMoss (Science Applications International Corporation): This paper represents a significant new base of reliability data for the most popular types of small generating units and will be a valuable resource for intelligent decisions between diesel- and gas-turbine-powered generation. The authors' component coding approach is excellent and would be a good basis for a standardized "component taxonomy" for diesel and gas-turbine generators.

The paper raises some questions for which answers would be valuable to system and reliability engineers contemplating similar projects, and we would appreciate the authors' comments on them.

First, as the authors remark, failure to start is the predominant failure mode of units of both types in "standby" service. (Independently developed reliability statistics on both nuclear-plant standby diesels and utility peaking gas turbines tend to confirm this observation.) As the authors also imply, the distinction between standby and continuous service is blurred in the industrial-commercial environment, because even the sets in nominally "continuous" duty typically operate cyclically, with many more starts than a base-loaded generating unit. Since starting reliability seems to be a critical RAM parameter, why are failure rates calculated exclusively in terms of failures per unit-year rather than failures per demand? The level of detail of the failure analysis in the paper suggests that the raw data were sufficient to distinguish between time- and demand-related failures and allow both failure rate per-unit time and failure-probability per demand to be calculated.

Second, although the RAM data were not conclusive and judgments about the relative merits of diesels versus gas turbines probably were outside the scope of this study, did the authors develop any insights into the optimum selection for various industrial, commercial building, and institutional applications?

P. F. Albrecht (General Electric Company): A key parameter for standby units is starting reliability. The text mentions starting reliability but does not give any statistics. I cannot determine how starting failures were treated. I assume they were counted as forced outages.

Another important event is "failed while not running." This is not discussed at all. These could be failures discovered by periodic testing or inspection. Thus, test frequency may be a very important parameter in determining operating availability. It does not appear that this factor was considered in the survey.

Basically, the authors have analyzed the data using a conventional two-state model approach. They have expressed results on both a period-hour and operating-hour base to suit a "variety of applications." In fact, a two-state model is not very useful for standby units, and the results presented are therefore difficult to use.

Pat O'Donnell (El Paso Natural Gas Company): The reliability survey data on diesel and gas-turbine generators collected by ARINC Research Corpo-

ration appear to provide an excellent data base for meaningful reliability studies on important equipment types. The results reflect an obvious intense and praiseworthy effort in assembling a well-organized and complete data base for its intended purpose. Personal plant visits, as reported in the paper, especially add to the credibility of the results. Although particular details on applications and circumstances of use are not listed, the number of plants, the number of power systems, periods of time, and the number of events counted are impressive and reflect very credible results.

As with any reliability survey, a given set of results always leads to questions and concerns related to any user's given experience background, and usually further manipulation and analysis of the data are required. My intent is to point out some questions and concerns that, hopefully, will lead to additional analyses. In many industries, economic studies comparing gas turbine/generators with reciprocating engine/generators are usually straightforward and simple, with the exception of reliability comparisons and the effects of reliability on economics. Hopefully, these new data will add a missing link and allow more meaningful and accurate comparisons to be made.

An important concern in evaluating the categories surveyed is the speed of the diesel engine. Typically, continuous duty units are designed and applied to run at slower speeds than standby units. "High-speed" reciprocating engines (e.g., 1200 r/min and higher) require frequent maintenance and predictable repair downtime compared with slow-speed units that simply do not experience the same mechanical stress. One would expect a higher failure rate or higher frequency or maintenance, or both, for high-speed engines than for slower speed engines. Will the data allow speed ranges to be identified and corresponding reliability comparisons to be made?

Starting reliability is an important concern, especially for standby or emergency applications. It is unclear if the failures shown for "starting systems" also mean "failures to start." The data might, in some cases, reflect component failures even though the generator set successfully started. Actual "failures to start" would be beneficial in comparing diesel engines with gas turbines, since there are many who believe there is a significant difference. Whether a unit is locally or remotely started normally requires an assessment of reliability in starting. The impact of a failure to start is obviously different when personnel are on site to address a problem immediately as compared with when personnel must travel to a site to address a problem.

Another concern that is important to reliability is the type of starter used. It appears from the data presented here that air and electric motors are two types of starters used. In the natural gas industry, expansion gas turbines are commonly used for starting turbines and definitely are much more reliable than electric starters, primarily because of the available gas supply. Can a closer analysis be made comparing the air systems with the electric motors?

Also regarding starting, the results reflect significant difference in failure rates between "continuous" and "standby" diesel units and "continuous" and "standby" gas turbines and state that this may be related to differences in actual in-use hours. One would also expect that the frequency of starting is different and might impact failure rates. Can this analysis be made?

The fuel system appears to be a significant contributor to failures. It is interesting that on "continuous" gas turbines, the fuel system is the least reliable part of the package. It would be beneficial if reasons could be identified. Are different types of fuels involved? If so, will the data collected allow comparing failure rates for each type?

The tabulated results in Appendix II, Tables VIII and IX, of the report suggest that possibly not all diesel units are truly packaged type (e.g., cooling towers, water heater). Can the data be refined further to identify which units are truly self-contained?

A last point of concern regards maintenance. A reciprocating engine is expected to be more demanding in routine maintenance requirements than a gas turbine. To qualify this statement, this is to say that it is easier to leave a gas turbine unattended, once it is running, than it is a reciprocating engine, especially if they are running continuously. There are various reasons why, some of which are the way the units are typically instrumented for protection and the number of moving parts and wear. If the MTBCM data include scheduled maintenance cycles, a comparison of failure rates for different cycles would be meaningful.

The results here reflect an excellent collection of data and should be very beneficial in making comparisons of these equipment types. In the application of reliability data an inevitable concern is the reason for differences in reliability between equipment types and applications. One obvious practical benefit is to be able to identify what corrective actions are encouraged by

TABLE XII
COMPARISON OF DIESEL AND GAS-TURBINE STARTING RELIABILITY STUDIES

Source	Number of Units	Start Attempts	Failed Starts	Starting Reliability
Gas-Turbine Starting Reliability Studies				
ARINC Research Corporation[1]	7	3,555	17	0.9952
Booz, Allen & Hamilton[2]	34	12,316	80	0.9935
Kongsberg Dresser Power[3]	38	17,749	141	0.9921
AT&T[4]	28	13,644	106	0.9922
Diesel Starting Reliability Studies				
ARINC Research Corporation[1]	—	—	—	0.97
Electric Power Research Institute (EPRI)[5]	155	22,320	83	0.9963
Consumers Power Company—Big Rock Point[6]	2	669	12	0.9821
Northeast Utilities—Millstone[6]	3	652	3	0.9954
Northeast Utilities—Connecticut Yankee[6]	2	642	2	0.9969
Commonwealth Edison Company—Zion[6]	4	1,693	30	0.9823
Consolidated Edison Company of New York, Inc.—Indian Point[6]	6	424	4	0.9906
Institute of Nuclear Power Operations (INPO)[7]	←———— Data not available ————→			0.9120
EPRI[8]	←———— Data not available ————→			0.9829

[1]ARINC Research Corporation. *Final Report—RAM Study of Diesel and Gas-Turbine Generator Sets.* Publication 4219-03-01-4803, October 1988.

[2]Booz, Allen Applied Research. *Small Gas Turbine Start Investigation.* April 1970.

[3]Kongsberg Dresser Power. Internal Study Comparing Diesels with Gas-Turbine Engines (unpublished), 1984.

[4]AT&T. Internal Study for Gas-Turbine Reliability (unpublished), 1980.

[5]Electric Power Research Institute. *Reliability of Emergency Diesel Generators at U.S. Nuclear Power Plants.* NSAC 108, September 1986.

[6]U.S. Nuclear Regulatory Commission. *Nuclear Computerized Library for Assessing Reactor Reliability (NUCLARR).* NUREG/CR-4639 EGG-2458, Volume 5, RX, June 1988.

[7]Institute of Nuclear Power Operations. *Nuclear Plant Reliability Data System. 1982 Annual Report.* 1983.

[8]Electric Power Research Institute. *Diesel Power Reliability at Nuclear Power Plants: Data Preliminary Analysis.* NP-2433, June 1982.

a user and which are encouraged by a manufacturer. Hopefully, additional analyses will be made addressing the concerns of this discussion and other similar concerns stimulated by the results presented here.

Closure

The authors appreciate the thorough review and the many constructive comments and recommendations offered in the preceding discussion. While space limitations prohibit addressing all of the suggestions offered, a response to some of the more frequently cited comments is provided in the following paragraphs.

Obtaining data on unit starting reliability was one of the objectives of the study. However, most of the plants surveyed did not record data necessary to determine starting reliability. While it was often possible to identify start failures through interpretation of the maintenance event descriptions, the number of start attempts was typically not retrievable. In addition, our discussions with plant personnel indicated that many start failures were corrected through minor adjustments that were usually not documented in maintenance or operating records. Because of the limited data available, starting reliability statistics were not presented in the paper.

Some information on starting reliability was obtained during the study. These data are presented in Table XII. Seven gas-turbine units provided data on start attempts and start failures during periodic testing. To obtain estimates of diesel starting reliability, we surveyed plant managers of four of the standby diesel plants to estimate the number of start failures in 100 attempts. We then averaged these estimates to obtain an estimated diesel starting reliability. Table XII also shows a comparison of values for diesel and gas-turbine starting reliability.

With regard to maintenance, data were categorized on the basis of the na-

ture of the individual maintenance task performed for each event. The maintenance codes do not refer to the cause of failure or the overall maintenance program for the plant. Additional reduction and analysis of the collected data would be required to investigate these issues.

An important feature of the computerized data base developed in this survey is the ability to sort and arrange the data to analyze specific issues regarding plant configuration, design, or operation. The preceding discussions have provided several beneficial suggestions for additional analyses. The results of additional data analyses or data collection activities under this program will be discussed in subsequent papers.

REFERENCES

[1] *Government-Industry Data Exchange Program Remote Terminal Users Guide,* GIDEP Operations Center, Oct. 1987.

[2] *Power Reliability Enhancement Program (PREP) GO Reliability/Availability Data Bank Development Effort,* Illinois Inst. Technol. Res. Inst. Reliability Analysis Center, Aug. 1984.

[3] *IEEE Recommended Practice for Design of Reliable Industrial and Commercial Power Systems,* ANSI/IEEE Standard 493, 1980.

[4] *IEEE Trial-Use Standard Definitions for Use in Reporting Electric Generating Unit Reliability, Availability, and Productivity,* ANSI/IEEE Standard 762, 1980.

[5] *Nuclear Power Reliability Data System (NPRDS) Data Retrieval Manual,* Inst. Nuclear Power Operations, INPO 83-034, Rev. 4, Sept. 1986.

[6] *Major Fixed Command, Control, and Communications Facilities Power Systems Design Features Manual,* U.S. Army Corps of Engineers, Huntsville Division, HNDSP-82-043-SD, Ch. 9, Aug. 1986.

[7] *RAM Study of Diesel and Gas Turbine Generator Sets,* ARINC Res. Pub. 4219-03-01-4803, Oct. 1988.

Clayton A. Smith received the B.S. degree in aerospace engineering from the University of Maryland, College Park.

He is a Staff Engineer for ARINC Research Corporation. He is responsible for the development of data bases for many reliability, availability, and maintainability projects, including those for the Electric Power Research Institute (EPRI), the Gas Research Institute (GRI), petroleum refineries, electric utilities, and the Army Corps of Engineers. He also conducts analyses to investigate component criticality, cost optimization, and modification evaluations.

Michael D. Donovan (A'86) received the B.S. degree in mechanical engineering from Vanderbilt University, Nashville, TN, and the M.S. degree in industrial management from the Georgia Institute of Technology, Atlanta.

He is the Manager of the Industrial Process Group of ARINC Research Corporation and is responsible for the company's engineering services to clients in energy, petrochemical, and manufacturing industries.

Mr. Donovan is Chairman of Station Operations for the Energy Development and Power Generation Committee of the Power Engineering Society and is a Registered Professional Industrial Engineer in the State of Georgia.

Michael J. Bartos received the B.S. degree in mechanical engineering from Pennsylvania State University, University Park, in 1985.

He is currently a Mechanical Engineer in the Power Reliability Enhancement Program (PREP) with the U.S. Army Corps of Engineers, Engineering and Housing Support Center (EHSC). He has been with the PREP program since 1987. His experience with the program includes engineering support for operating power plants and efforts to enhance reliability, survivability, and operation/maintenance of critical command, control, communications, and intelligence facilities and utilities. He was with the U.S. Navy's Cruise Missile Branch of the Airborne Weapons Engineering Division where he served as a Project Engineer from 1985. His experience in the area was design verification, lot acceptance tests, and maintenance concerns. He is also involved in tutorial disadvantaged youth in the Washington, DC, area.

Appendix M

Part 1
Comprehensive Bibliography on Electrical Service Interruption Costs

By
R. Billinton, G. Wacker, and E. Wojczynski
IEEE Transactions on Power Apparatus and Systems
Vol. PAS-102, No. 6
Jun. 1983, pp. 1831–1837

Part 2
Statistical and Analytical Evaluation of the Duration
and Cost of Consumer Interruptions

By
D. O. Koval and R. Billinton
1979 IEEE Winter Power Meeting, New York City, NY, Paper No. A79-057-1

IEEE Transactions on Power Apparatus and Systems, Vol. PAS-102, No. 6, June 1983

Comprehensive Bibliography On
Electrical Service Interruption Costs

R. Billinton, G. Wacker, E. Wojczynski
Power System Research Group
University of Saskatchewan
Saskatoon, Saskatchewan
Canada S7N 0W0

ABSTRACT

This paper contains a comprehensive list of publications relating to the theory and results of research into the worth of reliability. The emphasis is on research and surveys concerned with determining the impacts and estimating the costs to customers of electrical service interruptions. The application of cost estimates in system design and other related topics are included. The first four categories in the bibliography are germane to the topic and the lists are intended to be exhaustive. The remaining categories are peripheral to the bibliography topic but are relevant in terms of power system planning or otherwise; hence their lists are more or less representative of existing knowledge rather than comprehensive.

INTRODUCTION

Increased attention is being given to numerical evaluation of the worth of reliability in regard to electric power supply. The material contained in this bibliography illustrates the state of the art in the collection of information on electrical service interruption costs and provides an awareness of the utilization of these data in the evaluation of power system reliability worth.

Short descriptions are provided for approximately one third of the publications listed in the bibliography. These references can be considered as the major publications in the respective categories and could therefore be studied first before proceeding if necessary with a more detailed examination.

1. IMPACTS OF INTERRUPTIONS

This section contains references investigating and describing the specific impacts to customers and society from electrical service interruptions and the more general impacts resulting from reduction in the reliability of supply. Part 1(a) references emphasize a qualitative description of the impacts to customers during interruptions and the possible response of customers and society to further interruptions. Part 1(b) references investigate in a qualitative and quantitative manner the impacts of blackouts in New York, while Part 1(c) investigates the impacts of blackouts in general.

83 WM 066-8   A paper recommended and approved by the IEEE Power System Engineering Committee of the IEEE Power Engineering Society for presentation at the IEEE/PES 1983 Winter Meeting, New York, New York, January 30-February 4, 1983. Manuscript submitted August 24, 1982; made available for printing November 10, 1982.

1(a) General Impacts of Interruptions and Reduced Levels of Reliability

1. D. McWilliams, "Users Needs: Lighting, Start-up Power, Transportation, Mechanical Utilities, Heating, Refrigeration and Production", IEEE Transactions on Industry Applications, vol. IA-10, no. 2, March/April 1974, pp. 199-204.

2. Technical Advisory Committee on the Impact of Inadequate Electric Power Supply, "The Adequacy of Future Electric Power Supply: Problems and Policies", Federal Power Commission, March 1976.

This report discusses the evaluation of electric power supply adequacy; methods of dealing with shortages and the effect of the shortages as they pertain to government and utility decision making in the U.S.A. The favoured short term method during shortages is rationing amongst customers on either an equal share or an economic basis. The effects of shortages are discussed mainly in a qualitative manner.

3. D. Myers et al., "Impacts from a Decrease in Electric Power Service Reliability", Bonneville Power Administration, June 1976.

This 49 page report attempts to qualitatively describe the impacts to customers of a reduction in their electrical service reliability. Methodologies using global economic indices (e.g. GNP/KWHR), as a means to estimate customer interruption cost are critically evaluated while the wages paid/KWHR approach is applied to the Pacific Northwest Region. The possible adaptive responses of customers to decreased reliability levels are discussed.

4. W. Turner, "Problems with Lifts during Power Cuts", Electrical Review, vol. 201, no. 20, November 18, 1977, pp. 57-58.

5. R. Bird, "Power Cuts ... the Cost to Industry", Electrical Engineer (Australia), December 1977, pp. 23-24.

6. J. Hooper, "Standby-Generating Systems, Power Cuts - Can Industry Afford Them?" Electrical Review, vol. 202, no. 15, April 1978, pp. 23-25.

7. T. Key, "Diagnosing Power Quality-Related Computer Problems", IEEE Transactions on Industry Applications, vol. IA-15, no. 4, July/August 1979, pp. 381-393.

8. K. Neilson, "Throwing a New Light on Security", Electrical Review, vol. 205, no. 11, September 21, 1979, pp. 47-50.

1(b) Impacts of Large Scale Interruptions (Blackouts) in New York

There are several studies not listed which investigate the New York area blackouts but their emphasis is on the engineering aspects of the causes of the event rather than on the impacts.

9. Federal Power Commission, "Northeast Power Failure, Nov. 9 and 10, 1965", U.S. Government Printing Office, Washington, D.C., December 1965.

10. G. D. Friedlander, "The Northeast Power Failure a Blanket of Darkness", IEEE Spectrum, February 1966, pp. 54-73.

11. Department of City Planning, City of New York, "Blackout Commercial Damage Survey", July 1977.

12. Joint Committee on Defense Production, Congress of the United States, "Emergency Preparedness in the Electric Power Industry and the Implications of the New York Blackout for Emergency Planning (Hearings)", U.S. Government Printing Office, Washington, D.C., August 1977.

13. A. Kaufman and B. Daly, "The Cost of an Urban Blackout, The Consolidated Edison Blackout, July 13-14, 1977", The Library of Congress Congressional Research Service, Washington, D.C., June 1978.

This study investigates the losses associated with the 1977 blackout which are estimated to consist of at least 172.7 M$ of economic costs and 136.8 M$ of social costs. Economic costs are estimated from the variation in regional business activity indices while social costs are estimated using figures from government agencies.

14. J. Corwin and W. Miles, "Impact Assessment of the 1977 New York City Blackout", U.S. Department of Energy, Washington, D.C., July 1978.

This study collects and investigates data on the blackout and defines a framework in order to estimate the value of reliability with knowledge of blackout impacts. The main distinction drawn between the impacts is between direct impacts (56 M$) and indirect impacts (290 M$). social and organizational impacts are qualitatively analyzed.

15. P. D. Blair, "Planning for Power Failures and the New York City Experience", Canadian Conference on Communication and Power, 1978.

16. W. Miles, J. Corwin and P. Blair, "Cost of Power Outages - The 1977 New York City Blackout", 1979 Reliability Conference for the Electric Power Industry, April 1979, pp. 193-197.

1(c) Impacts of Large Scale Interruptions (Blackouts) - General

17. Law Enforcement Assistance Administration Emergency Energy Committee, "Preliminary Report on Rolling Blackouts", U.S. Department of Justice, March 1, 1974.

18. D. Myers, "The Economic Effects to a Metropolitan Area of a Power Outage Resulting from an Earthquake", Earthquake Engineering Systems Inc., San Francisco, February 1978.

This report develops a methodology and an analytical model for determining the costs to consumers of an electric power outage. Regional economic data are used as the basis. Residential willingness to pay as a measure of cost is estimated by means of the residential price elasticity (yielding $.50/KWHR).

2. THEORY OF RELIABILITY WORTH

This section contains references dealing with the determination and use of reliability worth. Some of the more abstract economics papers dealing with this topic are included in Section 5. Papers reporting and discussing the results of surveys and not the underlying theory are included in Section 3.

19. S.M. Dean, "Considerations Involved in Making System Investments for Improved Service Reliability", Edison Electric Institute Bulletin, November 1938, pp. 491-498.

20. S. Lalander and U. Sandstrom, "Costs for Disturbances and Their Influence on the Design of Power Systems", CIGRE Proceedings, 1952, paper 329.

21. W.H. Dickinson, "Economic Evaluation of Industrial Power-System Reliability", AIEE Transactions, November 1957, pp. 264-271.

22. K. Goldsmith, "The Price of Reliability in Electricity Supply" IEE Conference on the Economics of the Security of Supply, 1967, pp. 50-52.

23. F.S. Brown, "Economics of Reliability of Bulk Power Supply in the United States", IEE Conference on Economics of the Security of Supply, Pt. 1, pp. 150-154.

24. H. Lindner, "Economic Evaluation of the Reliability of the Electricity Supply of Industrial Installations", Electricity Council Translation O.A. Trans. 482, from Energietechnik, vol. 18, August 1968.

25. C. Starr, "Social Benefit versus Technological Risk", Science, September 19, 1969, pp. 1232-1238.

26. R.B. Shipley, A.D. Patton and J.S. Denison, "Power Reliability Cost vs Worth", IEEE PAS, 1972, pp. 2204-2212.

This paper proposes a method to roughly estimate the worth of reliability by estimating the loss of GNP due to unsupplied electrical energy. A preliminary evaluation of the economics of the USA power system is made using the resultant estimate of $.60/KWHR (1967). The paper concludes that the USA system was probably being planned with too high a target reliability standard.

27. W.G. Watson, "Rapporteur's Report on Session 1. - System and Component Reliability", Cired 73 Pt. II, Record of Discussion, 1973.

28. M.L. Telson, "The Economics of Reliability for Electric Generation Systems", Massachusetts Institute of Technology Energy Laboratory Report MIT-EL 73-106, May 1973.

This 266 page report investigates reliability evaluation methods and indirectly discusses customer impacts from interruptions, attempts to develop a methodology to quantify these impacts as well as include the costs in an economic study of the system reliability. Wages paid during interruptions divided by the non-residential energy consumption yields an estimate of $1.13/KWHR for the New York area. The present system reliability was found to be roughly 100 times more stringent than the optimum.

29. A. Kaufman, "Reliability Criteria – A Cost Benefit Analysis", Office of Economic Research, New York State Public Service Commission, August 1975.

This report discusses a cost-benefit analysis of the New York Power Pool generation reserve level. The interruption cost was estimated using value added techniques to be $.77/KWHR. The results indicate that a 1 day/year LOLE target reliability level is optimum.

30. M. L. Telson, "The Economics of Alternative Levels of Reliability for Electric Power Generation Systems", The Bell Journal of Economics, Autumn 1975, pp. 679–694.

This paper discusses customer impacts from interruptions and the quantification of the economic costs incurred. GNP/KWHR is proposed to be a conservative upper bound while the reasonable upper bound of wages/KWHR yields a figure for the USA of $.57. The optimum reliability standard was determined to be much less stringent than the target level presently used.

31. W.B. Shew, "Costs of Inadequate Capacity in the Electric Utility Industry", Energy Systems and Policy, vol. 2, no. 1, 1977, pp. 85–110.

This paper studies the cost to the USA economy of insufficient capacity in the 1980s by means of an optpimization between planned rationing and reduction of reserve. The costs of rationing are estimated using average price elasticities while interruption costs are estimated by means of GNP/KWHR. The total losses are found to be large but relatively insensitive to variation in the interruption cost estimate.

32. T.W. Berrie, "What Quality of Electricity Supply Can We Afford", Electrical Review, vol. 202, no. 1, January 1978, and "How To Work Out What Quality of Electricity Supply We Can Afford", Electrical Review, vol. 202, no. 3, January 20, 1978.

33. H. Khatib, "Economics of Reliability in Electrical Power Systems", Technicopy Limited, England, 1978.

This 157 page book investigates power system reliability evaluation techniques and the impacts of interruptions on customers. Previous studies attempting to determine the worth of reliability are critically evaluated and the results of surveys by the author of residential and industrial customers reported. Methodologies are presented to determine time of day variation of the cost of interruptions and to aggregate the interruptions caused by generation and distribution systems.

34. M.P. Bhavaraju and R. Billinton, "Cost of Power Interruptions – A User's Viewpoint", EPRI Workshop Proceedings, WS–77–60, March 1978.

This paper critically overviews cost-benefits evaluation of electric power supply reliability. Published research and data on customer interruption costs is discussed. Present data are concluded to be applicable only to local small random interruptions rather than larger wide scale blackouts.

35. R.L. Sullivan, "Worth of Reliability", EPRI Workshop Proceedings, WS–77–60, March 1978, pp. 6–40 to 6–50.

36. D.R. Myers, "Two Examples and Some Research Needs of Reliability/Worth Tradeoffs", EPRI Workshop Proceedings, WS–77–60, March 1978, pp. 6–29 to 6–39.

37. M.E. Samsa, K.A. Hub, G.C. Krohm, "Electricial Service Reliability: The Customer Perspective", U.S. Department of Energy, September 1978.

This 110 page report critically reviews studies concerned with the quantification of customer valuation of power system reliability. A customer classification scheme is proposed together with a discussion of the internal/external interruption costs of customers in the classification. Options for adjusting the reliability of the customers are discussed and categorized.

38. M. Munasinghe, "The Costs Incurred by Residential Electricity Consumers Due to Power Failures", World Bank Research Study, December 1978.

This paper presents a theoretical methodology for measuring residential interruption cost and the results of a survey of urban residential customers in Cascavel, Brazil. The main interruption cost is hypothesized to be the loss of evening leisure time which can be evaluated at the household income earning rate.

39. J. Nordin, "A Subscription Method of Determining Optimal Reserve Generating Capacity for an Electric Utility", The Engineering Economist, vol. 24, no. 4, 1979, pp. 249–257.

This paper proposes a methodology to optimize generation reliability by establishing a reserve capacity subscription system for customers. The principle involved is that system reliability would be determined and paid for by the individual customer subscriptions who could then be supplied at the level of reliability they opted for. Subscribing customers would have priority in being supplied before non-subscribers.

40. G. Wacker, E. Wojczynski, and R. Billinton, "Cost/Benefit Considerations in Providing an Adequate Electric Energy Supply", Third Symposium on Large Engineering Systems, July 1980.

41. K. Powell, "Valuing Reliability for Utility Planning, 1980". Reliability Conference for the Electric Power Industry, 1980, pp. 338–342.

42. Systems Control Incorporated, "Analysis of 12 Electric Power System Outages/Disturbances Impacting the Florida Peninsula", U.S. DOE report number DOE/RG 106359-1, December 1980.

43. R. Billinton, G. Wacker, and E. Wojczynski, "Quantitative Assessment of Power Supply Reliability – Reliability Cost/Reliability Worth Considerations", Alternative Energy Sources and Technology Symposium: Modelling Policy and Economics of Energy and Power Systems, May 1981. (Sponsored by The International Association of Science and Technology for Development – IASTED.)

44. Jack Faucett Associates, "Power Energy and Capacity Shortages – The 1976–77 Natural Gas Shortage", EPRI EA-1215, Volume 1, November 1979.

45. Jack Faucett Associates, "Power Shortage Costs and Efforts to Minimize: An Example", EPRI EA 1241, December 1979.

46. Jack Faucett Associates, "Analytical Framework for Evaluating Energy and Capacity Shortages", EPRI EA-1215, Volume 2, April 1980.

47. R. Billinton, G. Wacker and E. Wojczynski, "Power System Reliability Indices and Their Application in the Assessment of Reliability Worth", IEEE Conference Mexicon 1981, Mexico, CEA Transactions, Vol. 21, 1982.

48. A.P. Sanghvi, "Customer Outage Costs In Investment Planning Models for Optimizing Generation System Expansion and Reliability", CEA Transactions, Volume 21, 1982.

49. A.K. Lee and J.K. Snelson, "A Reliability Criterion For Generation Planning", CEA Transactions, Volume 21, 1982.

3. SURVEYS OF CUSTOMER COSTS

This section contains references which report and discuss surveys of customers attempting to quantify the worth of reliability. Section 3(a) deals with surveys other than those by Ontario Hydro which are included in section 3(b).

3(a) General

50. Swedish Committee on Supply Interruption Costs, "Costs of Interruptions in Electricity Supply", The Electricity Council, O.A. Translation 450, December 1969.

This Swedish survey was the first major investigation of consumer costs of interruptions. Direct questioning was preferred but only industrial customers were so surveyed. Residential and other cost estimates were obtained by discussion with representative organizations and worked examples. Residential costs were as significant as industrial costs. The cost was investigated as a function of interruption duration.

51. L. Lundberg, "Report of the Group of Experts on Quality of Service from the Consumer's Point of View", International Union of Producers and Distributors of Electrical Energy, Report 60/D.1, 1972.

This report is a collection and discussion of the experience and survey results of European utilities concerning the consumers' costs of interruptions. Only Swedish, French, and British utilities had performed studies and surveys. The Swedish survey is discussed above (ref. 50).

52. A. Jackson and B. Salvage, "Costs of Electricity-Supply Interruptions to Industrial Consumers-Summary", Proc. IEE, vol. 121, no. 12, December 1974, pp. 1575-1576.

53. A. Jackson and B. Salvage, "Costs of Electricity-Supply Interruptions to Industrial Consumers", IEE Library, 1974.

54. IEEE Committee, "Report on Reliability Survey of Industrial Plants, Part I: Reliability of Electrical Equipment", IEEE TPAS, March/April 1974, pp. 213-235.

IEEE Committee, "Report on Reliability Survey of Industrial Plants, Part II: Cost of Power Outages, Plant Restart Time, Critical Service Loss Duration Time, and Type of Loads Lost Versus Time of Power Outages", IEEE PAS, March/April 1974, pp. 236-241.

IEEE Committee, "Report on Reliability Survey of Industrial Plants, Part V: Plant Climate, Atmosphere, and Operating Schedule, the Average Age of Electrical Equipment, Percent Production Lost, and the Method of Restoring Electrical Service after a Failure", IEEE PAS, July/August 1974, pp. 463-466.

These papers report on a detailed survey of the reliability of electrical supply to industrial plants in the USA and Canada. Thirty companies with a total of 68 plants in nine main categories responded. The objective was to obtain information on the characteristics of failures of electrical supply equipment in the plants and the impact and cost of interruptions to help aid in the design of industrial power distribution systems. The forms used and the detailed response data are included.

55. P. Gannon, "Costs of Electrical Interruptions in Commercial Buildings", IEEE I & CPS Conference, 1975, pp. 123-129.

This IEEE sponsored survey of the cost to customers of interruptions in commercial buildings and offices received responses for 55 buildings from 48 companies. The survey form used was a simplified version of the 1972 IEEE industrial survey (ref. 54) and is included in the paper. The cost per KWHR unsupplied energy, per sq. ft./hr., and per employee/hr. was surveyed as a function of duration (15 min., 1 hour, > 1 hour) and computer usage.

3(b) Ontario Hydro

56. L.H. Berk, "Report on the Study of the London Power Interruption", OH Report No. PMA-76-2, March 1976.

A number of reports in this section each concern a segment of the Ontario Hydro customer market and have much in common in the way of objectives, approach and report features.

The surveys (refs. 57, 59, 61, 62, 63 and 64) were carried out in response to a suggestion of the Ontario Energy Board as expressed in their report on the hearings into the Ontario Hydro system expansion program. These surveys are part of a series of studies to investigate, amongst other topics, "the evaluation of alternate levels of reliability of power and energy supply from the viewpoint of the particular classes of customers and of the Province of Ontario as a whole." The purpose and scope of the surveys are stated to be:

1) To obtain customer estimates of costs and other effects of electric supply interruptions, voltage variations, frequency variations and rationing.
2) To gather data on customer groups for use in planning and operating the Ontario Hydro system.
3) To obtain information for use in seeking the cooperation of customer groups to reduce the adverse effects of operating problems which might occur in the future.

The customers in each sector (with the exception of residential customers) are classified into categories appropriate with their functions and their sensitivity to interruptions. Cost estimates for various durations of interruptions are reported on a $/KW of peak demand basis. Preferences for frequent but short as opposed to less frequent but longer interruptions are reported. Results concerning standby generation, variation of cost with time of season or day, non-monetary impacts (especially hazards), effects of voltage variation and advance warning and other factors are also reported. Each survey concerns the customer sector as the title of the reports suggest.

57. "Ontario Hydro Survey on Power System Reliability: Viewpoint of Large Users", OH Report No. PMA 76-5, April 1977.

58. L.V. Skof, "Customer Interruption Costs Vary Widely", Electrical World, July 15, 1977, pp. 64-65.

59. Market Facts of Canada Ltd., "Research Report: Residential Monitoring Survey of Energy Conservation: Extract on Supply Reliability", Ontario Hydro Customer Viewpoint Committee on Supply Reliability, August 1977.

60. L.H. e⁻k, "Up-Date on Ontario Hydro Surveys on Supply Reliability: Viewpoint of Selected Customer Groups", CEA Meeting, March 1978.

61. "Ontario Hydro Survey on Power System Reliability: Viewpoint of Small Industrial Users (Under 5000 KW)", OH Report No. R & U 78-3, April 1978.

62. "Ontario Hydro's Surveys of Power System Reliability: Viewpoint of Government and Institutional Users (Pilot Survey in the City of Guelph)", OH Report No. R & U 78-1, May 1978.

63. "Ontario Hydro Survey on Power System Reliability: Viewpoint of Farm Operators", OH Report No. R & U 78-5, December 1978.

64. "Survey of Power System Reliability: Viewpoint of Customers in Retail Trade and Service", OH Report No. R & U 79-7, July 1979.

65. "Energy Use in the Food and Beverage Industry in Ontario", OH Report No. R & U 79-4, July 1979.

66. "Ontario Hydro's Survey of Power System Reliability: Viewpoint of Customers in Office Buildings", Ontario Hydro Report No. R & U 80-5, March 1980.

67. "Ontario Hydro .Survey of Power System Reliability: Viewpoint of government and Institutional Users", Ontario Hydro Report No. R & U 80-6, March 1980.

68. "Ontario Hydro Survey on Power System Reliability: Summary of Customer Viewpoints:, Ontario Hydro Report No. R & MR 80-12, Dec. 1980.

4. OPTIMAL DESIGN AND PLANNING USING WORTH OF RELIABILITY

This section contains references mainly concerned with the application of cost of interruption data in system planning and design. Engineering and economic considerations are delved into. Some of the references attempt to determine the optimum level of reliability using the cost estimates.

Section 4(a) considers the application to generation systems, 4(b) to transmission and distribution systems, and 4(c) to power systems in general.

4(a) Optimal Design and Planning using Worth of Reliability - Generation Systems

69. D. Farrar et al., "A Model for the Determination of Optimal Generating System Expansion Patterns", M.I.T. Energy Lab Report MIT EL 73 009, February 1973.

70. M.G. Webb, "The Determination of Reserve Generating Capacity Criteria in Electricity Supply Systems", Applied Economics, 1977, pp. 19-31.

71. D. Dees et al., "The Effect of Load Growth Uncertainty on Generation System Expansion Planning", Proceedings of the American Power Conference, vol. 40, 1978, pp. 1223-1232.

72. M.H. Bensky et al., "The Cost Benefits of Alternative Generation Reserve Levels", Proceedings of the American Power Conference, vol. 40, 1978, pp. 1323-1332.

73. L. Guth and H. Zellner, "Costs and Benefits of Systems Reliability", Proceedings of the Sixth Annual Illinois Energy Conference, September 1978, pp. 121-137.

74. E.G. Cazalet, "Costs and Benefits of Over/Under in Electric Power System Planning", Proceedings of the Sixth Annual Illinois Energy Conference, September 1978, pp. 106-120.

75. Decision Focus Inc., "Costs and Benefits of Over/Under Capacity in Electric Power System Planning", EPRI EA-927, October 1978.

76. Ontario Hydro, "The SEPR Study; System Expansion Program Reassessment Study", Six Interim Reports and the Final Report, February 1979.

This series of 7 reports presents a summary of an extensive study carried out by Ontario Hydro to re-evaluate possible generation system expansion plans. The effect of changes in load growth on Ontario Hydro and its customers, and of nuclear-coal capacity mix, generating unit size and amount of reserve generation capacity are investigated. The cost of interruptions to customers as determined from surveys (see Section 3(b)) is used to estimate the impact of interruptions.

4(b) Optimal Design and Planning Using Worth of Reliability - Transmission and Distribution Systems

77. H.J. Sheppard, "The Economics of Reliability of Supply - Distribution (Great Britain)", The Economics of Reliability of Supply, IEE Publication 34, 1967 pt. 1, pp. 248-266.

78. W. Dickinson, "Economic Analysis of Reliability in Industrial Power Plant Power Systems", IEEE I & CPS Conference, May 1971, pp. 27-31.

79. P. Gannon, "Fundamentals of Reliability Techniques as Applied to Industrial Power Systems - Part 5 Assumed Costs", IEE I & CPS Conference, May 1971, p. 26.

80. C.R. Heising, "Reliability of Electric Power Transmission and Distribution Equipment", 1974 ASQC Technical Conference Transactions - Boston, pp. 314-319.

81. P. Gannon, "Costs of Interruptions; Economic Evaluation of Reliability", IEEE 1976 I & CPS (Industrial and Commercial Power Systems) Conference, pp. 105-110.

82. E. Dahl and J. Huse, "The Level of Continuity of Electricity Supply and Consequent Financial Implications", CIRED 77, 1977, pp. 138-141.

83. L. Lunden and L. Hammarson, "The Level of Continuity in Urban Distribution Systems and the Consequent Financial Implications", CIRED 77, 1977, pp. 142-145.

84. H. Persoz, et al., "Taking into Account Service Continuity and Quality in Distribution Network Planning", CIRED 77, 1977, pp. 123-126.

85. R. Billinton and D.O. Koval, "Evaluation of Reliability Worth in Distribution Systems", PSCC, vol. 1, 1978, pp. 218-225.

86. R.N. Allan et al., "Reliability Indices and Reliability Worth in Distribution Systems", EPRI Workshop Proceedings, WS-77-60, March 1978, pp. 6-20 to 6-28.

87. M. Munasinghe and W. Scott, "Long Range Distribution System Planning Based on Optimum Economic Reliability Levels", IEEE Summer Meeting, 1978, Paper A 78 576-1.

88. D.O. Koval and R. Billinton, "Statistical and Analytical Evaluation of the Duration and Cost of Consumers Interruptions", IEEE Winter Power Meeting, Paper No. A79 057-1, February 1979.

89. R.S. Tsai, "Reliability Worth Guides Distribution System Design", IEEE Transactions on Industry Applications, vol. IA-15, no. 4, July/August 1979, pp. 368-375.

4(c) Optimal Design and Planning Using Worth of Reliability - Power Systems in General

90. B. Mattsson and J. Nuder, "Simplified Use of Failure Statistics for Optimizing System and Equipment Design", CIGRE 1972, Paper 31-06.

91. W.G. Watson and K.R. Jarret, "Reliability Investment Strategies for HV and EHV Networks Based on an Incremental Approach", 1977 CIRED, pp. 134-137.

92. L. Markel, N. Ross, and N. Badertscher, "Analysis of Electric Power System Reliability", California Energy Resources Conservation and Development Commission", October 1976.

93. M. Munasinghe, "A New Approach to Power System Planning", IEEE 1979 PES Winter Meeting, Paper F 79 155-3.

5. ECONOMIC THEORY CONCERNING WORTH OF RELIABILITY AND POWER SYSTEM PLANNING

This section contains references in the economics field pertaining to Power System Planning and Worth of Reliability. There are many others but the ones most relevant were included. Many economics publications dealing with peak pricing discuss reliability but it is only of late that economics research considers reliability as a variable to be optimized rather than as an outside constraint.

94. M.L. Visscher, "Welfare-Maximizing Price and Output with Stochastic Demand: Comment", American Economic Review, vol. 63, 1973, pp. 224-229.

95. E.J. Mishan, "Cost-Benefit Analysis", Allen and Unwin, London, Second Edition, 1975.

96. E.K. Smith, "Reliability in the Demand for and Supply of Electricity", Ph.D. Dissertation, Texas A & M University, December 1975.

97. J. Stabler and L. St. Louis, "A Cost - Benefit Model for the Evaluation of Public Power Development Proposals", Western Regional Science Association Meeting, San Diego, February 1976.

98. L. Higgins, "Load Forecast Error as an Element in Planning Optimal Capacity in an Electrical Supply System", Public Utilities Forecasting Conference at Bowness-on-Windermere, March 1977.

99. B. Fischoff, "The Art of Cost-Benefit Analysis", The U.S. Department of Commerce, National Technical Information Service, February 1977.

100. D.T. Nguyen, "Public Utility Pricing with Stochastic Demands: A Note", Applied Economics, vol. 10, 1978, pp. 43-47.

101. M.A. Crew and P.R. Kleindorfer, "Reliability and Public Utility Pricing", The American Economic Review, vol. 68, no. 1, 1978, pp. 31-40.

102. M. Munasinghe and M. Gellerson, "Optimum Economic Power Supply Reliability", World Bank Staff Working Paper no. 311, January 1979.

103. M. Munasinghe and M. Gellerson, "Economic Criteria for Optimizing Power System Reliability Levels", The Bell Journal of Economics, vol. 10, no. 1, spring 1979, pp. 353-365.

104. J. Tschirhart and F. Jen, "Behaviour of a Monopoly Offering Interruptible Service", The Bell Journal of Economics, vol. 10, no. 1, Spring 1979, pp. 244-258.

6. COST OF RELIABILITY AND COST EFFECTIVENESS ANALYSIS

This section contains papers mainly concerned with determining some estimate of the relative cost and merit of design schemes. No estimate of the cost of interruptions is used.

This is only a partial bibliography, only a few of the pertinent papers on this topic are included to give some indication of the work in this area.

105. G.F.L. Dixon and H. Hammersley, "Reliability and Its Cost on Distribution Systems", IEE Conference on the Economics of Reliability of Supply, October 1967, pp. 81-84.

106. W. Cartwright and B.A. Coxson, "Reliability Engineering and Cost-Benefit Techniques for use in Power System Planning and Design", Electricity Council Research Memorandum ECR/M966, October 1976.

107. N.E. Chang, "Evaluate Distribution System Design by Cost Reliability Indices", IEEE PAS, vol. 96, no. 5, September/October 1977, pp. 1480-1490.

108. S. Cluts, M.K. Ravindra, and E. Filstein, "Transmission Reliability – What's it Worth?", 1979 Reliability Conference for the Electric Power Industry, pp. 23-29.

7.   MISCELLANEOUS PAPERS RELATED TO THE EVALUATION OF WORTH OF RELIABILITY

109. W.D. Rowe, "An Anatomy of Risk", John Wiley & Sons Ltd., New York, 1977.

This book discusses the determination of acceptable levels of risk for technological systems and programs.

110. G. Schofield and W. Marsh, "The Effect of Reliability on Utility Financial Results", Public Utilities Fortnightly, 1970, September 24, pp. 15-20.

111. H. Legler, "Modelling the Impact of Electricity Supply on Economic Growth and Employment", Progress in Cybernetics and systems Research, vol. IV, pp. 278-292.

112. A. Levis et al., "Large Scale System Effectiveness Analysis", U.S. Department of Energy, September 1978.

This report discusses the initial results of a theoretical study into the analysis of large scale effectiveness.

113. H.A. Cavanaugh, "How Much Will a KW Shortfall Cost?", Electrical World, January 15, 1979, pp. 43-46.

CONCLUSIONS

In compiling a bibliography of related publications it is not possible to include all the relevant material. The authors would like to apologize for any errors or omissions and hope that additional relevant publications will be advanced as discussion to this publication.

ACKNOWLEDGEMENT

The authors would like to thank the Canadian Electrical Association for providing the financial support for this work.

STATISTICAL AND ANALYTICAL EVALUATION OF THE DURATION AND COST OF CONSUMER INTERRUPTIONS

D.O. Koval,   R. Billinton
Power System Research Group
University of   Saskatchewan
Saskatoon, Sask. Canada

Abstract - This paper presents a method of expressing the duration of consumer interruptions in terms of statistical distributions. Discussions and illustrations of the results of recent publications by various governmental agencies and boards concerning consumer estimates of the worth associated with various levels of service reliability is presented. A procedure for combining the statistical distributions of interruption durations with non-linear reliability costing functions is illustrated by the application of various techniques to a practical single phase distribution underground lateral configuration. The approaches presented in this paper can be used in manual calculations or included in a digital computer program to enable the distribution engineer to evaluate the statistical distributions associated with the duration of consumer interruptions and the associated costs.

INTRODUCTION

Reliability evaluation of individual customer service continuity levels is an important consideration which can significantly affect the design and operating characteristics of distribution circuits. As utilities maintain and expand their existing distribution systems and plan future systems they are confronted with the task of appraising service continuity levels and comparing these levels with either a set of norms pertaining to the average frequency and duration of interruptions of various types of customers (i.e., industrial, commercial and residential) or with some economic criterion (e.g., reliability worth-reliability cost).

A thorough knowledge of the present quality of service reliability in any given specific distribution circuit configuration and the effectiveness of improvement schemes to this circuit may not be quantitatively known by utilities. Consistent and comprehensive reliability procedures presented in the literature [1-9] provide a means of evaluating the quality of service in existing distribution circuits and the effectiveness of any improvement measures in advance of their implementation.

In recent years considerable attention by various governmental agencies and boards has been directed at determining the worth to the customer of a particular level of service reliability and comparisons made with the cost of supplying this service [10-16]. The reliability worth to a customer has been appraised by evaluating the expected costs that would be incurred by a customer during an interruption. These various cost assessment associated with the various levels of reliability as determined by the utility and its customers provides a basis for reliability worth - reliability cost studies.

The majority of reliability techniques presented in the literature evaluate the average frequency and duration of consumer interruptions. The costs of consumer interruptions are then evaluated from these average values in the economic analysis of reliability cost. This paper will illustrate the application of the Monte Carlo simulation technique and the Pearson Method of Moments to evaluate the underlying distributions associated with the duration of consumer interruptions. A method of combining non-linear and/or discontinuous cost of interruption functions with the distribution of the duration of interruptions of a single phase distribution underground lateral. These total costs are compared with the total costs evaluated from average values and the errors presented and discussed.

ECONOMIC EVALUATION OF CONSUMER LOSSES DUE TO INTERRUPTIONS

One of the first comprehensive surveys of consumer losses due to system outages was conducted in Sweden in 1969 [11]. Their basic procedure was to first classify the consumers into the following categories: industrial, commercial, agricultural and domestic consumers. The estimates of the costs of interruptions were obtained by direct questioning of the various consumer groups. A similar method was used the Reliability Subcommittee of the Industrial and Commercial Power systems Committee of the IEEE. in 1973 to estimate the costs of supply interruptions to industrial plants [15]. In 1974 the Ontario Energy Board conducted hearing on the Ontario Hydro system Expansion Program and requested Ontario Hydro to study: "The evaluation of alternative levels of reliability of power and energy supply from the viewpoint of the particular classes of customers and of the Province of Ontario as a whole" [13,14]. The majority of published studies have been directed at evaluating industrial losses due to interruptions. Residential and commercial loss studies by various governmental agencies are still being planned or are in the process of being completed.

INDUSTRIAL ESTIMATES OF INTERRUPTION COSTS

The results of the Ontario Hydro survey [14] are shown in Figure 1. It is interesting to note that a momentary outage will result in an average cost to the industrial customer of $0.68 per kW of load interrupted while an interruption lasting one hour will cost $2.69 per kW. A significant point to note is that an outage of any substantial duration can be quite costly.

The results of a similar study conducted by the IEEE [15] for industrial plants in Canada and the United States is shown in Table 1. The basic costing units $/kW and $/kWh are related to the frequency of interruptions and the duration of interruptions, respectively. It is interesting to note that the smaller industrial plants have a higher per unit cost of interruptions than the larger industrial plants. Given these costing estimates, the reliability worth to industrial customers can be aggregated and compared with the reliability cost of the utility.

Table 1. Average cost of supply interruptions for industrial plants in U.S.A. and Canada

PLANTS	$/kW	$/kWh
All plants	1.89	2.68
Plants with maximum demand > 1000 kW	1.05	0.94
Plants with maximum demand < 1000 kW	4.59	8.11

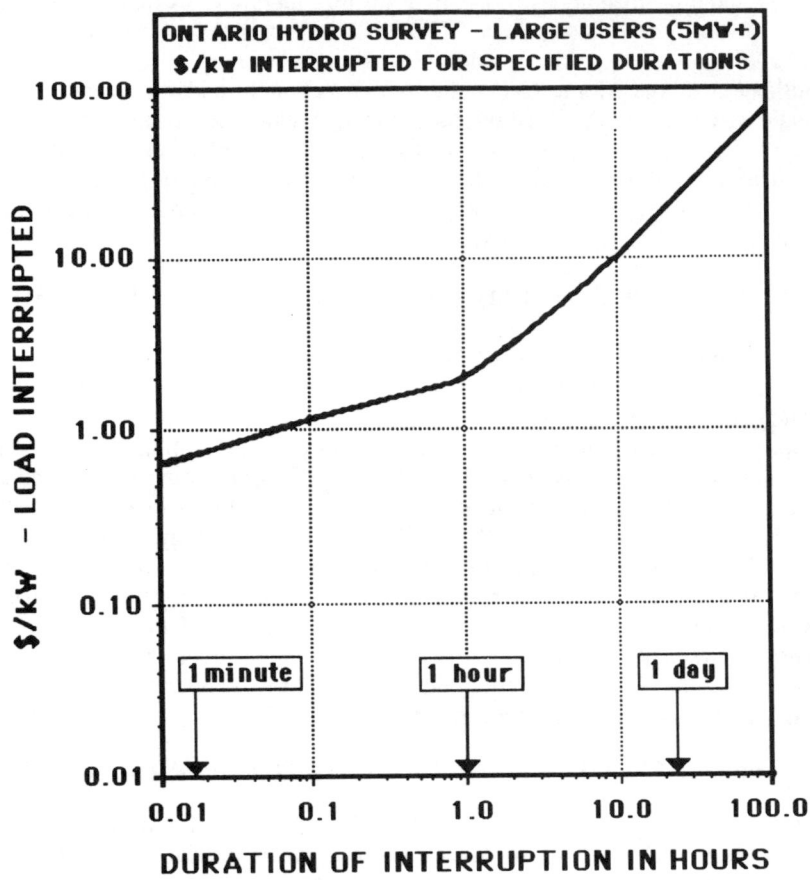

Figure 1. Ontario Hydro user estimates of interruption costs

RESIDENTIAL LOSSES DUE TO INTERRUPTIONS

Residential consumers can experience severe economic losses with unreliable circuit configurations and operating practices. Studies are presently being conducted by Ontario Hydro to determine the cost of supply interruptions to residential and commercial customers. Actual financial losses resulting from specific outages have been well documented in various newspapers (e.g., New York Times, August 31, 1973). Many of the reported financial losses to residential consumers have been dependent upon the duration of the outage, the time of day and the outdoor temperatures.

A study conducted in Great Britain [16] indicated that people value their leisure time at about the same level as their earning rate.   The value to the residential customer of undelivered energy ranged from \$0.50 to \$1.50   per kWh based on this assumption and an average home occupancy of 2 people with an average demand of 1 kW per household.

The Swedish study [11] in 1969 established a table of costs of losses to residential customers as a function of the duration of interruptions as shown in Table 2.

Table 2. Swedish residential cost of interruptions

IF THE INTERRUPTION LASTS FOR (hours)	FIXED COST AT START OF RANGE		VARIABLE COST WITHIN RANGE	
	(Skr/kW)	($US/kW)	(Skr/kW)	($US/kW)
0 - 1	0	0	4	0.80
1 - 2	4	0.80	4	0.80
2 - 8	8	1.60	3.5	0.70
8 - 24	29	5.80	2.5	0.50
24 - 48	69	13.80	6	1.20

A study conducted by the State of California[16] approached the cost of interruptions to residential customers in a slightly unique manner. Their criteria for estimating the cost of interruptions was stated as follows:

"As a first approximation, it is proposed here to look at appliance investment costs on the basis that when electrical energy is not available to the residential customer, his assets that have been purchased on the assumption of supply being available becomes useless."

By aggregating the consumer's domestic load including the electrical house wiring, a value of 10¢ /kWh was obtained.

A summary of the results of interruption costs for various classes of electric consumers conducted to date is shown in Figure 2.

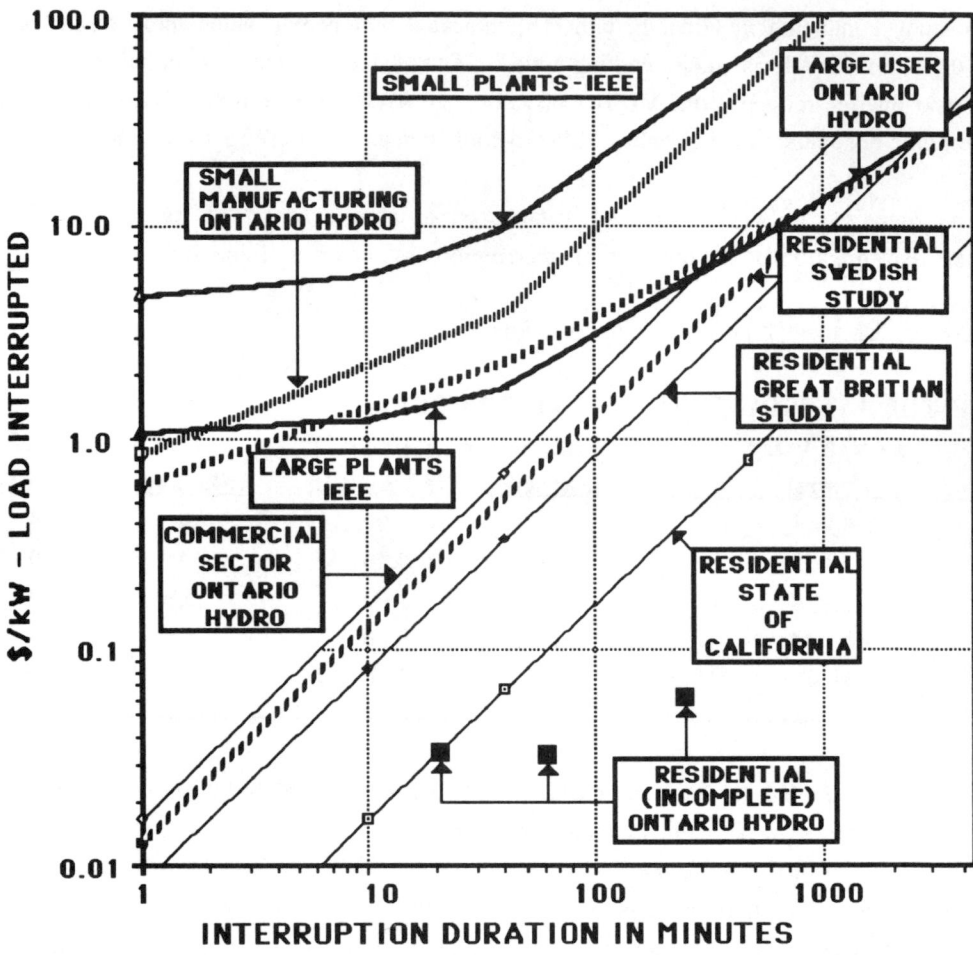

Figure 2.   Summary of interruption duration versus interruption costs for various classes of consumers.

FUNCTIONAL RELATIONSHIPS BETWEEN THE COST AND DURATION OF INTERRUPTIONS

The approximate relationships y(t) of the cost per kW of load interrupted and the duration of the interruption (t in minutes) for the various studies and classes of consumers is shown in Table 3.

The total cost (Y_I) of a single interruption of duration t expressed in minutes to an individual customer with a load of X kW is given by the following expression:

$$Y_I = y(t)[X] \tag{1}$$

Depending upon the class of customer, the appropriate function as illustrated in Table 3 is substituted in equation (1) to obtain the cost of a single interruption. To evaluate the annual average cost of interruptions (Y_A) as seen by an individual customer, equation (1) is multiplied by the average frequency of interruptions per year (f) as follows:

$$Y_A = f\big[(y(t)\big] X \tag{2}$$

Table 3. Functional relations between the cost of interruptions $y(t)$ in \$/kW and the duration of the interruption (t) in minutes.

TYPE OF CONSUMER		FUNCTION $y(t)$
RESIDENTIAL		
1. State of California		$y = t/600.0$
2. Great Britain		$y = t/120.0$ (lower value)
3. Swedish	(i)	$y = t/75.0$
	(ii)	$y = 0.2 + 0.0116t$
ONTARIO HYDRO		
4. Large user	(i)	$y = 0.60t^{0.36462509}$
	(ii)	$y = 0.253357795t$
5. Small manufacturer	(i)	$y = 0.85t$
	(ii)	$y = 0.090643129t^{1.00990454}$
6. Commercial	(i)	$y = 0.0165553704t^{1.00880454}$
	(ii)	$y = 0.013613460t^{1.056641667}$
IEEE		
7. Large plants		$y = 1.05 + 0.94t/60.0$
8. Small plants		$y = 4.59 + 8.11t/60.0$

389

The annual interruption cost can be estimated from the average interruption duration per failure $(E(t))$ provided the function $y(t)$ is a linear function of the interruption duration, for example, assume:

$$y(t) = a + bt$$

Then the annual average cost Y_A is:

$$Y_A = \frac{1}{n} \sum_{}^{n} [a + b\,t]\ f\ [X] \qquad (3)$$

$$= [a + b\,E(t)\,]\,f\ X$$

where: n - number of interruptions of duration t sampled over an integer number of years

If the costing function $y(t)$ is a non-linear function of the duraiton of the interruption, for example:

$$y(t) = a + b\,t + ct^2$$

then the average annual cost Y_A is:

$$Y_A = \frac{1}{n} \sum_{}^{n} [a + b\,t + c\,t^2]\ f\ [X]$$

$$Y_A = [a + b\,E(t) + c\,E(t^2)]\ f\ [X] \qquad (4)$$

If the underlying distribution of the interruption duration is exponential, then $E(t^2) = 2E(t)^2$. Then the average annual cost of interruptions reduces to:

$$Y_A = [a + b\,E(t) + 2\,c\,E(t)]\ f\ [X] \qquad (5)$$

The results of equation 5 are significantly different from the results obtained if the average value of t (i.e., $E(t)$) is substituted in the equation of the costing function as follows:

$$Y_A = [a + b\,E(t) + c\,E(t)^2]\ f\ [X] \qquad (6)$$

If the costing function in non-linear, then the simple substitution of the average value of the duration of the interruption into the costing expression may yield erroneous costing results. Knowledge of the underlying distribution of the interruption is required to evaluate the average annual cost of interruptions per customer. For example, if the underlying density function is a gamma function characterized by α and β, then the average annual cost of interruptions is simply:

$$Y_A = [a + b\ E(t) + c\ (\beta\ E(t) + E(t)^2)]\ f\ [X] \tag{7}$$

It has been illustrated by equations 5 and 7, that in some cases the non-linear costing functions can be modified to include the average duration of interruptions and the parameters of the underlying distribution. The costing studies for industrial and residential customers provide an indication of the range of the costs of interruptions within the study areas reported. A utility must assess its own consumer losses via questionnaires and interviews, however, before proceeding to utilize the general reliability worth-reliability cost methodology [13].

NON-LINEAR COST OF SUPPLY INTERRUPTIONS

The reliability levels presented in many papers have been average values. These average levels can be combined with linear costing functions to assess the total average cost of interruptions to various classes of customers. However, if the interruption costs are non-linear functions and/or discontinuous functions of the duration of interruptions, then the underlying density function of the interruption duration must be considered in the analysis. If the interruption costs are evaluated strictly from average values with non-linear and/or discontinuous cost functions, the results may be quite erroneous.

Initially, for the purpose of illustrating the magnitude of the costing error, the cost of interruptions will be evaluated from the average reliability levels of the customers being serviced by transformers T1 to T4 shown in Figure 3. Two isolation-restoration procedures will be studied, namely: the bisectionalizing and forward sequential and the evaluation will be based on the data and explanations given in Reference [9]. The respective outage activity procedural flow charts for the isolation-restoration procedures is shown in Figures 4 and 5.. Based on the zone branch concepts [7,9], the FZB and DTA arrays can be formulated for each isolation procedure as shown in Figures 6 and 7.

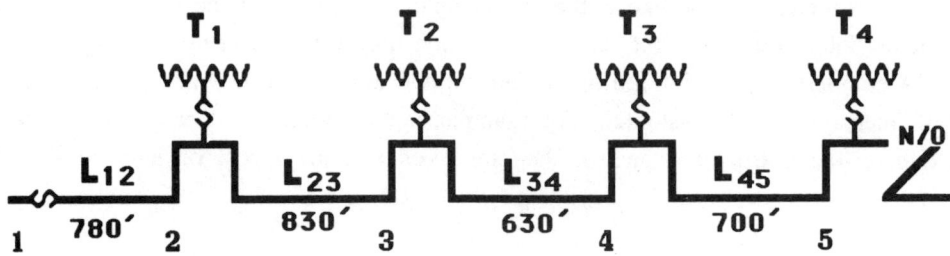

Figure 3. Single phase fused underground lateral-looped.

The average frequency of interruptions experienced by these customers on the single phase lateral is:

$$= \lambda_{\text{supply}} + \lambda_{\text{sub}} + \sum^{4} FZB(j) + \lambda_{\text{trasformer}} \quad \text{(defective + overloaded)} \tag{8}$$

$$= 0.10 + 0.025 + 2490(0.0685)/5280.0 + 0.004 + 0.002$$

$$= 0.0169086 \quad \text{interruptions per annum}$$

The average annual interruption duration experienced by any transformer location (**i**) is given by the following expression:

$$= (\lambda \ r)_{\text{supply}} + (\lambda \ r)_{\text{sub}} + \sum^{4} FZB(j)[DT \ A(i,j)] + \lambda_t \ r_t \tag{9}$$

A summary of the average annual interruption duration and the average duration per interruption is shown in Table 5.

Table 5 Summary of individual customer interruption durations for the bisectionalizing
and forward sequential isolation-restoration procedures.

TRANSFORMER NUMBER	BISECTIONALIZING PROCEDURE		FORWARD SEQUENTIAL PROCEDURE	
	minutes/year	minutes/failure	minutes/year	minutes/failure
1	11.650	68.900	10.924	64.606
2	10.924	64.606	12.076	71.419
3	11.527	68.172	12.679	74.986
4	11.754	69.515	12.906	76.328

The annual interruption costs for residential customers (e.g., Swedish study) is based on
the average annual interruption duration (U_A) for the single phase lateral with n_i customers,
and a load of 2 kW per customer, being served from each transformer is given by:

$$Y_T = \sum^{n} (2.0) \, n_i \, (U_A/75.0) \qquad (10)$$

The results of the total interruption cost for the lateral and each transformer location by the
Swedish Study and the State of California are given in Table 6.

Table 6. Annual cost of interruptions for a single phase lateral.

TRANSFORMER NUMBER	BISECTIONALIZING PROCEDURE		FORWARD SEQUENTIAL PROCEDURE	
	SWEDISH	STATE OF CALIFORNIA	SWEDISH	STATE OF CALIFORNIA
1	$ 8.70	$ 1.07	$ 8.16	$ 1.02
2	7.28	0.91	8.16	1.01
3	5.53	0.69	6.09	0.76
4	9.09	1.14	9.98	1.25
TOTAL COST FOR LATERAL	$30.60	$ 3.83	$32.28	$ 4.04

STATISTICAL DISTRIBUTION OF INTERRUPTION DURATIONS

The evaluation of non-linear or discontinuous interruption cost functions requires a knowledge of the statistical distributions associated with the duration of interruptions. The knowledge of the type of statistical distributions associated with the duration of interruptions has not been considered in any detail in the existing literature. This section of the paper will attempt to classify the type of distribution that will statistically represent the duration of interruptions for each individual transformer located in the single phase underground lateral of Figure 1 for two outage activity procedures as shown in Figures 4 and 5. The cost of interruptions based on these statistical distributions will be evaluated and compared with the cost of interruptions based solely on the expected values of the interruption durations.

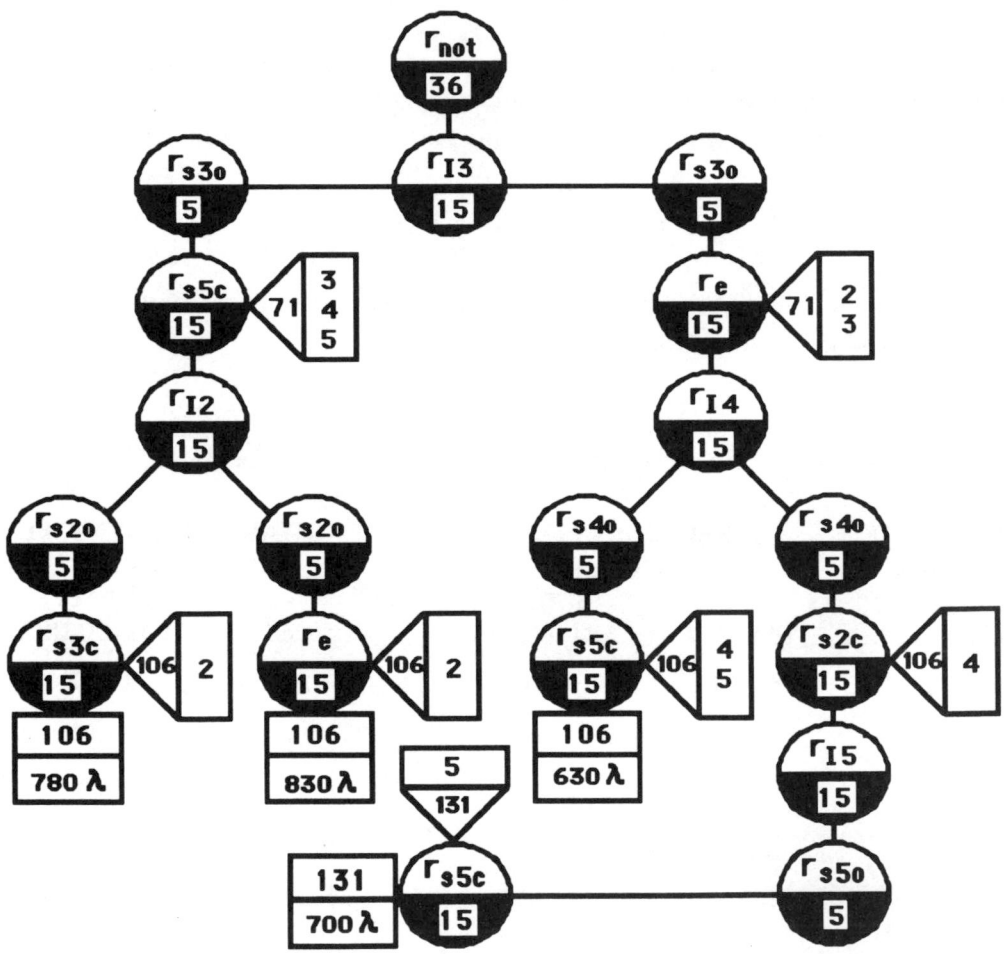

Figure 4. Outage Activity Procedural Flow Chart - Bisectionalizing Procedure

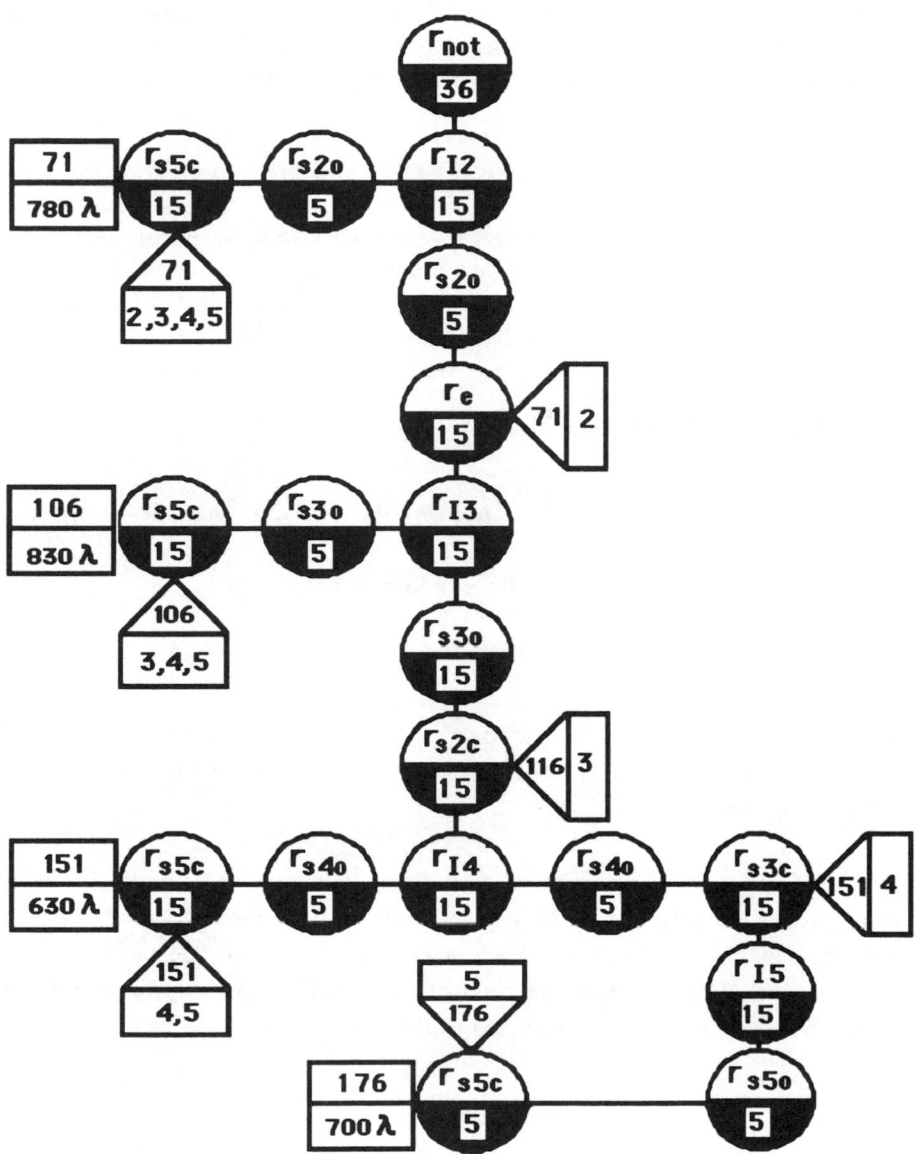

Figure 5. Outage Activity Procedural Flow Chart - Forward Sequential isolation restoration procedure

Each outage activity shown in Figures 4 and 5 for the various sectionalizing procedures is assumed to be exponentially distributed. With reference to Figure 1, the expected interruption duration of customers at bus 3, 4 and 5 (2,3,4 elements of the DTA array) for an outage in the 780 foot cable section is:

$$DTA(2,1) = DTA(3,1) = DTA(4,1) = E(r_{not}) + E(r_{I3}) + E(r_{s3o}) + E(r_{s5c})$$

$$= 36.0 + 15.0 + 5.0 + 15.0$$
$$= 71.0 \text{ minutes}$$

If the outage activities are assumed to be independent, then the variance of the interruption duration for the above example is:

$$V_{[DTA(2,1)]} = V_{[DTA(3,1)]} = V_{[DTA(4,1)]}$$

$$= V[r_{not} + r_{I3} + r_{s3o} + r_{s5c}]$$

$$= 36.0^2 + 15.0^2 + 5^2 + 15.0^2$$

$$= 1411.0$$

Given only the expected value and the variance of an interruption duration contained in the DTA array, it is not possible to deduce the type of distribution that will statistically represent the interruption durations.

The Monte Carlo simulation method was used to simulate the various interruption durations contained in the DTA arrays of Figures 6 and 7 from the individual outage activity durations contained in Figures 4 and 5. Each interruption duration element contained in the DTA array is the sum of the random independent generated variables t_i associated with the outage activities. The random variable $dta(i,j,k)$ is given by the following expression:

$$dta(i, j, k) = \sum_{i}^{m} t_i \tag{11}$$

where: k - kth sample

m = number of sequential activities associated with data(i,j,k)

780 λ	830 λ	630 λ	700 λ

FAILED ZONE BRANCH ARRAY FZB(j)

106 [1886]	106 [1886]	71 [1411]	71 [1411]
71 [1411]	71 [1411]	71 [1411]	71 [1411]
71 [1411]	71 [1411]	106 [1886]	106 [1886]
71 [1441]	71 [1411]	106 [1886]	131 [2161]

DOWN TIME ARRAY DTA(i,j)

where: [ ] - variance of DTA(i,j)

Figure 6. FZB and DTA arrays - Bisectionalizing procedure

780 λ	830 λ	630 λ	700 λ

FAILED ZONE BRANCH ARRAY FZB(j)

71 [1411]	71 [1411]	71 [1411]	71 [1411]
71 [1411]	106 [1886]	116 [2086]	116 [2086]
71 [1411]	106 [1886]	151 [2561]	151 [2561]
71 [1441]	106 [1886]	151 [2561]	176 [2836]

DOWN TIME ARRAY DTA(i,j)

where: [ ] - variance of DTA(i,j)

Figure 7. FZB AND DTA arrays - Forward sequential procedure

397

The frequency of occurrence f_k of the random variable $dta(i,j,k)$ is evaluated during the sampling process. The mean of the distribution for the element $DTA(i,j)$ is evaluated as follows:

$$\overline{DTA(i,j)} = \sum_{k}^{n} f_k\, [dta(i,j,k)] / \sum_{k}^{n} f_k \tag{12}$$

where: $\overline{DTA(i,j)}$ - sample mean

   n - total number of samples

The moments of the distribution of $DTA(i,j)$ are calculated as follows:

$$\mu_i = \sum_{k}^{n} f_k\, [dta(i,j,k) - DTA(i,j)]^i / \sum_{k}^{n} f_k \tag{13}$$

where: μ_i - the ith moment of the observed or simulated distribution

The Pearson Method of Moments [17] was considered as a method of selecting the type of distribution associated with the interruption durations. This method consists of essentially four basic steps as follows:

1. calculate the mean and the first four moments of the sample;
2. evaluate β_1 and β_2 coefficients as follows:

$$\beta_1 = \mu_3^2 / \mu_2^3 \tag{14}$$

$$\beta_2 = \mu_4 / \mu_2^2 \tag{15}$$

3. evaluate the kapa (κ) criterion as follows:

$$\kappa = \frac{\beta_1(\beta_2 + 3)^2}{4[4\beta_2 - 3\beta_1][2\beta_2 - 3\beta_1 - 6]} \tag{16}$$

4. determine which member of the Pearson Family of Curves best represents the interrupiton duration from the κ criterion and other criteria.

The results of the simulation studies for a sample size of 1000 for each element of the DTA arrays is shown in Table 7.

Table 7. Characteristic data for the Pearson Family of Curves

EXPECTED DURATION minutes/ failure	SIMULATED EXPECTATION minutes/ failure	β_1	β_2	κ	GAMMA DISTRIBUTION CRITERION $2\beta_2$ = $3\beta_1 + 6$	
71.0	70.6224	1.730	6.439	1.109	12.88	11.19
106.0	105.5164	1.239	5.902	0.591	11.80	9.72
116.0	115.8317	0.893	5.214	0..474	10.43	8.68
131.0	130.7071	1.095	5.295	0.805	10.59	9.20
151.0	150.2417	0.636	4.298	0.807	8.60	7.91
176.0	174.6800	0.472	3.694	3.454	7.39	7.42

From the above table, it appears that the "gamma" distribution may represent the interruption durations according to the "gamma distribution criterion". The characterizing parameters (α, β) of the theoretical "gamma" density function $(f(t))$ representing the interruption duration (t) are calculated from the expected value $E(t)$ and the variance $V(t)$ of the simulation study as follows:

$$\beta = V(t)/E(t) \tag{17}$$

$$\alpha = \frac{E(t)^2}{V(t)} - 1.0 \tag{18}$$

$$f(t) = \frac{t^{\alpha} e^{-t/\beta}}{\beta^{\alpha+1} \Gamma(\alpha+1)} \tag{19}$$

A typical frequency histogram for the theoretical and simulated values of the interruption duration (t) is shown in Figure 8.

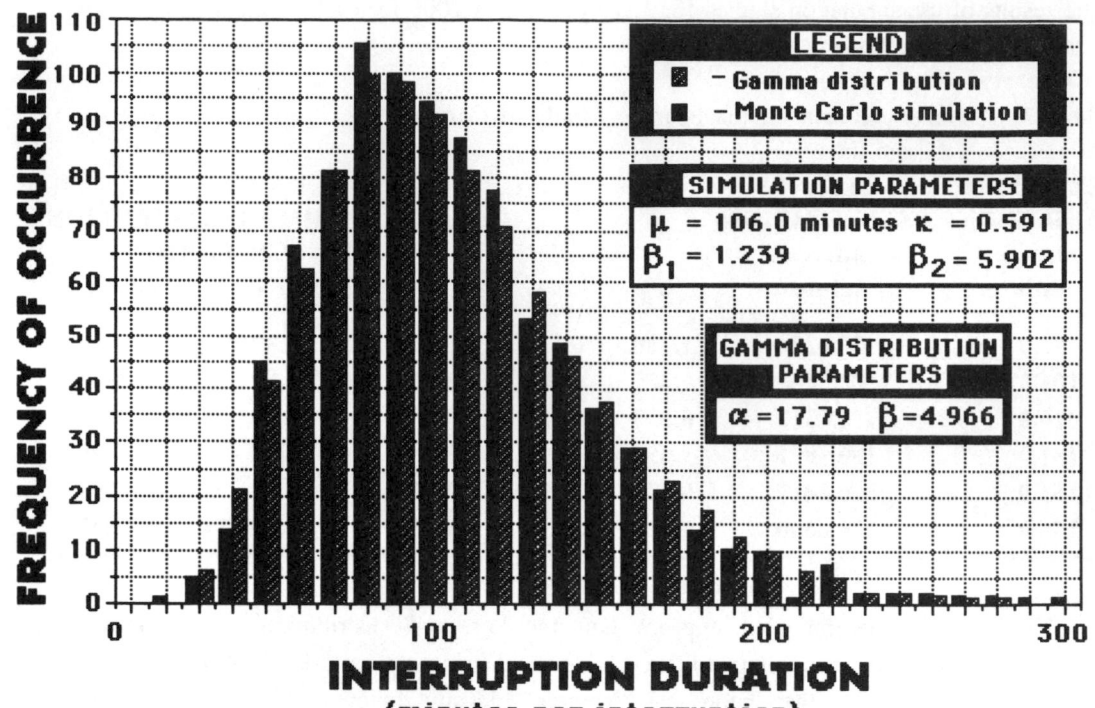

Figure 8. Interruption duration histogram

Given the theoretical frequencies of the "gamma" distribution for each interruption duration element of the DTA array, it is possible to test if the theoretical frequency distribution represents the actual frequency distribution generated by with Monte Carlo simulation by using the Chi-squared test for goodness of fit.

The Chi-squared value is estimated as follows:

$$\chi^2 = \frac{(f_0 - f_e)^2}{f_e} \qquad (20)$$

where: f_e - gamma theoretical frequency

f_0 - simulated actual frequency

A significance level of 0.05 was assumed. The Null hypothesis is made which suggests the actual and the theoretical distributions are drawn from the same population. If the calculated Chi-squared value is found to exceed the actual value at the 0.05 level of significance for the appropriate number of degrees of freedom, then the Null hypothesis is rejected. The two distributions are then said to be significantly different and the gamma distribution fails to represent the actual data. The results of the Chi-squared test for each interruption duration shown in Table 7 is shown in Table 8.

Table 8. Chi-squared results for the theoretical gamma distribution

EXPECTED DURATION min./failure	DEGREES OF FREEDOM	CALCULATED CHI SQUARED	CHI SQUARED 0.05	CHI SQUARED 0.01	GOODNESS OF FIT YES/NO
71.0	17	16.2390	27.587	33.409	YES
106.0	21	15.1765	32.671	38.932	YES
116.0	22	20.4252	33.924	40.298	YES
131.0	22	28.4106	33.924	40.298	YES
151.0	24	12.6553	36.415	42.980	YES
176.0	27	33.8617	40.113	46.963	YES

It appears from an examination of Table 8 that the "gamma" distribution adequately represents the duration of interruptions per cable section outage contained in the DTA array of Figures 6 and 7. Given the "gamma" distribution $f_{ij}(t)$ for each element of the DTA array and the expected costs in \$/kW of each element (Y_{ij}) can be evaluated as follows:

$$E(Y_{ij}) = \int_{0}^{t1} f_{ij}(t)\, y(t)\, dt + \int_{0}^{t2} f_{ij}(t)\, y(t)\, dt + \dots \tag{21}$$

where: $f_{ij}(t)$ - gamma distribution of the interruption duration associated with the ij th element of the DTA array

$y(t)$ - interruption duration costing function

$t1, t2$ - break points or discontinuities in the costing function

The above integrals are difficult to evaluate analytically. If the selected class width is very small in comparison to the standard deviation of the distribution, then equation 21 can be approximated by:

$$E(Y_{ij}) = \int_{t0}^{t0 + n1\Delta t} f_{ij}(t_m) \, y(t_m) \, \Delta t + \int_{t1}^{t1 + n2\Delta t} f_{ij}(t_m) \, y(t_m) \, \Delta t + \ldots \qquad (22)$$

where: Δt - class width

$n1$ - number of subintervals in the time range t_{i-1} to t_i

$y(t_m)$ - value of cost function at the mid point of the class width

$f_{ij}(t_m)$ - ordinate of the distribution function at the mid point of the class width

$t0, t1 \ldots$ - initial class mid points of the first interval in the time range t_{i-1} to t_i

Based on the Swedish cost of supply interruptions, the expected cost of interruptions per kW of load interrupted ($E(Y_{ij})$) associated with the elements of the **DTA** arrays of Figures 6 and 7 as shown in Table 9. If the average value of the duration is substituted directly into the costing equation, then the average cost per kW of load interrupted will be obtained as shown in Table 9.

Table 9. Expected cost of interruptions in $/kW for various duration elements of the DTA arrays in Figures 6 and 7.

ELEMENT IN DTA ARRAY	DISTRIBUTION E($/KW)	$/kW USING AVERAGE VALUE DIRECTLY	% ERROR
71.0	0.930162	0.9460	1.7
106.0	1.375657	1.4130	2.7
116.0	1.504034	1.5460	2.8
131.0	1.698030	1.7283	1.7
151.0	1.927739	1.9616	1.8
176.0	2.209981	2.2530	2.0

The average cost of interruptions (Y_I) due to primary cable outages that are only asssociated with each distribution transformer shown in Figure 1 can be evaluated as follows:

$$Y_i = \sum_k^{n_k} \sum_i^j kW_k \, (FZB(j)) \, (E(Y(i,j))) \tag{23}$$

where: n_k - number of services connected to a single distribution transformer

kW_k - kW load interrupted per customer

$FZB(J)$ - Element of failed zone branch array

$E(y(i,j))$ - expected cost of an interruption associcated with an interruption duration corresponding to the element in the $DTA(i,j)$ array

The expected costs of interruptions due to primary cable outages is shown in Table 10 for the bisectionalizing and forward sequential isolation-restoration procedures.

Table 10.   Cost of individual transformer customer interruptions due to primary cable outages only

TRANSFORMER NUMBER	COST OF INTERRUPTIONS (PRIMARY CABLE OUTAGES ONLY)	
	FORWARD SEQUENTIAL	BISECTIONALIZING
1	$1.983883	$1.606742
2	2.505206	1.771318
3	2.066557	1.552677
4	3.396251	2.669478

The interruption costs due to outages in the supply, substation and an individual transformer can be evaluated from equation number 23. The average costs are shown in Table 11.

Table 11.   Cost of individual transformer customer interruptions due to supply, substation and individual transformer outages

TRANSFORMER NUMBER	SUPPLY-TRANSFORMER	INDIVIDUAL TRANSFORMERS	TOTAL COST
1	$5.60	$0.27	$5.87
2	5.00	0.24	5.24
3	3.60	0.17	3.77
4	5.80	0.28	6.08

The average total cost of residential interruptions for each individual distribution transformer derived from the underlying distributions of the interruption durations and the expected interruption durations are shown in Table 12.

Table 12.   Summary of the total cost of interruaptions for individual transformers and their associated residential customers.

TRANSFORMER NUMBER	BISECTIONALIZING PROCEDURE		
	DISTRIBUTION	AVERAGE VALUE	% ERROR
1	$ 8.37	$ 8.70	3.94
2	7.01	7.28	3.85
3	5.32	5.53	3.95
4	8.75	9.09	3.89
TOTALS	$ 29.45	$ 30.60	3.90

TRANSFORMER NUMBER	FORWARD SEQUENTIAL PROCEDURE		
	DISTRIBUTION	AVERAGE VALUE	% ERROR
1	$ 7.85	$ 8.16	3.95
2	7.75	8.05	3.87
3	5.84	6.09	4.28
4	9.46	9.98	5.27
TOTALS	$ 30.92	$ 32.28	4.40

The error in the cost values derived from the underlying interruption duration distributions and the expected values can be quite significant. In these examples, the use of the expected values to estimate the costs of interruptions yield consistently higher values than when the underlying distributions are employed to estimate the costs. This consistency is due to the Swedish cost structure which has a slightly lower slope in the duration range 120-480 minutes which essentially weights the costs of the interruptions slightly downward.

It is difficult to generalize the magnitude and the direction of the costing errors.   The direction and magnitude of error is largely dependent upon the costing function for interruption durations and where the breakpoints occur with respect to the distribution location within the costing function range.   The error is also significantly dependent upon the circuit configuration and the component failure and repair rates.

CONCLUSION

This paper has presented the results of studies concerned with the cost of interruptions to various classes of consumers that have been conducted by various utilities and governmental agencies.   Functional relationships between the cost of interruptions and the duration of the interruption for the various studies were developed.

This paper illustrated the application of the Monte Carlo simulation technique and the Pearson Method of Moments to evaluate the underlying distributions associated with the duration of interruptions.   A method of combining non-linear and/or discontinuous cost of interruption functions with the distribution of the duration of interruptions is presented to assess the total cost of interruptions in a single phase distribution underground lateral configurations.   These total costs are compared with the total costs evaluated from average values and the errors presented and discussed.

REFERENCES

1. Koval, D.O., "Evaluation of Distribution System Reliability", Ph.D. Thesis, College of Graduate Studies, University of Saskatchewan, Saskatoon, Saskatchewan, Canada, June 1978.

2. Capra, R.L., Gangel, W.M., Lyon, S.V., "Underground Distribution System Design for Reliability", 1969, IEEE Trans., Vol. PAS-88, pp. 834-842.

3. Gangel, W.M., Schultz, N.R., Simpson, Jr., J.W., "Predicting Underground Distribution System Availability", 1970 IEEE Trans., Vol. PAS-89, pp. 268-274.

4. Mann, R.F., Hopkins, D.L., "A Yardstick for Evaluating Underground Distribution Designs", 6901-PWR, Conference Record Special Technical Conference on Underground Distribution, May 12-16, 1969, pp. 164-172.

5. Easley, J.H., "The Influence of Service Availability on the configuration of Undeerground Three-Phase Primary System", Paper 71 C42-PWR, presented at the Underground Distribution conference, Detroit, Mich., 1971.

6. Chang, N.E., "Evaluate Distribtuion System Design By Cost Reliability Indices", 1977 IEEE Winter Power Meeting, F77 167-0, New York, N.Y.

7. Koval, D.O., Billinton, R., "Evaluation of Distribution Circuit Reliability", 1977 IEEE Winter Power Meeting, F77 067-2, New York, N.Y.

8. Billinton, R., Koval, D.O., Grover, M.S., "Calculation of Reliability Worth", Canadian Electrical Association, CEA Transactions, Vol. 16, Part 3, Paper No. 77-SP-143.

9. Koval, D.O., Billinton, R.,"Evaluation the Effects of Isolation-restoration Procedures on Distribution Circuit Reliability Indices", 1978 Summer Power Meeting, Los Angeles, California, A78 512-6

10. Billinton, R., Koval, D.O., "Evaluation of Reliability Worth in Distribution Systems", Power Systems Computation Conference, PSCC Conference VI, Aug. 21-25, 1978, Darmstadt, Germany.

11. "Costs of Interruptions in Electric Supply", Swedish Report from Committee on Supply Interruption costs (English Edition), September 1969.

12. Shipley, R.B., Patton, A.D., Denison, H..S., "Power Reliability Cost vs. Worth", IEEE Transaction on Power Apparatus and systems, Vol. PAS-91, 19722, pp. 2204-2212.

13. Ontario Hydro Survey on Power System Reliability: Viewpoint of Large Users Report No. PMA 76-5.

14. Berk, L.H., Mackay, E.M., "Ontario Hydro Survey on Supply Reliability: Viewpoint of Large Users", Canadian Electrical Association,, March 21, 1977.

15. "Report on Reliability survey of Industrial Plants, Part 2 - cost of Power Outages, Plant Restart Time, Critical Service Loss Duration Time, and Type of Load Loss vs. Time of Power Outages: Report by Reliability Sub-Committee on Industrial and Commercial Power System Committee, May, 1973.

16. "Analysis of Electric Power System Reliability", Prepared for: Energy Assessments Division of the California Energy Resources Conservation and Development Commission, by System Control Inc., 11801 Page Mill Road, Palo Alto, California 943043.

17. Koval, D.O., Billinton, R., "Determination of Transmission Line Ampacities by Probability and Numerical Methods, 1970 IEEE Trans. Vol. PAS-89, pp. 1485-1492.

Index